Cultural Geographies

Cultural Geographies

An Introduction

John Horton

University of Northampton

and

Peter Kraftl

University of Leicester

LONDON AND NEW YORK

First edition published 2014
by Routledge
2 Park Square, Milton Park, Abingdon, Oxon OX14 4RN

and by Routledge
711 Third Avenue, New York, NY 10017

Routledge is an imprint of the Taylor & Francis Group, an informa business

British Library Cataloguing in Publication Data
A catalogue record for this book is available from the British Library

Library of Congress Cataloging in Publication Data
A catalog record for this book has been requested

ISBN: 978-0-415-74016-6 (hbk)
ISBN: 978-0-273-71968-7 (pbk)

Typeset in Minion Pro
by Jouve India

Printed in Great Britain by Bell & Bain Ltd, Glasgow

For mums and dads. For Faith, Juliet and Emily.

Contents

Preface

Cultural geography is a major, vibrant, diverse and ever-changing subdiscipline of geography: in fact, it is so varied that many geographers (as we do in this book) tend to speak of cultural geograph*ies*, in the plural. Cultural geographers have taken part in some of the key debates in the history of academic geography – from changing views of landscape, to the role of artistic and written representations in understanding social inequalities, and from debates about emotion, performance and materiality to the experience of identities and subcultures.

Cultural geographies excite us, as authors, and ignited our passion for studying, researching and teaching geography. The over-riding aim of this book is to communicate some of the diversity, dynamism and controversy that continues to stimulate us and many other geographers. More than that, this book aims not only to provide an overview of key approaches to cultural geographies, and an introduction to some key case studies of research by cultural geographers, but also to encourage you to think critically and creatively about cultural geographies. It seeks to encourage you to raise questions about cultural geographies that matter to you and your studies, to the place where you live, and to other places both similar to and different from those with which you are familiar. In so doing, it aims not only to support your learning about cultural geographies, or your development of an essay or research project, but, we hope, to provoke you to think about how, when and where cultural geographies *matter*.

The book takes a broad view of culture, which can refer to (usually collective) ways of doing life, ways of thinking, ways of representing the world, ways of behaving, ways of believing, ways of feeling, ways of creating material objects, and ways of identifying with or against others (see Chapter 1 for further discussion). It takes a similarly broad view of cultural geographies, looking not simply at how 'culture' is somehow expressed 'over space' or 'in place', but also how cultural processes – ways of doing, thinking, etc. – are part and parcel of the processes that make the world look and feel as it does.

This all means that, while the book pays somewhat greater attention to recent trends in cultural geographies (where our own research and teaching is situated), we aim to provide a sense of some of the continuities and overlaps with earlier approaches. Rather than represent a somehow chronological biography of (some) cultural geographies, the book is therefore structured around three kinds of idea, which may either be read separately or in conjunction with other parts of the book.

Part 1 focuses on cultural practices and politics, outlining how geographers have explored consumption, production, identities, subcultures, resistance and activism. Part 2 considers cultural materials, texts and media, looking at cultural geographies of landscape, architecture, sport, fiction, dance, visual art and much more besides. Part 3 examines cultural theories that have developed (and continue to develop) in the work of cultural geographers – from theories of materiality and everydayness to theories of space, place, emotion and embodiment.

In order to meet the above aims, the book contains several features. The introductory chapter provides an overview of definitions of, and approaches to, cultural geographies, as well as a 'how-to' guide for using the book. Each part of the book contains a brief introduction that sets the scene for the following chapters. Every chapter contains a chapter summary, an introductory exercise to get you thinking about a particular concept or problem, concluding reflections, and an annotated list of further reading. While you may wish to read each chapter in a stand-alone way (for instance, if you want an overview of cultural geographic research about architecture or consumption), we include cross-references to other chapters in **bold**, which encourage you to consider the important ways in which specific ideas or examples link elsewhere in the subdiscipline, and beyond.

The concluding chapter offers some final – and in some cases personal – reflections that briefly address questions about the future of cultural geographies, and of where cultural geographies might take *you*.

John Horton
Peter Kraftl

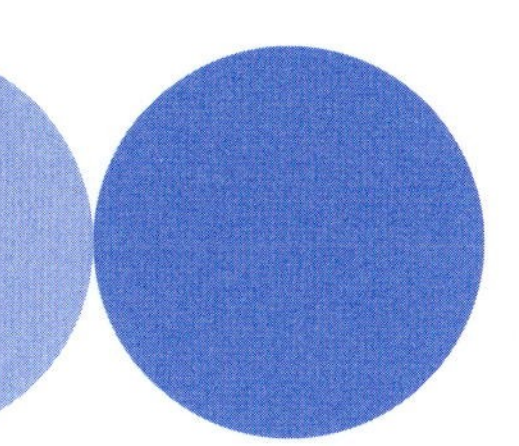

Authors' acknowledgements

Writing a book about cultural geographies – especially one that fulfils the aims we have outlined above – is no small endeavour. This book could therefore not have been written without the support of many people over a period of several years. We are hugely grateful to Andrew Taylor at Pearson for initially proposing and supporting this book. We also thank Professor Chris Philo for important words of support and encouragement at the outset. Andrew Taylor, Rufus Curnow, Helen Leech and Patrick Bond have consistently provided outstanding editorial support and guidance. The generous, detailed and constructive comments of five anonymous peer reviewers have helped us to revise and develop each and every chapter. Thanks are also due to all at Pearson who been involved in the production, formatting, illustration, copy-editing, proofreading and indexing of the book. Any remaining errors and shortcomings are, of course, our own.

John would like to thank Professor Paul Cloke for sparking an interest in social and cultural geographies. The University of Bristol's School of Geographical Sciences, 1995–2003, was an outstanding place in which to study and develop such an interest. Among many friends, colleagues and tutors from there and then, special thanks must go to Sergei Shubin and Emma Jane Roe for enduring friendship and ongoing discussions about the merits (or otherwise) of contemporary social and cultural geography. At The University of Northampton, the mentorship of Professor Hugh Matthews was greatly valued, and Professor Andrew Pilkington supported the writing of this book. Many thanks are due to undergraduate students who have studied GEO2028 *Issues in Human Geography*, GEO2029 *Approaches to Research in Human Geography* and human geography dissertations at Northampton between 2005 and 2012: much of the material in this book began life as lecture and seminar materials for these modules, and students have provided valuable feedback on earlier drafts of this book's chapters. Many people have made The University of Northampton a supportive and interesting workplace during the writing of this book: particular recognition must go to outstanding geography postgraduate students Michelle Pyer, Lee Dunkley, Matthew Callender and Rebekah Ryder and outstanding postdoctoral researchers Victoria Bell and Sophie Hadfield-Hill. Finally, as always, thank you to Mum, Dad and Faith for keeping me going during the writing of this book.

Peter would like to thank Pyrs Gruffudd and Marcus Doel for inspiring an interest in cultural geographies in all their richness and diversity. During my time at The University of Northampton, several people contributed directly or indirectly to my passion for communicating geographical research to students in a variety of ways that I had never anticipated; but special thanks in this regard go to Faith Tucker and Greg Spellman. I should also like to acknowledge the mentorship of Hugh Matthews. My thinking and writing on cultural geographies have flourished during my time at the University of Leicester. For this I have to thank several colleagues who are ever-supportive: Jenny Pickerill, Gavin Brown, Martin Phillips, Clare Madge and Katy Bennett. Several chapters have developed in tandem with lectures and seminars given since 2003 at Swansea University, The University of Northampton, and, especially, the University of Leicester. In particular, I should like to thank University of Leicester students taking *Geographies of Youth Cultures*, *Social & Cultural Geographies* and *Spaces of Identity* for helping me to refine some of the explanations, examples and pedagogic exercises that now form part of this book. Finally, thanks to Mum, Dad, Martin, Juliet and Emily.

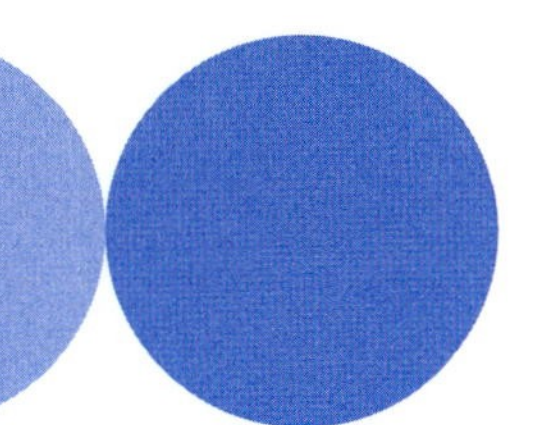

Publisher's acknowledgements

We are grateful to the following for permission to reproduce copyright material:

Figures

Figure 1.1 after *Handbook of Cultural Geography*, illustrated, abridged, Sage, London (Anderson, K., Domosh, M., Pile, S. and Thirft, N. (eds.) 2003) 1–35, A rough guide pp. 1–6, Reproduced by permission of SAGE Publications, London, Los Angeles, New Delhi and Singapore, Copyright (© 2003); Figure 3.1e after *Doing Cultural Studies: The Story of the Sony Walkman*, 1 ed, Sage, London (P. du Gay 1997) Introduction, Circuit of Culture p. 3, Reproduced by permission of SAGE Publications, London, Los Angels, New Delhi and Singapore, Copyright © 1997; Figure 4.1 from Building in wood in the eastern United States, *Geographical Review*, 56, pp. 40–66 (Kniffen, F. and Glassie, H. 1966); Figure 6.1 adapted from Lying with Maps, *Statistical Science* 20, pp. 215–222 (Monmonier, M. 2005); Figure 7.1 from An event of geographical ethics in spaces of affect, *Transactions of the Institute of British Geographers*, 28, pp. 488–507 (McCormack, D. 2003), p. 492, Reproduced with permission of John Wiley & Sons Ltd.; Figure 7.2 from An event of geographical ethics in spaces of affect, *Transactions of the Institute of British Geographers*, 29, pp. 488–507 (McCormack, D. 2003), p. 497; Figure 9.2 from Steps to an Ecology of Place, *Human Geography Today*, p. 303 (Thrift, N. 1999), Blackwell, Oxford, Permissions granted by Nigel Thrift.

Screenshots

Screenshot 10.6 from 1950s visitor map of London Zoo, Zoo Guide © Published in 1958 by The Zoological Society of London.

Tables

Table 2.1 from Creative Economy Report (UN 2008, 2011), United Nations, pp. 302–303; Table 2.2 from Creative Economy Report (UN 2008, 2011), United Nations, pp. 319–320; Table 2.3 from dePropris, L., Chapain, C., Cooke, P., MacNeill, S. and Mateos-Garcia, J. (2009) The Geography of Creativity, NESTA, London, p. 19., JCIS/ABI and ONS (2007); Table 8.1 adapted from Office for National Statistics, http://www.ons.gov.uk/about-statistics/classifications/archived/ethnic-interim/presenting-data/index.html, Contains public sector information licensed under the Open Government Licence (OGL) v1.0. http://www.nationalarchives.gov.uk/doc/open-government-licence/open-government licence.htm

Text

Poetry on page 138 from Eshun G. 01/09/2010, Student, Department of Geography, University of Leicester http://www2.le.ac.uk/departments/geography, courtesy of Dr Gabriel Eshun; Box 8.1 adapted from Australian Government, Department of Immigration and Citizenship.

Picture Credits

The publisher would like to thank the following for their kind permission to reproduce their photographs:

(Key: b-bottom; c-centre; l-left; r-right; t-top)

Alamy Images: Aardvark 238, Adrian Arbib 75, Alamy Celebrity 5br, Alex Segre 61, 251, Colin Palmer Photography 62, Corbis Super RF 174, Fancy 58, Guillem Lopez 167b, Ivy Close Images 112, Jeffrey Blacker 144, John Peter Photography 97, Kevin Britland 259b, Mike Dobel 96, Peter Greenhalgh (UKpix.com) 167t, Photo Alto 253, Picturesbyrob 59, Richard Baker 231, Robert Harding Picture Library 259tl, Travelscope 146; **Pat Bond**: 213; **Corbis**: Bettmann 92; **Adam Crowley**: 5tr; **Dr Faith Tucker**: 182, 186, 187, 193, 259tr; **Dr Sophie Hadfield-Hill**: 98, 113; **FLPA Images of Nature**: D P Wilson 217; **Getty Images**: 71, 242, AFP 257, 258, Alvis Upitis 152, Bongarts 149, Chen Hanquan 273, DEA Picture Library 134, Tim Graham 65; **Gwynedd Archaeological Trust**: 45; **John Frost Historical Newspapers**: 122; **John Horton**: 202, 206, 214, 235, 240; **Peter Kraftl**: 48t, 48b, 51, 103, 111, 239, 269; **Press Association Images**: PA Archive 106; **Rex Features**: Invicta Kent Media 74, Per Lindgren 227; **Shutterstock.com**: Alexander Raths 5tl, Claudio Zaccherini 185, Dainis Derics 282, Kjersti Joergensen 47, Lakov Filimonov 69, Pincasso 270, Radio Kafka 5bl, Vasaleki 246, Withgod 32; **Stoke Potteries Museum**: 218

Cover images: *Front:* **Alamy Images:** Adrien Veczan

In some instances we have been unable to trace the owners of copyright material, and we would appreciate any information that would enable us to do so.

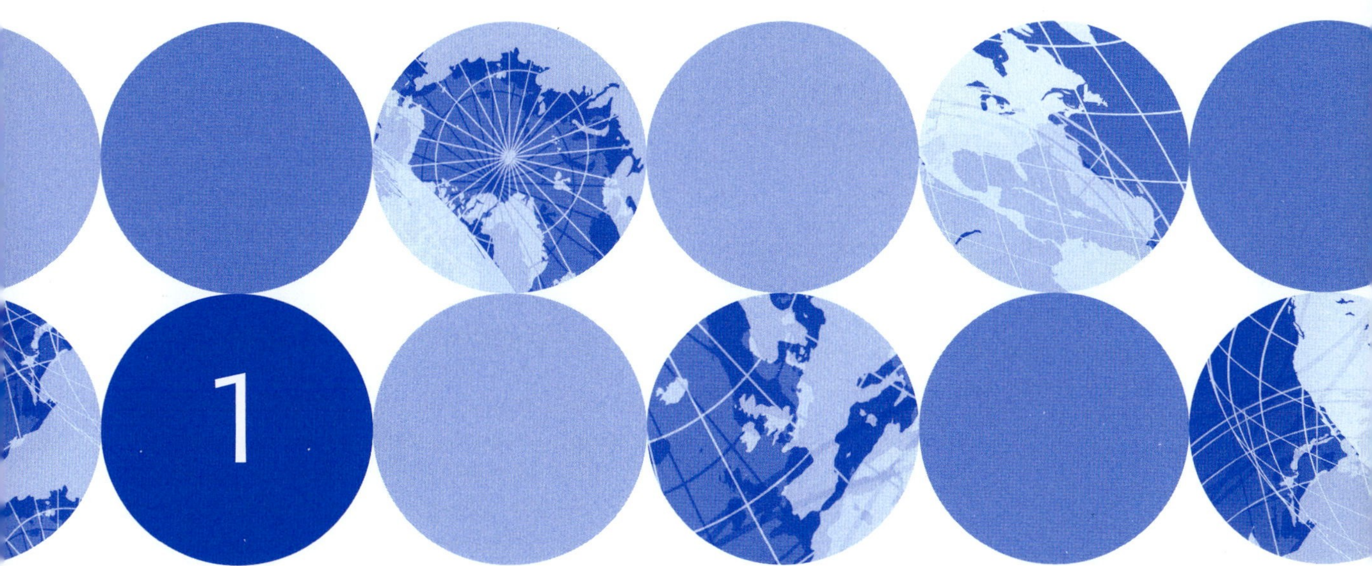

Introduction

Before you read this chapter . . .

Think about what you want to get out of this book.

- If you are looking for information about specific topics studied by cultural geographers read Sections 1.1 and 1.2, and then use Box 1.1 to explore the topics and case studies in this book.
- If and when you need more in-depth background information about the concept of culture, the cultural turn, and the history of cultural geography, read Sections 1.3 and 1.4.

Chapter Map

- 'Cultural geography': where to begin?
- Starting points: using this book
- Multiple meanings of 'culture'
- Multiple versions of 'cultural geography'
- Cultural geographies now

1.1 'Cultural geography': where to begin?

Cultural geography is a major, vibrant area of research and scholarship within the discipline of human geography. This book aims to provide an introduction to this remarkably diverse, exciting, fascinating, frequently controversial, sometimes infuriating body of work. This introductory chapter has the following sections.

- Section 1.2 provides a user's guide to this book, and makes some introductory propositions about how you should approach its later chapters, and indeed cultural geography in general. Start here if you would appreciate some direct, applied guidance on making use of this book and studying cultural geography.
- Section 1.3 introduces the concept of 'culture', and Section 1.4 introduces 'cultural geography' and provides a brief history of work by cultural geographers. Start with these sections if you require some background information about the rather complex history of the subdiscipline of cultural geography.

1.2 Starting points: using this book (or, we love cultural geography?)

So, we have been thinking for a long time about how to start this book. Seriously, writing a book is not an easy process. We have been trying to get straight what exactly it is that we want to tell you about cultural geography. And it comes down to this . . . We love cultural geography . . . at least sometimes . . . but it is hard to explain what it is.

So, there it is. Let us begin by explaining that sentence, one bit at a time.

We love cultural geography. Our starting point for writing this book is a conviction that cultural geographers have done some of the most important, exciting, interesting, refreshing, thought-provokingly zesty work in human geography over the last half-century. Like many human geographers, we have been fascinated and inspired, in all kinds of ways, by the work of cultural geographers. Our aim, in writing this book, is to convey a sense of why so many human geographers find the subdiscipline of cultural geography inspiring and interesting. After working through this book, perhaps you too will come to love cultural geography – at least some of it, at least a little.

Or at least we *sometimes* love cultural geography. As we outline in Section 1.4, there are some important reasons why many geographers are critical of much work in cultural geography. Quite frankly, many critics find cultural geography rather infuriating and problematic and, quite honestly, we have to concede that their critiques sometimes hit home. So, although we write from a position of loving cultural geography, throughout this book we also acknowledge and reflect upon critical responses to the work of cultural geographers. After working through this book, you may conclude that cultural geographers could do more, or do better. Indeed, this book is designed to help you develop skills of critical reflection and analysis that are crucial when reading and doing cultural geography.

Having said that we love cultural geography, we need to explain that *it is hard to explain what cultural geography is*, exactly. The term 'cultural geography' has meant – and continues to mean – quite different things to different people. As we explain in Sections 1.3 and 1.4, cultural geography is a complex, diffuse and fragmented subdiscipline. It comprises multiple traditions, and countless forms of research and scholarship about countless topics; it can seem unclear how or whether 'it' constitutes a coherent subdiscipline at all. This can make cultural geography seem inaccessible, especially for those trying to 'get into' it for the first time. In this book, rather than dwell on the complexity of defining cultural geography, we have tried to provide some 'ways in' to studying the subdiscipline, by focusing upon some key concepts, debates and topics studied by cultural geographers. The anxiety about how to define cultural geography is always there in the background (and laid out in some detail in Sections 1.3 and 1.4). However, in writing this book we have opted to foreground some things that cultural geographers do well – *what cultural geography is good for* or, again, the reasons why one might love cultural geography – rather than dwell upon definitional and conceptual anxieties (which are fascinating and important to us as academic human geographers, but perhaps less immediately useful to readers).

So, we love cultural geography . . . at least sometimes . . . but it is hard to explain what it is. We suggest that cultural geography is important, useful and inspiring – despite its critics, and despite uncertainty and anxiety over what 'it' is – for at least three big reasons.

- Cultural geography – in its various forms – presents us with a rich body of research and evidence-based theory about geographies of cultural practices and politics in diverse contexts.
- Cultural geography offers a major resource of concepts and in-depth research exploring the geographical importance of cultural materials, media, texts and representations in particular contexts.
- Cultural geography has been, and continues to be, one of the most theoretically adventurous, open-minded and avant-garde subdisciplines within human geography. Through the work of cultural geographers, many lines of social and cultural

theory have become important within the wider discipline of human geography.

We have used these three points as the basis for this book's structure. So the book has three core sections, as follows.

- *Part 1 – Cultural processes and politics.* This set of chapters explores the geographical significance of cultural processes and politics. We shall focus on the key concepts of **cultural production** (Chapter 2) and **cultural consumption** (Chapter 3), and explore how these processes are important to the production of cultural spaces, groups and identities. We shall also explain how cultural production and consumption are always interrelated (although this has sometimes been overlooked in academic research, which has tended to focus upon *either* cultural production *or* cultural consumption).
- *Part 2 – Several cultural geographies.* This set of chapters explores some examples of phenomena studied by cultural geographers. We shall introduce geographical work on **buildings** (Chapter 4), **landscapes** (Chapter 5), **texts** (Chapter 6), forms of **performance** (Chapter 7) and **identities** (Chapter 8). In the process, we shall consider the importance of cultural texts, media, representations and materials in everyday spaces.
- *Part 3 – Key concepts for cultural geographers.* This set of chapters introduces geographers' engagements with key lines of social and cultural theory. We focus upon the importance for cultural geographers of theories of **everydayness** (Chapter 9), **materiality** (Chapter 10), **emotion and affect** (Chapter 11), **bodies and embodiment** (Chapter 12) and **space and place** (Chapter 13).

Through these sections, we introduce diverse key topics, debates, concepts and case studies drawn from the work of researchers working in diverse contexts. As you read the chapters that follow, you will encounter the work of geographers, past and present, working within different traditions of cultural geography (Section 1.4), and working with different concepts of 'culture' (Section 1.3). You will also encounter many other geographers and social scientists whose work contributes to key areas of work in cultural geography, even though those individuals may not define themselves as 'cultural geographers' at all. The chapters that follow are intended to provide something of an introduction to the subdiscipline by focusing on a selection of important lines of work addressing some key issues. We hope that each chapter will offer you a 'way in' to each body of work, and equip you with some key concepts, understanding, vocabulary and skills. To support you as you read, every chapter contains the features and activities listed in Box 1.1.

1.3 Multiple meanings of 'culture'

Now, it would make all our lives easier if we could begin with a single, clear, simple, indisputable definition of 'cultural geography'. But it would be misleading and problematic to offer this kind of definition, for several reasons.

For starters, there is that word 'culture'. 'Culture' – and therefore 'cultural' – has multiple meanings, and is a notoriously difficult concept to pin down with any precision. It is one of those words that is very familiar, widely used and taken for granted, but actually quite difficult to define. Indeed, the cultural theorist Raymond Williams (1976: 87) famously describes 'culture' as

> one of the two or three most complicated words in the English language . . . partly because of its intricate historical development in several European languages, but mainly because it has now come to be used for important concepts in several distinct intellectual disciplines and in several distinct and incompatible systems of thought.

To illustrate the first of these points, Williams attempts to trace the historical development of the word 'culture' through various European literary and archival sources. He demonstrates that 'culture' has never had a single, stable definition; rather, the concept of 'culture' has constantly and complexly evolved from multiple linguistic roots over at least 1000 years, and has therefore meant quite different things to different people in different historical and geographical contexts. For example, he notes that, since the eighteenth century, at least four broad, often overlapping meanings of 'culture' have been in common usage,

Box 1.1
Features of this book

You will encounter the following features as you read this book. These features are designed to support your understanding, and to help you to engage with each chapter's key points.

- *Before you read this chapter* – Each chapter begins with a few questions or short activities. We would encourage you to take a few moments to consider these before launching into the chapter. They have been designed to give you a 'way in' to the chapter, to get you thinking about something relevant (often rooted in your everyday life).
- *Case studies* – Each chapter contains boxes that highlight key case studies, or provide a focused discussion of a key concept. The boxes are referred to in the text, but they could also be used as stand-alone, bite-sized examples. We have selected case studies that ground key concepts in everyday spaces in different parts of the world.
- *Making connections* – Many of the issues covered in this book are interrelated. Within each chapter we have used **bold** text to point you towards relevant sections of other chapters.
- *Chapter summary* – Each chapter concludes with a bullet-point summary of major issues covered in the preceding pages. Use this to check that you have 'got' each key point.
- *Some key readings* – At the end of each chapter, we direct you to a selection of key texts that will help you to enhance and extend your understanding of key issues. We provide a brief 'user's guide' to each reading, to help you choose which will be most useful to you.

simultaneously, in the English language. These four versions of 'culture' are illustrated in Photograph 1.1, and are described in the following bullet points.

Williams (1976) identifies four meanings of the word 'culture', as follows.

- Culture as a (somewhat obscure, archaic or technical) noun relating to the literal "tending of something, basically crops or animals" (p.87). As in: 'a bacterial culture' (micro-organisms stored and reproduced in a laboratory), 'a tissue culture' (animal or plant cells grown in laboratory conditions), 'a crop culture' (a system, space or set of technologies for maximising yield of a particular crop), and indeed 'agri*culture*' and '*culti*vation'.
- Culture as an idea or ideal describing "a general process of intellectual, spiritual and aesthetic development" (p.91). As in: 'she is very cultured', 'they are not very cultured', 'a cultured civilisation', 'a footballer with a cultured left foot'. Note that this version of 'culture' frequently involves making an implicit but problematic value judgement about who, what or where is 'cultured' (or *not*).
- Culture as a distinct grouping of people: "a particular way of life, whether of a people, a period, a group, or humanity in general" (p.91). As in: 'this is my culture', 'English culture', 'European culture', 'medieval culture', 'fan culture' 'organisational culture', 'the cultures of East Africa', 'indigenous cultures'. Note that this version of culture is frequently pluralised: there are many *cultures*. Note, too, that this version of 'culture' often entails implicit but problematic assumptions about who *belongs* (or does not belong) to which 'culture'.
- Culture as "the works and practices of intellectual and especially artistic activity" (p.91). As in: 'cultural objects', 'high culture', 'going to a gallery to get some culture', 'the Ministry of Culture', 'culture vulture'. Note that this version of 'culture' has often relied on implicit but problematic value judgements about what *kinds* of 'intellectual and artistic activity' are valued as 'culture', and which are not. Countless prefixes have been coined to denote different forms of 'culture' (and perhaps signal their difference from 'normal' culture): e.g. 'low culture', 'popular culture', 'local culture', 'folk culture', 'indigenous culture', 'subculture'.

Already, then, we have at least four definitions of 'culture', which suggests that 'cultural geography' could mean at least four different things, depending on which version of 'cultural' is being used at the time! And indeed it is possible to trace how, at different

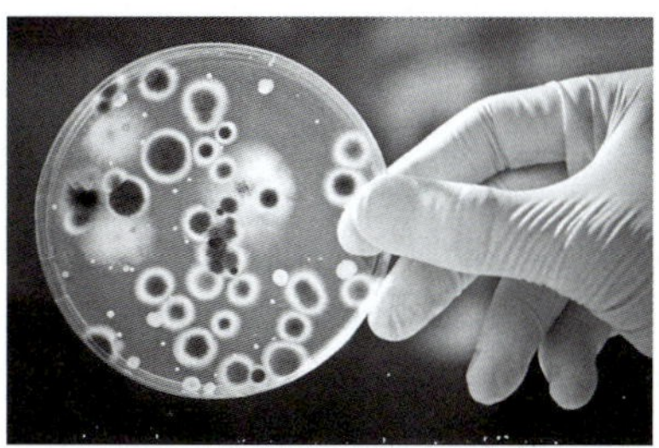

(a) A Petri dish containing a 'bacterial culture'

(b) The Athens Parthenon: an icon of 'cultured civilisation'

(c) England football fans: an expression of 'English culture'

(d) Ballet: an example of 'high culture'

Photograph 1.1 Four uses of the word 'culture'.

Source: (a) After Williams (1976: 87–93) Shutterstock.com/Alexander Raths; (b) Alamy Images/Adam Crowley; (c) Shutterstock.com/Radio Kaflea; (d) Alamy Images/Alamy Celebrity.

times, various groups of 'cultural geographers' *have* been specifically preoccupied with one or several of the above-listed notions of 'culture' (Cloke *et al.* 2004: 139–141). So already it should be clear that the apparently singular term 'cultural geography' actually encompasses multiple lines of geographical work relating to multiple notions of culture.

Let us return to Williams' assertion that 'culture' is "one of the two or three most complicated words in the English language": in the latter part of that quotation, he notes that the (already complex) notion of 'culture' has been further complicated by its usage in diverse "intellectual disciplines and . . . systems of thought". To briefly illustrate this point, Williams notes how academics from two ostensibly similar academic disciplines used the word 'culture' as a specific technical term within their work during the 1970s. Thus, for European archaeologists, 'studying culture' referred to recording the material artefacts of a specific archaeological site, whereas for European historians (perhaps working down the corridor at the same university), 'studying culture' referred to interpreting the historical significance of symbols, artworks and images. In each case, the term 'culture' had been adopted within an academic discipline to mean something very specific. As well as differing from each other, these meanings differed from (or at least represented a very specific, subtle twist upon) everyday uses of the word 'culture'.

As Williams' examples suggest, there is no single, unified academic definition or theory of 'culture'. On the contrary, there are multiple working definitions and concepts of 'culture', which can refer to very specific aspects of particular academic disciplines, in particular geographical and historical contexts. To give one estimate, Crang (1998: 2) reckons that more than 150 different, specific definitions of 'culture' can be found in academic literature published during the 1950s. Confronted with so many versions of 'culture' (some no doubt closely interrelated, others profoundly different), we might wonder: what exactly *are* cultural geographers talking about when they talk about 'culture'?

Anderson *et al.* (2003a) provide a significant and useful reflection upon this question. Surveying the major topics and debates that have been central to cultural geography over the last century, they suggest that the term 'culture' has been used by geographers to refer to some diverse, specific 'geographical problems'. That is, they suggest, cultural geography has never been a singular endeavour dealing with a singular concept of 'culture'; rather, cultural geography has encompassed multiple, disparate, 'far-flung' forms of research and scholarship, relating to multiple aspects and forms of 'culture'. To illustrate this multiplicity, they identify five understandings of 'culture' that can be found in the work of 'cultural geographers' past and present. That is, Anderson *et al.* (2003a: 3–6) argue that a geographer whose work focuses on 'culture' may actually be interested in one or more of the following things:

- Culture as distribution of things: how "cultural artefacts, from . . . everyday personal items . . . like furniture and clothing, to the larger-scale and more public artefacts such as buildings and roads . . . [relate to] the values, livelihoods, beliefs and identities of the cultures who have produced them . . . [and their] social, economic and political dynamics."
- *Culture as a way of life*: "the assortment of practices that constitute people and place, life and landscape . . . [t]he values, beliefs, languages, meanings and practices that make up people's 'ways of life'".
- *Culture as meaning*: "how and why landscapes become embedded with individual and cultural meaning and in turn create new meanings".
- *Culture as doing*: "attempting to reinstate the richness of the world which is so often rendered in conventional academic accounts as 'just' everyday life", for example by exploring practices such as performance, protest and play.
- *Culture as power*: "the ways in which space, place and nature are implicated in – and constitutive of – unjust, unequal and uneven power relations" in all manner of cultural contexts.

Or, to put it another way, five different sets of questions could be asked about any given cultural object – such as the textbook you are reading (see Figure 1.1).

Anderson *et al.* (2003a) thus pinpoint five specific, distinctive notions of 'culture' that can be found in the work of cultural geographers; and, indeed, you will see these concepts recur in this book's many case studies. Moreover, they emphasise that cultural geography should be understood as an open, living, constantly unfolding intellectual terrain, in which new questions relating to the above-listed concepts of 'culture' (and new concepts of 'culture') constantly emerge and interrelate.

1.4 Multiple versions of 'cultural geography'

So, to recap, we have noted that: (i) the notion of 'culture' is notoriously complex; and (ii) multiple, quite specific and distinctive, notions of 'culture' have been central to the work of cultural geographers. To further complicate matters, it is important to recognise that the term 'cultural geography' itself has meant – and continues to mean – some quite different things to different people. Consider Box 1.2, for example: here we have collated some definitions of 'cultural geography', written in different contexts. Look at these diverse definitions before reading on. And note that we are not trying to present any of these definitions as definitively right or wrong; rather, we want to give an impression of how the term 'cultural geography' has multiple meanings, and has been continually (re) defined and contested.

The quotations juxtaposed in Box 1.2 should suggest several characteristics of cultural geography. First, it should be clear that cultural geography has meant (and continues to mean) different things to different people: for example, the quotation by Sauer is referring to something quite different from the quotation by McDowell. As we outline in this section, there is no single cultural geography; rather, we should think in terms of multiple coexisting, overlapping versions or traditions of research in cultural geography. Second, you may note that several of the quotations critically reflect upon older or other

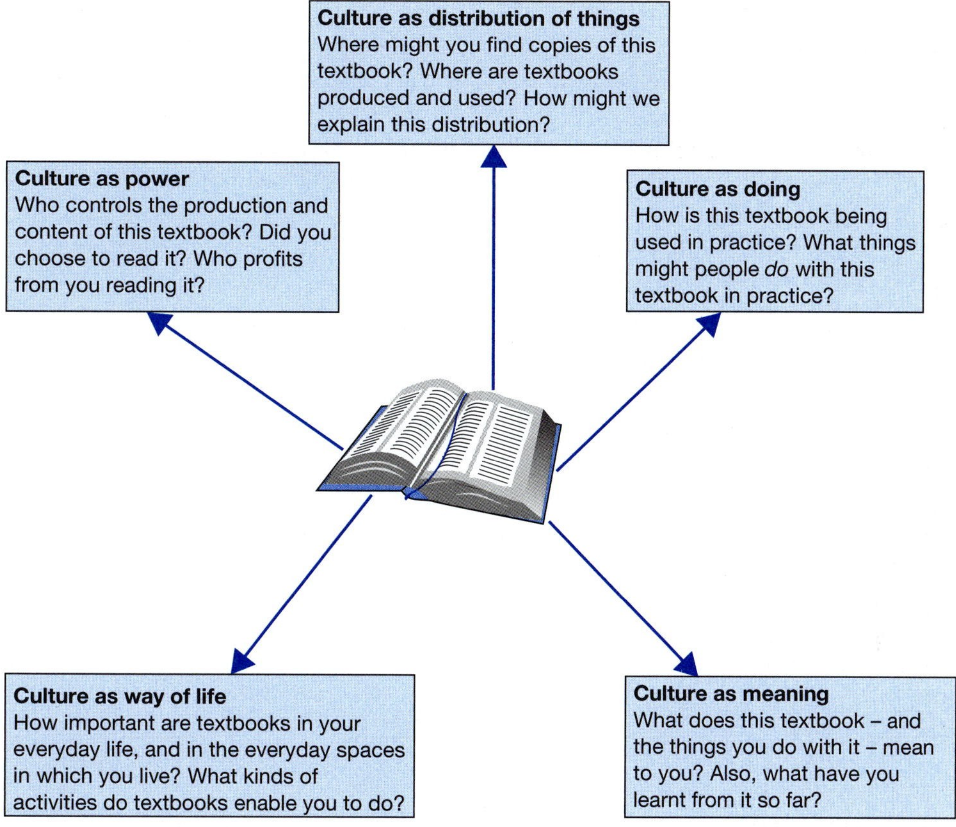

Figure 1.1 Five sets of questions we could ask about this textbook.

Source: After *Handbook of Cultural Geography*, 1, Sage, London (Anderson, K., Domosh, M., Pile, S. and Thrift, N. (eds) 2002) pp 1–6, Dec 24, 2002.

work in cultural geography, and implicitly call for cultural geographers to change or develop their work accordingly: for example, see Jackson's satisfaction that cultural geographers have begun to overcome an 'aversion to theory'. As we outline in the following pages, cultural geography has developed and changed over time through a process of internal critique and contestation. Indeed, third, Shurmer-Smith and Hannam's sense of cultural geography as 'very much up for grabs' suggests a subdiscipline that is – perhaps excitingly, perhaps dauntingly – undergoing constant change and renewal. Fourth, (GIS expert) Openshaw's forceful comments illustrate how cultural geography has frequently been controversial and much maligned by other geographers, not least for its perceived 'squelchy soft' research and 'second-rate philosophical thinking'. Fifth, it should be clear from the quotations by Mitchell and Lanegran that cultural geography is now characterised by its remarkably and dauntingly broad agenda (which may, indeed, be 'all over the place').

To get to grips with the remarkably diverse array of work encompassed by cultural geography, it is useful to have some understanding of several key traditions of geographical research and scholarship. Looking back over the last century, it is possible to identify six significant, but very diverse, 'versions' of cultural geography, four of which remain influential for many (different groups of) cultural geographers in the twenty-first century. These six versions of cultural geography are shown in the timeline in Figure 1.2, and are outlined in the following subsections.

Box 1.2
'Cultural geography is . . .'

Consider the following definitions of 'cultural geography', written in different places, at different times. Try to relate these quotations to the six versions of cultural geography described in the following pages.

> Cultural geography is . . . concerned with those works of man that are inscribed onto the earth's surface and give to it characteristic expression.
>
> (Sauer 1962[1931]: 32)

> Cultural geography has begun to assume a more central position in the current rethinking of human geography. Cultural geographers are now experimenting with a range of new ideas and approaches, their aversion to theory now firmly overcome.
>
> (Jackson 1989: 1)

> Cultural geography is one of the most exciting areas of geographical work at the moment. Ranging from analyses of everyday objects, views of nature in art or film, to studies of the meaning of landscapes and the social construction of place-based identities, it covers numerous issues.
>
> (McDowell 1994:146)

> Cultural geography is currently in a very exciting stage and many people believe it is at the cutting edge not only of human geography, but the way that all geography is constituted . . . [C]ultural geography is very much up for grabs at the moment . . . [and] has only begun to scratch the surface of the possibilities available to it.
>
> (Shurmer-Smith and Hannam 1994: 215–216)

> [Cultural geography is an example of] "unstable theoretical speculation and descriptive storytelling . . . squelchy soft qualitative research . . . second-rate philosophical thinking" (Openshaw 1991: 623); "the Catherine Cookson approach" to research
>
> (Openshaw 1996: 761).

> It could now be said that cultural geography is *the* agenda, but it is an agenda without any clear focus. Cultural geography . . . rushed off in a million new directions . . . [It] is now simply 'all over the place'.
>
> (Mitchell 2000: xiv)

> [We] cannot convey the depth and scope of research and writing on cultural geography . . . Indeed, some writers assert that recent developments in cultural geography imply that it will soon encompass *all* of human geography.
>
> (Lanegran 2007: 181)

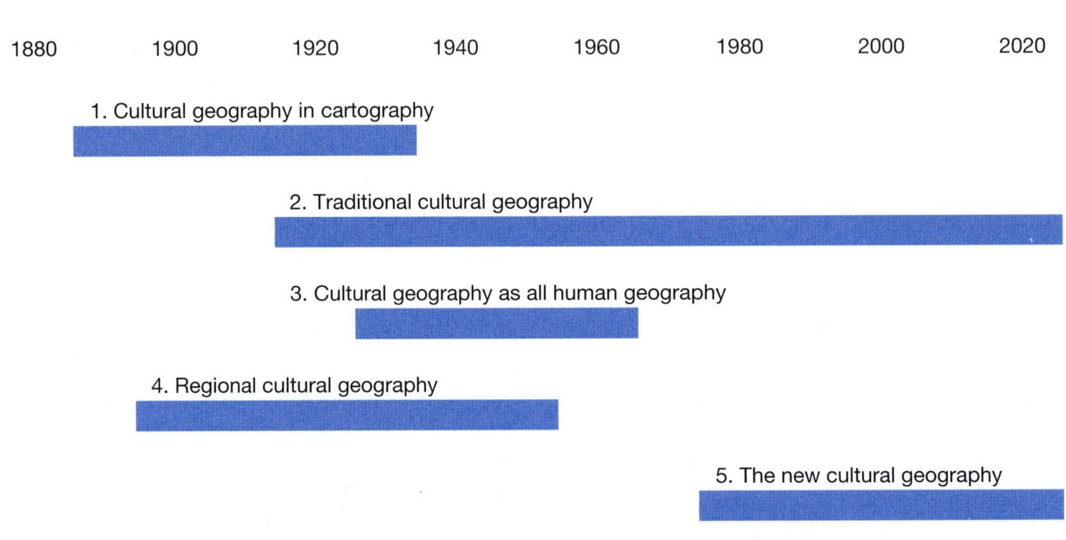

Figure 1.2 Different versions of 'cultural geography': an approximate timeline.

Version 1: Cultural geography in cartography (North America, c.1890–1930)

In North American military cartography, and topographic maps produced by the United States Geological Survey during the first decades of the twentieth century, 'cultural geography' was used as a specific technical term to describe the layer of symbols "designating the 'works of man' . . . in contrast with the land and water forms of nature. On these maps, 'culture' is printed in black – as compared with nature, represented in brown for relief . . . and blue for water features" (Platt 1962: 35). Note that we *shall not* be using the term 'cultural geography' in this sense in this book.

Version 2: Traditional cultural geography (North America 1925–present)

Key readings: Sauer (1925), Wagner and Mikesell (1962), Denevan and Mathewson (2009)

This major field of human geography in North America was pioneered by Carl Sauer (1889–1975; arguably the most influential American geographer of the twentieth century) and his colleagues and students at the University of California at Berkeley (the so-called 'Berkeley School' of geographers). These cultural geographers sought to map "the impress of the works of man [sic]" (Sauer 1925: 38) upon the landscape, and thus chronicle "the capacity of man [sic] to alter his natural environment" (Sauer 1956: 49). Five key 'building blocks' of this version of cultural geography are outlined in Figure 1.3, and are explored in the following discussion.

As Figure 1.3 suggests, traditional cultural geography can be understood as a particular approach to mapping the impact of human activity on landscapes. Sauer's most famous essay, *The Morphology of Landscape* (1925), presented a kind of conceptual and methodological template for cultural geographers, which was influential in shaping the direction and concerns of many (especially North American) cultural geographers for at least six decades. In the most widely cited passage from *The Morphology of Landscape*, Sauer (1925: 46) argued that cultural geography should involve mapping how:

> the cultural landscape is fashioned out of a natural landscape by a culture group. Culture is the agent, the natural area is the medium, the cultural landscape is the result. Under the influence of a given culture, itself changing through time, the landscape undergoes development, passing through phases, and probably reaching ultimately the end of its cycle of development. With the introduction of another, that is to say alien, culture, a rejuvenation of the natural landscape sets in, or a new landscape is superimposed on remnants of an older one. The natural landscape is of course of fundamental importance, for it supplies the raw materials out of which the cultural landscape is formed. The shaping force, however, lies in the culture itself.

Thus this form of cultural geography attempted to record how the 'natural landscape' (as shaped by geology, climate, hydrology, chemistry, weathering, erosion, vegetation, etc.) has been altered by different 'cultures', resulting in diverse, successive 'cultural landscapes' as

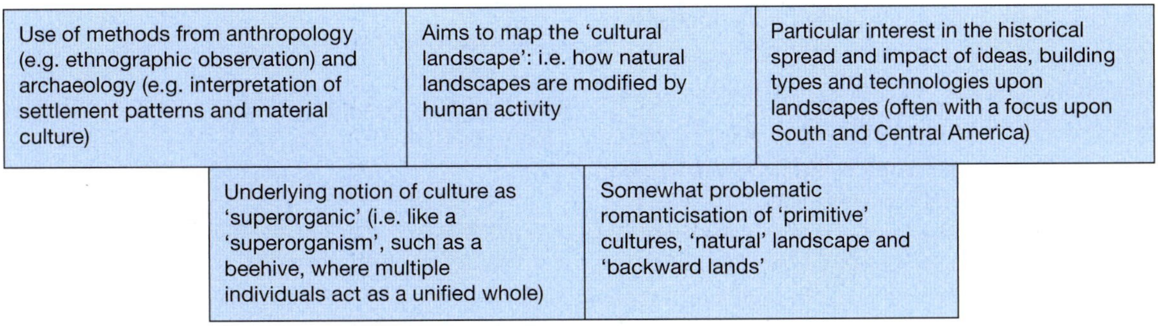

Use of methods from anthropology (e.g. ethnographic observation) and archaeology (e.g. interpretation of settlement patterns and material culture)	Aims to map the 'cultural landscape': i.e. how natural landscapes are modified by human activity	Particular interest in the historical spread and impact of ideas, building types and technologies upon landscapes (often with a focus upon South and Central America)
Underlying notion of culture as 'superorganic' (i.e. like a 'superorganism', such as a beehive, where multiple individuals act as a unified whole)	Somewhat problematic romanticisation of 'primitive' cultures, 'natural' landscape and 'backward lands'	

Figure 1.3 Five building blocks of traditional cultural geography.

> the works of man [sic] express themselves in the cultural landscape. There may be a succession of these landscapes with a succession of cultures. They are derived in each case from the natural landscape, man expressing his place in nature as a distinct agent of modification.
>
> (p.43)

In practice, Sauer pioneered and advocated the use of methods derived from anthropology (e.g. ethnographic observation of 'traditional' and 'vernacular' cultural practices, and interpretation of objects of material culture) and archaeology (e.g. the interpretation of settlement patterns and traces of human activity in landscape morphology) to interpret the development of the 'cultural landscape' in particular places (Sauer 1925). Sauer and his colleagues, students and followers developed these methods through numerous studies of the transition from 'natural' to 'cultural' landscapes in diverse contexts, often with a focus on Latin America prior to European colonisation. For example, members of the Berkeley School investigated the impact of building types, settlement creation and population movements (Sauer, 1934), and the spread of certain ideas (e.g. the use of fire; see Sauer 1950) and technological innovations (e.g. relating to plant cultivation and animal domestication; see Sauer 1965), upon landscapes in particular historical contexts. In so doing, they developed several concepts that remain influential, and still inform the work of geographers. These include the notions of 'cultural hearths' (places where particular ideas, technologies, norms or forms of artefact originate), 'cultural diffusion' (how artefacts, ideas, technologies and norms spread and are transmitted from their cultural hearth, and then from place to place over time) and 'culture areas' or 'culture regions' (distinctive areas that share an identifiable 'culture' at a given point in time). These investigations and concepts tended to be united by a particular notion of culture as a kind of 'superorganic thing' in itself, neatly outlined by one of Sauer's protégés in the following quotation:

> We are describing a culture, not the individuals who participate in it. Obviously, a culture cannot exist without bodies and minds to flesh it out; but culture is also something both of and beyond the participating members. Its totality is palpably greater than the sum of its parts, for it is superorganic [i.e. like a 'superorganism', such as an ants' nest or beehive, where a collective entity is like a 'thing' in itself] and supra individual [i.e. more than a group of individuals, and bigger than any individual] in nature, and entity with a structure, set of processes and momentum of its own.
>
> (Zelinsky 1973: 40–41)

Although nearly a century old, Sauer's description of cultural landscapes remains elegant and thought provoking, and the Berkeley School's investigations remain fascinating in their detailed theorisation of historical geographical patterns. Indeed, this form of cultural geography remains significant for many (especially North American) geographers. Case studies of work from this form of cultural geography appear throughout this book: in particular, see Chapters 4 and 5. However, it should be noted that this form of cultural geography has latterly been subject to significant critique, not least because of the 'superorganic' concept of culture outlined in the preceding quotation. See version 5 below for a discussion of this critique.

Version 3: 'Cultural geography' as a term for all human geography (North America, c.1930–70)

Such was the dominance of 'traditional' cultural geography in North America that it became somewhat common to use 'cultural geography' to refer to *all* human geography. 'Cultural geography' was quite often used in this sense in North American textbooks and course and university department/faculty names during the middle decades of the twentieth century. We *shall not* be using the term 'cultural geography' in this sense in this book; but note that some sources (especially non-academic and online texts) do retain this usage.

Version 4: Regional cultural geography (Continental Europe, c.1900–60)

Key readings: Vidal de la Blache (1908, 1926), Sorre (1962[1948])

A range of work by European regional historians and geographers during the first half of the twentieth century is often described as a key, early form of cultural geography. Of particular note was the work of Paul Vidal de la Blache (1845–1918), the so-called 'father of French geography', and his students and colleagues in Parisian universities. Their work focused on the complex, intimate, mutual relationships that exist between *paysan, paysage et pays*: that is, between people (or 'peasants' or 'folk'), landscapes and regions (Vidal de la Blache 1913; Gregory 2009). Five key building blocks of this version of cultural geography are outlined in Figure 1.4, and are explored in the following discussion.

As Figure 1.4 suggests, regional cultural geography was a set of approaches concerned with exploring connections between landscapes, regions and the lifestyles of folk living within them. Vidal de la Blache (1908: 6) observed that *paysan, paysage et pays* were invariably complexly and intimately "intermingled", "combined" and "networked" in mutual, ongoing relationships, such that "man [sic] modifies and humanises the landscape through his works, but he is also shaped by the landscape, and so becomes part of that landscape." Although Vidal de la Blache did not tend to refer to himself as a 'cultural geographer', his work is now celebrated for its relatively sophisticated understanding of interactions between cultural groups and landscapes – an understanding that was arguably more sophisticated than the Berkeley School's notion of 'cultural landscape'. For:

> [a]ccording to Vidal, it is unreasonable to draw boundaries between natural and cultural phenomena; they should be regarded as united and inseparable . . . the animal and plant life of France during the nineteenth century, for example, was quite different from what it would have been had the country not been inhabited for centuries by human beings . . . Each community adjusts to prevailing natural conditions in its own way, and the result of the adjustment may reflect centuries of development. Each single small community, therefore, has characteristics that will not be found in other places, even in places where the natural conditions are practically the same. In the course of time, humanity and nature adapt to each other like a snail to its shell. In fact, the relationship . . . becomes so intimate that it is not possible to distinguish the influence of humanity on nature from that of nature on humanity.
>
> (Holt-Jensen 2009: 68)

Vidal de la Blache used the term *un genre de vie* ('a way of life') to describe the set of habitual behaviours, cultural forms and social interactions that characterise a given cultural group in/of a particular landscape. For Vidal de la Blache and his followers, a *genre de vie* had three key characteristics. First, any *genre de vie* consists of a complex and heterogeneous array of interrelated features, encompassing "individual behaviour, determined by personality, social status and professional mores, and the habits of groups" *and* the specific features and "atmospheres" of local and regional landscapes *and* "visible material elements . . . non-material or spiritual elements, and . . . of course . . . social elements" (Sorre 1962[1948]: 411, 400). Second, each *genre de vie* has unique, local features: as Holt-Jensen notes in the quotation above, each has distinctive characteristics that will not be found elsewhere, even in places that appear to be very similar. Moreover, *genres de vie* are not permanent, but subject to change. For:

> as assemblages of techniques, genres de vie are active expressions of the adaptations of human groups to their natural settings. The specialisations and stability of this adaptation depend on the specialisation, stability and durability of the genre de vie, and local changes in the former are expressed in variations in the latter.
>
> (Sorre 1962[1948]: 401)

Use of archaeological and archival research methods to 'discover' the lives and geographies of 'ordinary' people and cultures	Aim to understand connections between *paysan, paysage* and *pays*: i.e. between people, landscapes and regions	Particular interest in the geographies and lifestyles of preindustrial peasant and agricultural communities (often with a focus upon French regions)
Notion of *genre de vie* ('way of life'): the habitual lifestyles and cultural practices that characterise a particular group in their landscape	Somewhat problematic romanticisation of peasants, folk culture and 'humble', 'authentic' expressions of regional identity	

Figure 1.4 Five building blocks of regional cultural geography.

Third, the importance of *genres de vie* had, problematically, been overlooked by previous European geographers and historians. For, Vidal de la Blache (1908: 4) argued,

> [t]he geographers of the past were concerned to explain the position of the largest cities, ports and centres of government. It did not even cross their minds to focus their attention on the villages, or more humble forms of groupings.

In practice, Vidal de la Blache advocated research focusing on geographically bounded areas with distinctive *genres de vie*. His own work attempted to explore *genres de vie* in pre-industrial peasant and agricultural communities in different regions of France, and trace the impacts of industrialisation and emergent communications technologies upon these distinctive cultures and landscapes. His work should be understood as a pioneering example of a broader turn, among many contemporary continental European historians and geographers, to investigate local and regional histories and geographies of 'ordinary' people and cultures in different regions of Europe. Of particular note was the so-called '*Annales* School' of historians, named after the foundation of the journal *Annales d'Histoire Économique et Sociale* in 1929. As Holt-Jensen (2009: 67) describes, the *Annales* School aimed to

> reform the discipline of history by focusing on the history of ordinary people through such means as archaeological and other physical evidence and private archives, rather than through the traditional way of recounting history based on a study of written sources that stressed political events, military confrontations and the lives of kings, emperors and ministers. The 'new' history of the *Annales* School sought to discover the fate of the lower classes, the ways in which they were organised, and the techniques they used in farming or producing tools and other artefacts.

The work of Vidal de la Blache and the *Annales* School prompted a substantial body of geographical and historical research exploring the distinctive geographies and *genres de vie* of diverse regions of Europe, including French cities, Belgium, Poland and the Mediterranean (Braudel and Reynolds 1975[1948]). Much of this work remains influential for (mainly European) geographers and historians. Chapters 4 and 5 include some further discussion of research from this context.

Version 5: The 'new' cultural geography (UK, c.1985 - present)

Key readings: Jackson (1989), Philo (1991a), McDowell (1994)

The term 'new cultural geography' was coined by British human geographers to describe how their research had developed through, and as a consequence of, the so-called 'cultural turn' during the 1980s and 1990s. Figure 1.5 identifies some key components of this cultural turn. For a fuller discussion of the cultural turn, see Box 1.3; alternatively, read on for an introduction to the 'new' cultural geography.

The notion of a cultural turn – as laid out in Figure 1.5 and Box 1.3 – can feel a little abstract and obscure, so let us consider how the new cultural geography actually unfolded. In practice, a series of debates and conferences at British universities between 1988 and 1991, critically reflecting upon the state and future of geographical research into social issues, are often identified as central to the cultural turn in geography, and the development of the new cultural geography (see Philo 1991a, 2000). These debates explored the geographical importance of "'the cultural' – culture as the 'glue' of beliefs, values and ideals present in the collective heads and hearts of human communities" (Philo 1991b: 1). Here, it was argued that understanding of this complex, cultural 'glue' had been strangely absent from the work of human geographers, even those who were supposedly 'cultural geographers' (in the traditional sense outlined in version 2 above). The 'new cultural geography' is thus used to describe the work of (initially predominantly British) geographers who attempted to understand the importance of 'the cultural' in relation to geographical issues, in this context. Key building blocks of this version of cultural geography are outlined in Figure 1.6.

New cultural geographers were sometimes treated with suspicion by other geographers. They were caricatured as 'wishy-washy' practitioners with unimportant research interests, whose work directed attention away from really important issues; indeed, this caricature of cultural geographers still persists, to some extent. However, the scale and impacts of the new cultural

Critique of quantitative research methods, which produced limited understandings of individuals' everyday experiences and geographies

Critiques of Marxian theories of society, which tended to treat culture as insignificant

Many geographers sought new ideas and methods, and began to explore work in cultural studies, and the work of feminist, postmodern and postcolonial theorists

Many geographers inspired by, and active in, politicised social movements; anger at conservative social-political context of Thatcher–Reagan eras in UK/USA

A need to make sense of major social and cultural changes in capitalist economies

Figure 1.5 Five key components of the 'cultural turn'.

Box 1.3
The cultural turn

The cultural turn is a:

> catch-all description given to a set of intellectual shifts which took place from the mid-1980s. These were not confined to geography, but were felt across the humanities and the social sciences, and even influenced debate within the natural sciences.
>
> (Goodwin 2004: 69–70)

The phrase 'cultural turn' suggests a neat, single, clear-cut change of direction (i.e. *a* turn towards culture), but this is misleading; 'it' actually comprised a complex "series of sometimes interwoven strands" of activity, where many academics became more sensitised to previously overlooked issues and questions (Johnston 1997: 271). Some key strands of the 'cultural turn' were as follows.

- During the 1980s and 1990s, many social scientists began to investigate some apparently new, significant social and cultural changes in contemporary capitalist economies. For example, they noted: the fracturing and decline of traditional forms of identity (e.g. relating to religion, class or union membership) because of ongoing processes of economic and political restructuring; the growing economic importance of the cultural, creative and entertainment industries; the growing importance of popular, mass-produced, mass-marketed and electronic media and commodities in people's everyday lives; the increasing prevalence of globalised brands and commodities; and the emergence of new identities and lifestyles as a result of the preceding trends, including the formation of new forms of politicised activism that brought together very diverse people who sought to resist, critique and subvert these trends (Jackson 1989; Anderson and Gale 1999). It was argued that new lines of research, and new concepts of identity, lifestyle and 'cultural politics', were needed to make sense of all this.
- From the late 1960s onwards, an increasing number of academics were active in, or inspired by, politicised social movements such as women's rights, gay rights and anti-racist campaigns. Individually and collectively, these activists drew attention to the role of 'cultural norms' and media representations in perpetuating social exclusion and marginalising 'alternative' viewpoints from the cultural 'mainstream'. Arguably, anger about a "depressing political climate" in the UK and USA during the 1980s (characterised by the right-wing Thatcher and Reagan administrations, respectively)

Box 1.3 continued

fuelled this critique: for, in this context, conservative cultural norms were especially common in the mainstream media, and feminist, gay rights and antiracist campaigners were considerably antagonised and marginalised within many sections of contemporary popular culture (Jackson 1989: 7–8).

- In academic disciplines such as sociology and human geography, there was growing criticism of Marxian theories of society, which had been very influential during the 1970s. In short, Marxian theories were criticised for treating culture as insignificant, paying little attention to the actions and agency of individuals within societies, and focusing on social class to the exclusion of other forms of social difference. It was argued that Marxian theories of society needed to be expanded and refreshed with (or, for some, superseded by) new concepts, better able to address these problems. Moreover, critics of Marxian approaches called for new lines of sociological and geographical research to begin to move on from Marxian scholars' arguably limited exploration of the importance of culture in contemporary societies.
- During the 1960s and 1970s, human geography and other, similar academic disciplines were dominated by quantitative approaches to research. That is, research questions were often approached via scientific hypothesis testing, statistical techniques and computer-aided analyses of large datasets. Subsequently, a succession of critics argued that this kind of approach often lacked the capacity to understand individuals' motivations, and the complex details of everyday geographies. Different groups of these critics sought to develop new methods and approaches to research that would afford more in-depth, qualitative understandings of geographical issues.
- In the context of these debates and historical shifts, many social scientists sought new ideas and methods to develop new approaches to research, and move on from the conceptual and methodological problems listed above. Frequently, this entailed exploring concepts and literature from other disciplines. So, for example, the cultural turn in geography can be understood as a period when many human geographers began to engage with an exciting profusion of ideas beyond the traditional boundaries of their subject (e.g. exploring ideas from cultural studies, landscape studies and literary theory became very significant for many human geographers).

geography were ultimately substantial (reflecting the far-reaching impacts, and multiple 'strands', of the cultural turn as outlined in Box 1.3). This body of work established cultural geography as a major, reinvigorated, widely recognised subdiscipline of human geography in the UK and elsewhere. To give one example, in 1988 the Institute of British Geographers (the professional organisation for academic geographers in the UK) formally changed the name of its Social Geography Study Group to the Social *and Cultural* Geography Study Group. As Philo (2000: 28) notes, this change reflected a significant 'critical mass' of interest in 'the cultural' among many British geographers, inspired by "the remarkably rich arc of the cultural turn". For example, just after the change of title, Cosgrove (in Philo 1991b: 1) wrote:

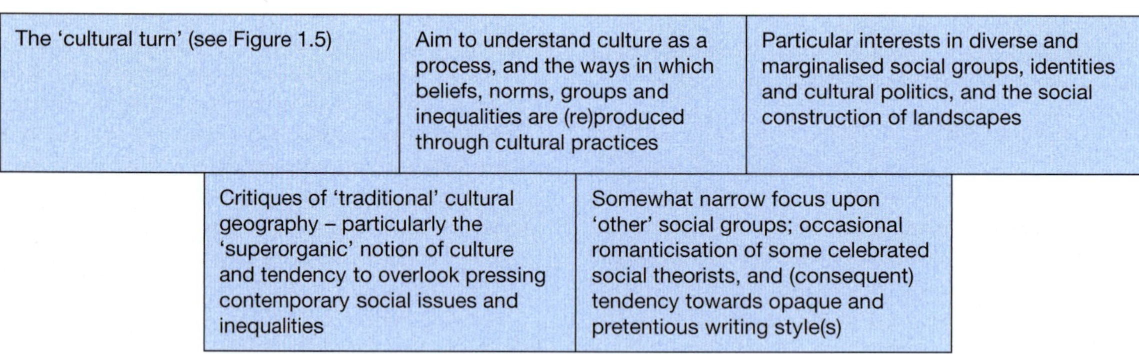

Figure 1.6 Five building blocks of the new cultural geography.

> the change in title . . . is an entirely welcome event for someone like myself who has always believed that human geography should celebrate the cultural diversity of our world and pay attention to the ways in which human beliefs, values and ideals continuously shape its landscapes. It is a change which signals a profound and, in some respects, an overdue change in geographical philosophy and methodology . . . [This demands geographical research with] a sense of history, and of human agency as the collective expression of real people with imagination, ideas, desires, beliefs and values – all those characteristics that are captured within that powerful but thankfully elusive term, 'culture'.

Let us focus on some of the key points from this quotation. The quotation maps out the following aims and concerns, which were central to the work of many new cultural geographers:

- a call for research exploring 'beliefs, values and ideals': i.e. the complex, emotional, qualitative 'glue' of communities described by Philo in an earlier quotation;
- a need to 'celebrate . . . cultural diversity': that is, to acknowledge the complex array of cultural groups, activities and identities that may coexist in a given space, and to resist and challenge the marginalisation of any of them;
- a desire to highlight the importance of practices, experiences and 'imagination, ideas, desires, beliefs and values' of individual 'real people' in particular geographical contexts;
- an understanding that these cultural practices 'continuously shape . . . landscapes' – that is, that cultural practices are constantly ongoing, and can be explored by investigating the ways in which landscapes are represented, mapped, understood and experienced;
- an acknowledgement that 'culture' is always a complex, contested concept, and hence it is a 'thankfully elusive term' as – again – we should celebrate the diversity of 'cultures' and understandings of 'culture' that exist;
- recognition that new methods and concepts were required to investigate 'culture', given all of the above.

Here and now, these understandings of culture may seem rather unproblematic and commonsensical, but there and then they were intended to present a somewhat radical challenge to many geographers' prevailing assumptions. Specifically, many new cultural geographers saw their work as profoundly challenging the assumptions, concepts and working practices of traditional cultural geography (see version 2). They thus called for "a new cultural geography deliberately removed from (and in many cases in opposition to) the preoccupations of Sauer and the Berkeley School" (Gregory 1994: 133). Jackson (1989: 9), for example, argued that [traditional] "cultural geography is in urgent need of reappraisal; its conception of culture is badly outdated and its interest in the physical expression of landscape is unnecessarily limited." Indeed, like many new cultural geographers, Jackson (1989:1) was strongly critical of the limited, "conservative" and "antiquarian" focus of the Berkeley School, arguing that their approach to culture – "limited to the interpretation of historical, rural and relict landscapes, and to a static mapping of the distribution of culture traits, from barns and cabins to field systems and graveyards" – was obsolete and basically irrelevant to pressing contemporary cultural issues.

Many new cultural geographers were especially critical of the 'superorganic' notion of culture (see version 2), which, they argued, had come to be taken for granted by many geographers via the influential work of the Berkeley School. It was argued that this had produced an approach to studying cultural geography that:

> (1) sees culture as a given, rather than a process; (2) . . . sees culture as externally causal . . . in terms of personality, art, economics or politics; (3) . . . postulates a spurious holism of culture which transcends the actions, interactions, interests, passions and interpretations of individuals, who are seen as the creations and carriers of their culture, not the creators of it; and (4) [has proceeded] . . . in the direction of description rather than interpretation.
> (Shurmer-Smith and Hannam 1994: 5)

Let us take each of these points in turn, and in so doing reiterate some key concerns of the new cultural geography. So, first, it was argued that traditional cultural geographers had imagined culture to be a given 'thing' in itself, overlooking the complex, continuous and dynamic processes and practices that constitute particular cultures. Thus new cultural geographers called for an understanding of

> culture as a *process* in which people are actively engaged . . . a dynamic mix of symbols, beliefs, languages and practices that people create, not a fixed thing or entity governing humans. [Culture] is not some evolutionary inheritance as it was for Sauer.
>
> (Anderson 1999: 4)

Second, moreover, it was argued that traditional cultural geographers understood culture as shaping individuals and social, historical and political contexts, but rarely vice versa. That is, they arguably often overlooked how cultures, cultural norms (and the meaning of 'culture' itself) may be shaped by individuals, and by diverse social factors, and by complex forms of cultural politics. Thus new cultural geographers sought to redress this imbalance by exploring the complex politics of cultures in practice, and the complex factors and forms of agency, power and conflict that shape these politics, in diverse contexts. Third, it was argued that traditional cultural geography often conceived of a 'culture' being essentially consistent over large landscapes or regions – irrespective of the 'actions, interactions, interests passions and interpretations of individuals' occupying such spaces. Thus new cultural geographers called for acknowledgement of how cultures are complexly differentiated in terms of individuals' age, gender, ethnicity, or social class, and "the *plurality* of cultures and the multiplicity of landscapes with which those cultures are identified" (Jackson 1989: 1).

Fourth, it was claimed that the 'toolkit' of methods and concepts used within traditional cultural geography was limited, overly descriptive, and inappropriate for addressing key contemporary issues relating to cultural geographies. New cultural geographers therefore advocated moving away from the description of 'cultural landscapes', and called for greater engagement with a wider range of literatures, concepts and methods to develop better understandings, analyses and interpretative concepts about such issues. This move was perhaps most visible in new cultural geographers' work on **landscape** (see Chapter 5). New cultural geographers advocated a radical break from traditional cultural geography's understanding of landscape. As noted previously, traditional cultural geographers tended to view landscapes as things that could be read and interpreted, fairly unproblematically, to find out about the impacts and activities of cultural groups. New cultural geographers argued that this approach was problematic, and overlooked the complex range of representations, discourses, symbols and processes through which landscapes are actually experienced and culturally constructed. Through their research on diverse landscapes, and associated processes, these geographers opened out a new concern with cultural **production** and **consumption** of **representations** (see Chapters 2, 3 and 6), fuelled by geographers' engagements with work from literary, artistic and cultural studies (Cosgrove 1984; Cosgrove and Daniels 1988).

In practice, as McDowell (1994: 156–168) identifies, work within the new cultural geography initially tended to crystallise around some broad research themes:

- *New landscape studies* (see Chapter 5) – i.e. studies of landscape in the wake of critiques of earlier approaches to 'cultural landscape', recognising that ideals about landscape reflect contemporary power relations, and that the way people see landscape is framed by cultural norms, discourses and representations such as painting, poetry and scientific exploration. Such studies explored the cultural construction of landscape, nature and countryside in, for example, art, literature, mass media, place promotion and regeneration literatures, and the heritage, tourism and leisure industries (Cosgrove and Daniels 1988; Wilson 1991; Duncan and Ley 1993).
- *Shared meanings and social identities* (see Chapters 3 and 8) – i.e. studies of links between social relations, cultural norms and spaces, via such case studies as changing lifestyles (Zukin 1995), social exclusions and inequalities (Davis 1990), oppositional social movements (Anderson and Gale 1999) or youth subcultures (Valentine *et al.* 1998). These studies often focused on contemporary urban, metropolitan settings (quite different from the rural, historic focus of much traditional cultural geography), drawing upon the work of urban social theorists.
- *Global cultures and sense of place* (see Chapters 8 and 10) – i.e. research "uncovering the connections between globalisation and local . . . [experiences], revealing the multiple senses of place held by inhabitants . . . recording the ways in which global trends are interacting with locally based customs and social

practices to create new layers of meaning" (McDowell 1994: 165). Such studies frequently focused on identities and cultural practices in globalised, multicultural or postcolonial contexts, for example through explorations of ethnicity, transnationalism, migration or diasporic communities (Anderson 1988).

Case studies of geographers' research relating to these themes can be found throughout this book. However, it is important to recognise that the new cultural geography resulted in more than just a body of literature on these topics. For the new cultural geography was characterised by a spirit of conceptual experimentation and adventure. We previously noted that the new cultural geography emerged from a series of critiques of traditional cultural geography, and its underlying concepts of culture and landscape. New cultural geographers sought new ideas and methods to move on from traditional approaches. In so doing, they explored other academic disciplines, including the following areas of social theory.

- The emerging discipline of cultural studies was important in: challenging elitist notions of 'high culture', and the assumptions implicit in popular understanding of 'culture'; emphasising the many practices that constitute 'culture' in everyday contexts; exploring the importance of cultural consumption for individuals' **identities** (Chapter 8); and theorising diverse forms and scales of **cultural politics** (Chapters 2 and 3) (Jackson 1989).
- Work from literary, heritage and artistic studies, and associated work in psychology, provided a vocabulary and set of concepts relating to the production and reception of cultural representations, which were particularly important for geographers exploring the representation of landscapes.
- Feminist critiques of society (and, within this, academic practices) were vital in: developing concepts with which to critique social and academic norms and representations; drawing attention to the experiences of diverse social and cultural groups who were marginalised within society, and often overlooked by academic researchers; highlighting and exploring the importance of identities, positionalities, **emotions** and **bodies** in social interactions and inequalities (see Chapters 11 and 12); and developing and popularising new, qualitative research methods within which to explore these kinds of issues (see Rose 1993; WGSG 1997).
- Postmodernist philosophers' theories of media, **texts,** meanings and urban **landscapes** (Chapters 4, 5 and 6), and poststructuralist philosophers' reflections on everyday lives, experiences, **emotions,** practices and **performances** (Chapters 7 and 11) became important influences for many human geographers.
- Postcolonialist critiques of literature and art became influential, providing: a theoretical framework for understanding the representation (or underrepresentation) of excluded groups and 'others' within contemporary culture (Chapter 2); understandings of the complex cultural aftermath of imperialism; a critique of the historical representation of 'exotic', 'native', 'uncultured' people and **landscapes** by colonialists, explorers and geographers (Chapter 5).

These encounters with new areas of theory and literature remain a key legacy of new cultural geography, and we return to these concepts at numerous points throughout this book.

Version 6: Critiques of new cultural geography (UK/USA c.1991 - present)

Key readings: Mitchell (1995, 2000), Thrift (2000b), Lorimer (2005, 2007, 2008)

The subdiscipline of cultural geography has developed rapidly, and in *all kinds* of directions, in the wake of the new cultural geography. Indeed, much recent work by cultural geographers might be understood as belonging to a somewhat different phase in the subdiscipline's development. Five key building blocks of this sixth version of cultural geography are outlined in Figure 1.7.

As Figure 1.7 suggests, the new cultural geography has prompted a range of emotions. There is considerable excitement at the range of work and new ideas made possible through the new cultural geography. However, there have also been a range of critiques of new cultural geography by cultural geographers. These critiques have often revolved around three broad sets of anxieties.

First, many Anglo-American cultural geographers have begun to be explicitly critical of the new cultural

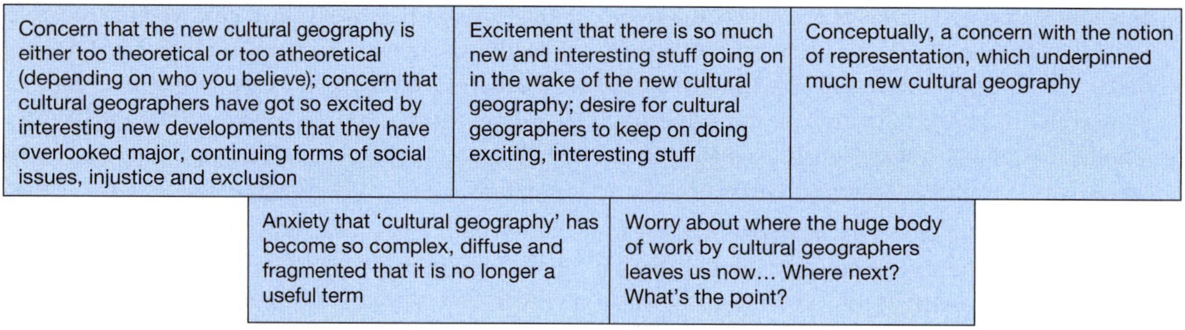

Concern that the new cultural geography is either too theoretical or too atheoretical (depending on who you believe); concern that cultural geographers have got so excited by interesting new developments that they have overlooked major, continuing forms of social issues, injustice and exclusion	Excitement that there is so much new and interesting stuff going on in the wake of the new cultural geography; desire for cultural geographers to keep on doing exciting, interesting stuff	Conceptually, a concern with the notion of representation, which underpinned much new cultural geography
Anxiety that 'cultural geography' has become so complex, diffuse and fragmented that it is no longer a useful term	Worry about where the huge body of work by cultural geographers leaves us now... Where next? What's the point?	

Figure 1.7 Five building blocks of critiques of new cultural geography.

geography, and work with a stated aim of moving on from its themes and assumptions. Even in the midst of the new cultural geography, some cultural geographers had begun to articulate this kind of critique (Rose 1991; Thrift 1991). Over the subsequent decade many Anglo-American cultural geographers appeared to be increasingly self-reflective and often critical of their own practice within the new cultural geography. Indeed, by 2000, several established cultural geographers appeared to be experiencing some kind of millennial angst about the state of the subdiscipline! Some were anxious about the (perhaps limited or unadventurous) concepts, methods and approaches that had become popular via the cultural turn:

> The problem is that cultural geography needs to be both more theoretical *and* more empirical . . . More theoretical certainly. Most cultural geographers are not trained in theory, and use it as though it were a technique (hence the bizarre calls for 'applying' particular theories . . .) . . . But cultural geography is not empirical enough either. Its range of methods is remarkably small and, underneath all the rhetoric, really quite conservative.
>
> (Thrift 2000b: 5)

Others questioned the (perhaps limited) set of research themes that had preoccupied geographers in the wake of the cultural turn, and suggested that these preoccupations had often led geographers away from contemporary social, political and cultural issues. For example, it was noted that new cultural geographers had ultimately had relatively little to say about issues of poverty, social exclusion and contemporary social and political inequalities. For instance:

> [in the wake of the 'cultural turn'] we need to keep an eye open to the processes . . . which are the stuff of everyday social practices, relations and struggles, and which underpin social group formation, the constitution of social systems and social structures, and the social dynamics of inclusion and exclusion. More concretely . . . it is to continue paying urgent attention to the mundane workings of families and communities . . . ; it is to register the battles to get by on a daily basis, to earn a crust, to keep the house warm, to cope with the neighbours, to walk down the street without being afraid . . . and so on and so on . . . I cannot help feeling that the cultural turn in human geography has risked emptying out much of this stuff from our lenses, from both the approaches that we adopt and the subject matters we tackle.
>
> (Philo 2000: 37)

Yet others critiqued the notion of 'culture', or suggested that a preoccupation with culture had prevented cultural geographers from understanding the issues and processes that were – actually – at the heart of their work (Mitchell 1995). For example:

> no decent cultural analysis can draw on culture itself as a source of explanation; rather culture is always something to be explained as it is socially produced through myriad struggles over and in spaces, scales, and landscapes.
>
> (Mitchell 2000: xvi)

Others called for new cultural geographers to be more politicised, and to *do* more than just study the cultural politics at the heart of their work. To quote Mitchell (2000: 294) again:

> cultural geography must be more than study and analysis. Since culture is politics by another name, cultural geography must become an intervention in cultural

> politics. No other cultural geography is sufficient to the work of cultural justice that everywhere remains to be done.

Critiques such as these called for new cultural geographers to 'move on', or at least to change their practices, in different ways: to become more theoretically and methodologically experimental *and/or* to (re)focus upon social and political issues *and/or* to 'get over' the concept of 'culture' *and/or* to become more politically active. And indeed, over the last decade, many cultural geographers *have* attempted to travel in each of these directions, producing a rapidly expanding and diversifying field of new work, loosely connected by a critical stance towards the new cultural geography, and older versions of cultural geography too.

Second, there is considerable anxiety about the coherence of cultural geography as a subdiscipline: many critics question whether 'cultural geography' is a useful term, or even whether cultural geography exists in any meaningful sense any more. These questions have emerged, in large part, because concepts, questions and methods developed during the new cultural geography have come to be remarkably influential in many other areas of human geography. That is, cultural geography is no longer done solely by 'cultural geographers'. Rather, the questions and practices of cultural geography have been adopted and taken elsewhere by many other human geographers: *all* "geography . . . has become . . . embodied by cultural discourse" to a greater or lesser extent (Cook *et al.* 2000: xii).Thus, in hindsight, Philo (2000: 28) suggests that the main achievement of the new cultural geography was

> to heighten our senses of how all things cultural might be raised to a much more prominent position in studies *throughout* . . . human geography (and not just one or two neatly parcelled-off subdisciplines) . . . [T]he amazingly rich arc of the cultural turn . . . has undoubtedly sent shockwaves throughout the length and breadth of human geography, leading to debates (more or less explicit, more or less heated) about the merits of a cultural turn within (say) economic geography, political geography, population geography, environmental geography and elsewhere.

There therefore exist countless pockets of what could legitimately be termed 'cultural geography' research going on in practically all areas of human geography, and even in some areas of environmental and physical geography. The case studies included throughout this book are intended to give a flavour of this remarkable breadth of work. However, many cultural geographers now caution that cultural geography has become too diffuse and intimidating: "a highly charged intellectual world [which] can be a daunting experience for students. Cultural geography's agenda is vast and challenging, its intellectual sweep audaciously ambitious" (Jackson 2003: 136).

A third set of anxieties – which we shall explore in detail in Part 3 of this book – relate to the methods, interests and underlying notions of representation which were central to the work of new cultural geographers. It is argued that new cultural geographers sometimes tended to underestimate the significance of **everydayness** (Chapter 9), **material objects** (Chapter 10), **emotions** (Chapter 11), **bodies** (Chapter 12), and **spatial** processes (Chapter 13) for cultural geographies. In thinking through the importance of these aspects of human geographies, some cultural geographers have turned to a body of concepts known as **non-representational** theories (see **Chapter 9** and Lorimer 2005, 2007, 2008).

1.5 Cultural geographies now

In this chapter we have identified six distinctive, coexisting forms of cultural geography, four of which (versions 2, 4, 5 and 6) are either still ongoing or remain influential. We have also tried to give a flavour of the diverse work ongoing within each version of cultural geography, and especially the substantial expansion, diversification and fragmentation of Anglo-American cultural geography in the wake of the cultural turn. So, where does this leave us? Or, perhaps, what *is* cultural geography now? As one kind of answer to this question, Box 1.4 lists the topics covered in three international academic cultural geography journals during 2009–10. Take a look at this remarkably eclectic list before reading on.

Box 1.4 provides a kind of snapshot of research by cultural geographers in 2009–10, and prompts three points. First, there is simply *so much* interesting, intriguing, thought-provoking, eye-opening work represented in there. Even in just two years' editions of just three journals there are *dozens* of new research

Box 1.4
Cultural geography, 2009–10: from A to Z

The following topics were explored in research papers published in three international academic cultural geography journals (namely *Cultural Geographies, Journal of Cultural Geography* and *Social and Cultural Geography*) during 2009–10. You may wish to compare this list with the contents of earlier or later versions of the same journals.

> African refugee resettlement in the US; airport security checkpoints; Angel Island Immigration Station; Antoine de Saint-Exupéry (1900–1944); Antonioni's *L'Avventura* (1960); art and climate change; artistic interventions and border security; autism, identity and disclosure; Barbudan open-range cattle herding; Berlin's Palace of the Republic; biodiversity offsets in New South Wales; boundaries in Coast Salish territories; Brazilian transnational migration; British West Indian travel narratives, 1815–1914; 'Burqinis'™; celebrities, image control and public space; children's popular culture; Christian identities in Singapore; commemorating Karelian evacuation in Finland; commemorating the Royal Ulster Constabulary; community radio in Knoxville; condominium development in Toronto; cultural imports in Cape Town; cultural representations of mental illness; 'cycling citizens'; Danish marriage laws; demotion of Platt National Park, Oklahoma (1976); Dubai's landscape mix; election campaigning in Mississippi; encounters with possums in suburban Australian homes; everyday encumbrance in the railway station; expatriate everyday life in Dar es Salaam; extraordinary geographies of the Hundertwasser-Haus; fear of violence in public space in Umeå; film and urban redevelopment; fire management in indigenous landscapes; food consumption and anxiety; food and identity in contemporary Indian women's writing; garden narratives; geographical imaginaries and EU enlargement; governmentality in a Brazilian favela; grief and belief; Gypsy and Traveller mobility; health and race in Louisiana, 1878–1956; Hispanic population change in Detroit; identity and geopolitics in Hergé's *Adventures of Tintin*; indigenous counter-mapping in Nicaragua and Belize; indigenous spatiality in Thailand; installation art; intellectual disability; inter-ethnic relations; Italian investors in Slovakia, Romania and Ukraine; John Wesley Powell (1834–1902); landscape and cinematography; leisure and Indonesian nation-building; low-paid migrant men in London; masculinity and Australian wilderness; media discourses in Romania; mega churches in the USA; men's public toilets as spaces of heterosexuality; methodologies for more-than-human geographies; military masculinities; mobile time spaces of quiescence; Moldova's place in the post-socialist world; morale and the 'war on terror'; myths, memory and traditional knowledge in north-eastern Namibia; naturism, nature and the senses; neo-Malthusianism as bio-political governance; Norwegian family reunification law; *Paraíso en el Nuevo Mundo* (1655); participatory mapping and GIS; performing wartime memories; place-marketing images; place names and emotional geographies; place symbolism and land politics in *Beowulf*; policing in Seattle; Polish nurses in Norway; postcolonial memory-makings in Cape Verde; post-smoking selves; public space in post-apartheid Cape Town; queer migration; racialized privilege in New York; reflections on making a film; regionalism/landscape dialectics in L'Horta de València; religious affiliation in Glasgow; responses to economic restructuring in Florida; rural gentrification in the Old and New Wests; same-sex parenting; Seattle's desegregated schools; *Scipione l'Africano* (1937); secular iconoclasm; sexuality and park space; sound artworks in Western Montana; South African queer politics; spiritual landscapes; surfing and sensuality on the Gold Coast; Swahili trading practices; synchronisation in urban landscapes; taboos, conservation and sacred groves in Tallensi-Nabdam; televangelical publics; the early Argentine sound film *Los tres berretines*; the 1992 Los Angeles riots and narrative-based geovisualization; thinking landscape relationally; Timothy Dwight's travel writing (1817); tourist souvenirs; *Traffic in Souls* (1913); tributes to Denis Cosgrove (1948–2008); Triestine perspectives on city-state relations; TV detective tours; Ultimate Frisbee; Vashti Bunyan; urban decay in Ankara; urban youth in Ghana; walking and suburban nature-talk; *WALL-E* and the philosophy of Alain Badiou; zombie geographies and the undead city.

papers that sound fascinating to us; hopefully there will be many that catch your attention, too. That *all* of this stuff is going on simultaneously is testimony to the remarkably rich, remarkably diverse present of cultural geography. But, second, the juxtaposition of all these topics might seem problematic, or even comical. We might wonder: what *does* all this stuff have in common? What on earth *do* African refugee resettlement, Barbudan open-range cattle herding, an early Argentinean sound film, surfing and sensuality, *WALL-E*, Alain Badiou's philosophy and zombies (or indeed *any* of the other topics listed in Box 1.4) have in common? The answer to this question is that all of these topics were the focus for researchers who were motivated by at least one of the versions of cultural geography outlined in Section 1.4, and who sought to investigate at least one of the versions of culture outlined at the end of Section 1.3. But even with this in mind, the list of topics in Box 1.4 can seem incoherent and dauntingly diffuse. Indeed, third, the sheer range of work going on under the umbrella of cultural geography can seem excessively diverse and profoundly difficult to keep up with, or even summarise. For Philo (2000: 29), for example,

> there are just so many good new things to read which look interesting in the literature, packing out the pages of even the supposedly more 'staid' geographical journals, that it is incredibly difficult to 'keep up': to retain either a sense of the overall intellectual landscape or a knowledge of the more detailed excursions within this landscape. What is also difficult is the pace of the changes occurring: the rapidity with which new possibilities for thought and practice come tumbling into the literature.

This kind of anxiety has led many cultural geographers to proclaim the impossibility of writing a neat summary or explanation of cultural geography. For Mitchell (2000: xiv), for instance,

> 'there is no way . . . that I can survey, let alone analyse, the breadth and depth of cultural geography in . . . [one] book. The field is too unwieldy, too chaotic, too diffuse.'

In this chapter, then, we have shown that the subdiscipline of cultural geography is remarkably diffuse, difficult to define, and difficult to summarise. Indeed, we have shown that cultural geography is *not* a single, coherent, neatly definable subdiscipline; instead, there are countless fields of work and scholarship by cultural geographers whose work is motivated by different traditions of cultural geography and different meanings of 'culture'. We realise that this may be discouraging. But we urge you not to be deterred by all this diffuse complexity. Indeed, we should like to end this introductory chapter by reminding you of the statement at the beginning of Section 1.2: we love cultural geography . . . at least sometimes . . . but it is hard to explain what it is. In short, cultural geography is worth the hassle, and in this book we shall focus on the reasons to love cultural geography. As a starting point, see Box 1.5. Here, we present several propositions about how you should approach the following chapters, and indeed the subdiscipline of cultural geography in general. On that note, we invite you to begin exploring the rest of this book. Bear the propositions in Box 1.5 in mind as you read . . .

Box 1.5
How to tackle cultural geography

- Be critical; when studying cultural geography, *your* opinions and reflections count. But be constructively critical; don't just dismiss ideas or approaches.
- In particular, be critical of anyone offering an easy, singular definition of 'cultural geography', or a definitive agenda for what cultural geography 'should' be.
- Work out what you think. Constantly ask yourself: 'where do I stand in relation to this?' Develop critiques of your own. But remain open-minded to anything that you feel negative about.
- Be open-minded. In particular, be open-minded to forms of research and scholarship that you may not have encountered before. And do not be dismissive of older forms of research and scholarship.
- Don't worry too much about the 'boundary' of cultural geography. Explore relevant ideas and literature that may not be written by cultural geographers. Explore relevant ideas and literature from other academic disciplines.
- Ask questions. It is always useful to ask 'why?' Or 'what's the point?' But don't dismiss anything until you have got an answer to these questions.
- Reflect on the 'point' – the usefulness and the application of research and concepts by cultural geographers. Ask yourself: in your opinion, what makes research worthwhile or useful? Ask: how, exactly, do the approaches and concepts of cultural geography extend and enhance our understanding of key issues?

Box 1.5 continued

- Read widely, and be adventurous and experimental in your reading. For example, there must be *something* in the long list in Box 1.4 that catches your interest. Find the paper; read it, and see where it leads you.
- Realise that cultural geography involves a set of specific, challenging academic skills and research methods. It is not just 'wishy-washy', but requires some key skills, academic rigour and understanding of demanding concepts. Keep working on these skills.
- Indeed, recognise that *doing* cultural geography may entail some forms of research and scholarship that are different from those you have previously encountered: it may include "looking, feeling, thinking, playing, talking, writing, photographing, drawing, assembling, collecting, recording and filming as well as the more familiar reading and listening" (Shurmer-Smith, 2002a: 4).
- Talk to people. Find a cultural geographer. Talk to them or email them, and find out how and why they 'got into' cultural geography.
- Know that cultural geography is an open-ended, evolving, ongoing subdiscipline; you can extend it through your own research and scholarship.

Summary

- Cultural geography is a major, vibrant area of research and scholarship within the discipline of human geography. While this book was being written, cultural geographers were studying everything from African refugee resettlement in the US to zombie geographies, and all points inbetween (see Box 1.4).
- Cultural geography provides us with a rich body of research about cultural practices and politics, and the importance of cultural materials, media, texts and representations, in diverse contexts. Cultural geography is also one of the most theoretically adventurous and avant-garde subdisciplines within human geography.
- 'Culture' and 'cultural geography' are complex terms to define: different geographers have different understandings of these terms.
- Looking back over the last century, we can identify multiple significant, but very diverse, versions of cultural geography. You should pay particularly close attention to the shifts from 'traditional cultural geography' through 'new cultural geography' to 'critiques of new cultural geography' discussed in Section 1.4.

Some key readings

There are many, many useful and fascinating texts relating to cultural geography. In the space below, we can include just a few examples of key readings that link to the discussion in this chapter.

Anderson, K., Domosh, M., Pile, S. and Thrift, N. (eds) (2003) *Handbook of Cultural Geography*, Sage, London.
An excellent collection of essays reflecting upon different thematic and conceptual areas of cultural geography. In effect, the book surveys the period when critics began to call for a move on from new cultural geography.

Cosgrove, D. and Daniels, S. (eds) (1988) *The Iconography of Landscape: Essays on the symbolic representation, design, and use of past environments*, Cambridge University Press, Cambridge.
As noted earlier, geographical research on landscape and representation was central to the new cultural geography. This is a classic text of this genre: a collection of studies of historical representations of landscapes.

Crang, M. (1998) *Cultural Geography*, Routledge, London.
An excellent, accessible textbook, which is particularly useful as an introduction to the body of work that arose from the new cultural geography.

Jackson, P. (1989) *Maps of Meaning*, Routledge, London.
A classic text of the new cultural geography, which provides a clear and readable argument calling for geographers to engage more with work from cultural studies and other social scientific disciplines.

Lorimer, H. (2005) Cultural geography: the busyness of being 'more-than-representational'. *Progress in Human Geography*, **29,** 83–94.

This punchy paper (see also Lorimer 2007, 2008) explores some recent critical and non-representational currents within cultural geography.

Nayak, A. and Jeffrey, A. (2011) *Geographical Thought: An introduction to ideas in human geography*, Pearson, Harlow.

This book introduces broad shifts in thinking within human geography that have contextualised the developments in cultural geography described in this chapter. Read in conjunction with the Shurmer-Smith book listed below.

Shurmer-Smith, P. (ed.) (2002) *Doing Cultural Geography*, Sage, London.

A very helpful book that introduces different approaches to cultural geography, as contextualised by broader shifts in geographical thought.

Zelinsky, W. (1973) *The Cultural Geography of the United States*, Prentice Hall, Englewood Cliffs.

An excellent resource that draws upon, and encapsulates, more 'traditional' forms of work in cultural geography.

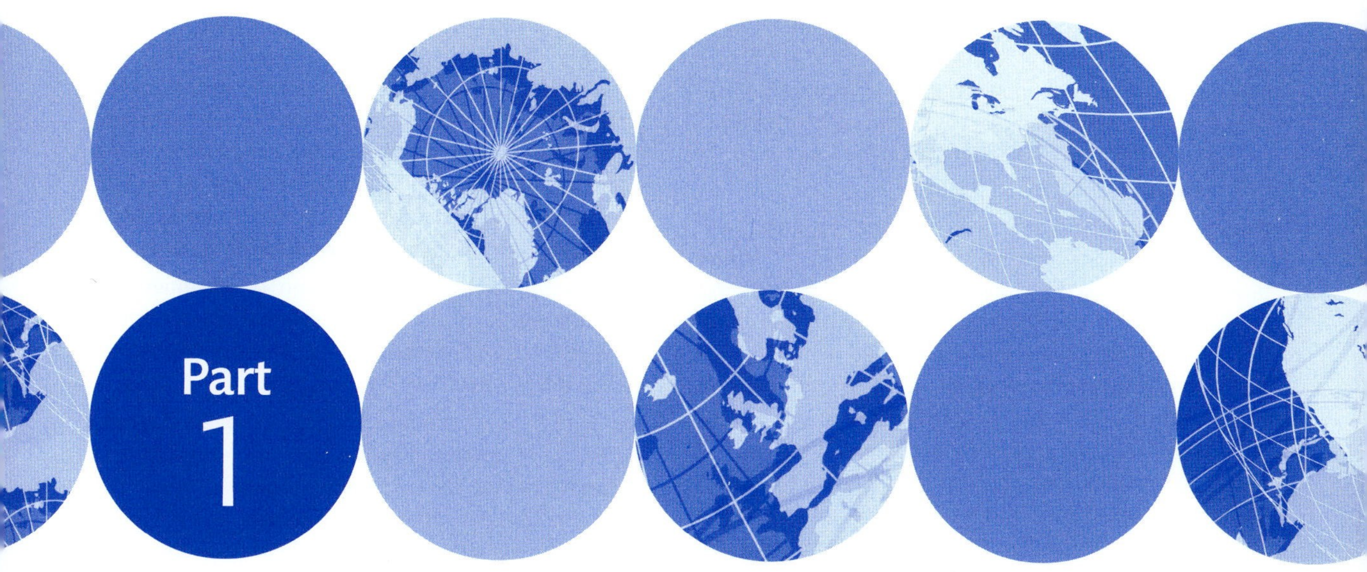

Cultural processes and politics

Part Map

- Chapter 2 - Cultural production
- Chapter 3 - Cultural consumption

In Chapter 1 we identified three major ways in which the work of cultural geographers is important:

- researching cultural processes and their complex geographies and politics;
- exploring the importance of cultural materials, media, texts and representations in all kinds of space and geographical context;
- engaging with new lines of social and cultural theory, and reflecting upon their importance for human geographers.

Part 1 deals with the first of these themes. The following chapters introduce some key concepts and geographical research exploring cultural processes and politics. We also introduce a broader context of ideas and research - particularly by sociologists and cultural studies scholars - that has inspired many geographers working in this context. As we shall explain, a great deal of this work has tended to focus on *either*:

- **cultural production** - the processes through which all manner of cultural objects, spaces, commodities, texts, representations and media are created; *or*
- **cultural consumption** - how cultural objects, spaces, commodities, texts, representations and media are encountered, purchased, and used by consumers.

We shall focus on each of these topics in turn. Chapter 2 introduces some key processes of cultural production, including: how and where cultural objects are literally manufactured; how meanings, norms and notions of 'good taste' are produced and attached to certain cultural objects in particular geographical and historical contexts; and how cultural spaces are produced and regulated. We note, too, how these forms of cultural production often effectively (re)produce particular and unequal forms of power relations. Chapter 3 introduces the notion of cultural consumption, with a particular geographical focus on: the substantial importance of consumption practices in everyday spaces; the development of spaces of/for consumption within 'consumer societies'; practices that involve the consumption *of* places; and the geographical complexity of consumption. We note how consumption is central to one's cultural identity, and can sometimes be central to politicised practices that seek to contest or subvert the forms of cultural power identified in Chapter 2.

Chapters 2 and 3 should help you get to grips with important bodies of research about processes of cultural production and consumption respectively. But - and this is a big *but* - it is crucial to bear in mind that these processes are not neatly separated in real life. Although much classic research in this context *did* tend to focus on *either* cultural production *or* consumption, more recent work within cultural geography and cultural studies has begun to recognise that this was deeply problematic. In Section 3.5 we outline why this 'either/or' approach was so problematic, and argue that the processes discussed in this section should actually be understood as always already closely and complexly interconnected.

After reading Part 1, several later chapters of this book should help you to extend and develop your understanding of cultural processes and politics in three key ways.

- Part 1 provides a kind of general 'sketch' of how cultural objects and spaces are produced and encountered. Several chapters in Part 2 provide a much more detailed and specific focus on particular kinds of cultural object. Now consider the complex geographies evident in the later chapters on **built spaces** (Chapter 4), images of **landscape** (Chapter 5), diverse cultural **texts** (Chapter 6) or forms of **performance** (Chapter 7).
- Follow up Chapter 3's discussion of identities and subcultures by exploring the extended discussion of concepts of **performativity** and **identity formation** in Chapters 7 and 8.
- In Part 3, we introduce notions of **everydayness** (Chapter 9), **materiality** (Chapter 10), **emotion** and **affect** (Chapter 11), and **embodiment** (Chapter 12), and discuss the inherent complexity of all human geographies. Consider how these concepts are relevant to the processes outlined in Chapters 2 and 3. It should become clear that *all* cultural production and consumption involves embodied, emotional, everyday practices in complex and material spaces. How does this realisation add to, or complexify, our understanding of cultural processes and politics?

Cultural production

Before you read this chapter . . .

Consider this book. Try to answer the following questions. (Note any gaps in your knowledge.)

- Where and how was it made?
- Who was involved in making it, and under what conditions did they work?
- Who decided its content?
- Who profits from the publication of the book?
- Think about the spaces in which this book can be found. How, and by whom, are these spaces produced?

Chapter Map

2.1 Introduction: producing a cultural geography textbook

Think, for a moment, about this book. How much do you know about this object? Where, when and how *was* it made? As authors we can shed light on *some* aspects of this book's production: the many hours spent writing (or not writing) on computers in the English East Midlands; the many notepads, pens and printer cartridges used in the process; the many meetings and emails between the authors and commissioning editors; the contracts, drafts, proofs and other paperwork signed, sealed and delivered; the whole process commissioned by a publishing company based in Harlow, England; the book itself printed in Gosport, England. Many other aspects of this process are rather more difficult to map. (Where *were* the raw materials used in the book's paper, cover and inks sourced? How *did* the book get from the printers to wherever you picked it up? *Who* was responsible for typesetting, inking, maintaining the production line, or driving the delivery lorries? What are their working lives like? What exchanges of finance, and what behind-the-scenes goings-on in the publishing industry, enabled the book to be commissioned? Who, ultimately, benefits from the production of this book?) As we shall suggest, cultural objects, texts and media are typically created via complex processes of production, although these processes frequently pass unnoticed by consumers of these objects.

This chapter introduces some of the ways in which geographers have explored processes of cultural production. As we shall explain, questions of cultural production have been important for many cultural geographers. To help you understand this importance, this chapter outlines:

- why cultural production has been a major concern, one way or another, for many cultural geographers;
- some key concepts that have been useful to geographers exploring cultural production;
- some key work by economic, urban and industrial geographers that explored the making of cultural objects, texts and media;
- some geographical research about the creation and maintenance of different kinds of cultural spaces.

2.2 Questioning cultural production

In Chapter 1, we noted that the word '**culture**' can be used to describe a number of things. It can refer to:

- texts, objects and spaces created by human beings: everything from small-scale, individually crafted artefacts through mass-produced commodities and globalised media to large-scale public spaces and architectures;
- practices, habits and lifestyles that relate to these cultural texts, objects and spaces;
- meanings, norms and value judgements that relate to all of the above;
- power relations, group identities and inequalities that relate to all of the above.

We also identified several lines of work by cultural geographers who have explored some spatial patterns, processes and consequences of all this. The question of 'cultural production' – how all these cultural objects, spaces, practices, meanings and relations are made and come to be as they are – has been a key issue for many cultural geographers. Indeed, we can think of cultural production as a foundational, but divisive, issue for many cultural geographers: on the one hand, many cultural geographers have been motivated to explore processes, politics and spaces of cultural production; on the other hand, many others have been motivated to react *against* a perceived predominance of work on cultural production. Either way, there is an acknowledgement that cultural production is a significant issue: for some, it *should* be a central and motivating concern for cultural geography; for others, it has been an overwhelming focus for *too much* geographical work in this context.

This chapter focuses on the making of cultural objects, meanings and spaces: here, we outline how studying cultural production reveals processes, and forms of cultural politics, that are fundamentally geographical. In Chapter 3, however, we introduce some important critiques of geographical work on cultural production. In particular, we recommend that you read Section 3.5 – **connecting cultural production and consumption** – when you have reached the end of this chapter.

Here are three key reasons why cultural production became a foundational concern for many cultural geographers. You may find that referring to Box 1.3, and the related commentary, helps you to place these points in a broader context.

- A good deal of early, classic work in human geography – particularly those lines of work named **traditional cultural geography** and **regional cultural geography** in Chapter 1 – explored connections between human activity and landscape morphology. As we outlined in Section 1.4, traditional cultural geographers, most notably the **Berkeley School,** conducted many studies of 'cultural landscapes': exploring, for example, how particular forms of material culture, manufactured objects, dwelling spaces, technologies and creative practices emerged and spread from particular 'cultural hearths', and resulted in changes to 'natural' landscapes. Somewhat similarly in spirit, early regional geographers such as the ***Annales* School** investigated how particular agricultural or manufacturing traditions shaped distinctive local and regional 'ways of life' as manifest in, for example, local artistic and folk cultures. Although these lines of geographical work have been criticised, and to some extent disowned, by later cultural geographers (see Section 1.4), their recognition that cultural artefacts, practices and lifestyles are closely linked to agricultural, manufacturing and landscape processes remains important. These early geographers were important in shaping expectations of what human geographers do; and their understandings of cultural production continued to shape and direct the work of cultural geographers for many decades.

- As noted in Section 1.4, a **new cultural geography** emerged from geographers' encounters with contemporary work from other disciplines, including cultural studies, during the 1980s and 1990s. Some of the most exciting and important work within cultural studies during the 1970s and 1980s had argued for research exploring the production – and producers – of popular cultural media such as newspapers, television programmes, movies and pop music. It was argued that such research would unveil the often-overlooked ideological content of such media, and the particular, exclusionary, inequitable processes and industries that produced them. This heady, politicised line of work inspired many geographers who encountered it during the 'new cultural geography'. See Section 2.3 for an outline of some key arguments about cultural production that originated in cultural studies and inspired many geographers at this time (but led others to critique an overwhelming focus upon cultural production – see Chapter 3).
- Cultural geographers were by no means the only or first geographers to explore processes of cultural production. Indeed, during the 1970s and 1980s, at least two major bodies of geographical work had focused upon these issues. First, many economic, urban and industrial geographers wrote extensively about cultural industries, means and chains of cultural production, the commodification of cultural goods, and linkages between cultural and economic production in diverse contexts. Second, many urban geographers and planners theorised the production and regulation of cultural activities, spaces, events and identities within diverse built environments. See Sections 2.4 and 2.5 for discussion of these two lines of work. In effect, then, 'new cultural geographers' found themselves working in a context that had, in certain respects, already been theorised and researched extensively. Some cultural geographers extended and appropriated these existing lines of work, but – again – some critiqued an exclusive focus upon cultural production (see Chapter 3).

2.3 Making meanings, discourses and taste: key concepts from cultural studies

We have already noted that concepts and research from the disciplines of cultural and media studies were fundamentally important in the development of the new cultural geography during the 1980s and 1990s. There and then, some of the most important and exciting work within cultural studies focused upon questions of cultural production, such as:

- *How* are cultural texts, media and objects made?
- *Who* makes them?
- Does this '*who*' affect the content and form of these texts, media and objects?

We should note, straight away, that this production-focused work was not the *only* exciting and important stuff going on within contemporary cultural studies; as we shall explore in the next chapter, another strand of cultural studies – which was especially influential for many cultural geographers – focused on the consumption of cultural products.

Let us consider why production-focused work was so important within cultural studies – and, then, so influential for cultural geographers. Cultural studies scholars set out to challenge some deep-seated, traditional assumptions about the making of cultural texts, objects and media. In particular, they argued that twentieth-century art critics and cultural commentators had popularised and idealised a centuries-old 'myth' of cultural creativity whereby:

- poets, artists, composers, novelists and other cultural producers, were understood as gifted, autonomous, talented creators or geniuses;
- the activities of these individuals were viewed as special, rarefied, high-status, creative, more-or-less spontaneous practices;
- these practices were autonomous and creative, ideally occurring independently of social, economic or political contexts (indeed, it was considered vulgar for creators to be evidently commercial or ideological);
- poems, artworks, musical scores, novels, and so on, can be read and evaluated as objects in their own right – independent of the work involved in their production and, again, independent of their social, economic or political context;
- certain kinds of cultural production ('high culture' – see below) were seen as inherently more special and high-status than others.

This myth of cultural creativity was considered problematic and inaccurate, in three key senses. First, it was argued that, by romanticising the spontaneously gifted creativity of individuals, it overlooked the actual work involved in cultural production. This realisation led to a major line of work in cultural studies, investigating the practices and techniques through which meanings are actually made. Inspired by linguistic and semiotic theorists, cultural studies scholars developed new research methods to explore the content, form and meanings of all manner of cultural texts and media, and the practices through which meanings are communicated therein. Often, this led to detailed consideration of linguistic assumptions and techniques. Consider the following questioning, by leading cultural theorist Stuart Hall (1997: 21), of how combinations of letters convey particular meanings in particular cultural contexts.

> The question, then, is how do people who belong to the same culture . . . know that the arbitrary combination of letters and sounds that makes up the word TREE will . . . represent the concept 'a large plant . . . '? One possibility would be that the objects in the world themselves embody and fix in some way their 'true' meaning. But it is not at all clear that real trees *know* that they are trees, and even less clear that they know that the word in English which represents the concept of themselves is written TREE whereas in French it is written ARBRE! As far as they are concerned, it could just as easily be written COW or VACHE or indeed XYZ. The meaning is *not* in the object or person or thing, nor is it *in* the word. It is we who fix the meaning so firmly that, after a while, it comes to seem natural and inevitable. The meaning is *constructed by the system of representation*.

Now, this kind of talk of TREEs, VACHEs, XYZs and systems of representation may initially seem rather eccentric and esoteric. However, in practice, cultural studies researchers used this sort of close reading of the minutiae of representation to explore how meanings were produced in all manner of cultural products: in popular music, cinema, television, advertisements, news media, to name a few. For example, cultural studies researchers applied methods of textual analysis – employing a close, systematic reading of language and images – to unpack the content of contemporary popular news reporting, revealing politicised and partial messages present in apparently 'neutral' reportage (see Box 2.1 and Chapter 6). This brand of radically critical reflection upon linguistic norms – exploring how ideas, images, sentences and, therefore, all cultural products are put together in practice – proved widely influential, not least among cultural geographers. As we shall see, methods of textual and discourse analysis developed within cultural studies at this time were taken up and applied, in new and exciting ways, in diverse geographical research: for example, see our later chapters on landscapes, texts and performance. More broadly, the realisation that taken-for-granted meanings (such as TREE) are demonstrably socially constructed and culturally specific (i.e. they are produced by groups of people, and they are not universally understood) has been profoundly important for many branches of cultural geography. As we shall outline in later chapters, this notion of socially constructed meanings has been a touchstone of geographers' work on topics such as **buildings** (Chapter 4), **landscapes** (Chapter 5), **identities** (Chapter 8), **emotions** (Chapter 11) and **bodies** (Chapter 12).

A second, related, objection to the myth of cultural creativity is that it overlooks connections between cultural production and wider social, economic and political contexts. In particular, it conceals how these wider contexts are important in determining who produces culture (and who does not), and how this affects the content and form of cultural texts and media. In the previous paragraph we introduced the idea that cultural texts, media and meanings are socially constructed. This concept led many cultural studies researchers to consider *who* is involved in this process: that is, who, exactly, is involved in the social construction of cultural texts, media and meanings? It was noted that not everyone participates in cultural production; more importantly, it was argued that not everybody is *able* to participate in the kinds of cultural practice celebrated by the myth of cultural creativity. In order to devote one's life to creative cultural practices, it is distinctly advantageous to have economic capital (money, economic security, or a financial backer) and what cultural theorists call 'cultural capital' (particular skills, knowledge, opinions, habits, or access to networks of power and influence). Cultural studies researchers critiqued the evidently exclusionary nature of cultural production. Historically, it was noted that prestigious cultural creators had overwhelmingly been white, educated, metropolitan, upper-class European or North American men, rich in economic and/or cultural capital. Likewise, it was argued that many social groups (notably women, ethnic and religious minorities, disabled people, and indeed those lacking in economic and cultural capital) were systematically underrepresented within the cultural industries (e.g. the worlds of publishing, media, art – see the next section), past and present. Or, more precisely, it was noted that particular aspects of cultural production (decision-making, commissioning work, creative practices themselves) were considered high status, and tended to be done by a narrowly elitist and exclusionary section of society. Meanwhile, many other forms of labour integral to cultural production (the printing of novels, the assembly of paintbrushes, the making of paints, the maintenance of performance spaces, for example) had tended to be overlooked as lower status than, and somehow separate from, the mythologised practices of cultural creativity. See Chapter 10 for examples of geographical work that has taken up this task of mapping the hidden labour behind geographies of cultural production.

A key achievement of cultural studies during the 1970s and 1980s was therefore to reveal how differential participation in cultural production could be mapped directly onto broader contemporary socio-economic inequalities: it generally remains true that, across most domains of cultural production, high-status roles tend to be occupied by those who are relatively privileged within society, whereas those who belong to socially excluded groups are overwhelmingly underrepresented. A related achievement of cultural studies was to explain how these inequalities in cultural production affected the form and content of cultural products. Cultural studies researchers argued that, in effect, all cultural texts and media can be read as ideological: implicitly reflecting, serving and promoting the interests of those who produce them (and, therefore, mostly serving the interests of particular social groups who are most often involved in cultural production). The term 'hegemony' was widely used in this body of work, and in subsequent work by cultural geographers. The Italian Marxist Antonio Gramsci is usually credited with coining this term, to describe the processes through which "consent is sought and won" by powerful groups within society (Hartley 1994: 59). Within cultural studies, this concept was applied to the study of cultural texts and media. It was argued that all manner of cultural products (including news media – see Box 2.1) should be understood as 'hegemonic' in that:

- they promote or serve the interests of those who produce them;
- they present the worldview of these cultural producers as commonsense, ideal and normal;
- in so doing, they actively disseminate and reproduce the norms and prejudices of cultural producers as natural and universal;
- despite this partial and narrow worldview, cultural products are presented as neutral or natural;
- they thus represent "existing social, political and economic arrangements as being natural and inevitable" (Crane 1992: 87);
- that is, they conceal inequalities in society, and cultural production, or present them as unproblematic and inevitable;
- they neutralise, or provide no room for, dissenting voices and worldviews.

Much cultural studies research during this period involved the unveiling of hegemonic norms and propagandist methods in seemingly apolitical cultural texts and media such as television dramas, popular music, cinema, advertisements and, again, news media (see Box 2.1).

This concern with hegemony prompted many researchers in cultural studies and other disciplines (including cultural geography) to consider connections between cultural production and social, economic and political power. Two twentieth-century philosophers – Michel Foucault and Edward Said – have been deeply influential for those seeking to understand these connections. The work of both of these key thinkers was wide-ranging, and Boxes 2.2 and 2.3 provide only brief introductions to their importance in the context of cultural production: see Chapter 6 for some extended discussion of these ideas.

Foucault's work on 'discourse' (see Box 2.2) prompted many social scientists to develop methods of discourse analysis that have been applied to a diverse range of cultural texts, objects and media. Within cultural geography, for example, one can find a very varied range of work interrogating discourses in all sorts of settings: in policy documents (Evans and Honeyford 2012), economic forecasts (Peet 2007), school buildings

Box 2.1
'Putting reality together': producing news media

Photograph 2.1 How does television news 'put reality together'?
Source: Shutterstock.com/Withgod.

Within cultural studies, much pioneering work on cultural production has focused on news media, whether in print, on television or online.

News media are interesting cultural products because: (i) they are typically presented, and often uncritically digested, as a factual, neutral record of current affairs; (ii) for many people, they are one of the most important sources of representations of the wider world in general, including political and controversial issues. Consider the following quotation by a researcher whose work was important in exploring how news media 'put reality together' for our consumption.

> The issue for us was this: if television puts 'reality' together, what is the nature of the product? The naturalistic quality of news presentation, . . . grounded in professional claims to objectivity and impartiality, led us to ask what kind of cultural artefact it is. One way of breaking into this . . . is to look

> at matters that are controversial . . . and examine how they are treated on television news. We can proceed to look at the news story as narrative: at the verbal and visual grammar in which it is expressed; at the use of graphics and other symbolic expressions; at the use of headlines; at who is interviewed, and the form and content of these interviews . . . , in short, at the way information is organised and at the implicit and explanations that are put before us.
>
> (Eldridge 1993: 4–5)

This kind of approach should prompt us to ask questions about the newspapers, headlines news broadcasts and online news feeds we encounter every day. You might like to look at one example of today's news media and consider the following questions.

- What techniques are used to 'put reality together'? (See Hartley 1994; Matheson 2005.)
- How do headlines, sequencing and running order structure the news and attach significance to particular events?
- How are language, phrases or scripting used to describe incidents in a particular way?
- How are images and iconography used in the presentation of the news?
- Are any clichés or conventions used?
- Whose perspectives are represented? (And whose are not?)
- What techniques are used to present this news medium as neutral, authoritative and factual?
- How is the same news story represented in two different news sources? What is learnt from this kind of comparison?
- How are news media shaped by the way in which they are produced? (See Herman and Chomsky 1988.)
- Who makes decisions about how this news medium is produced?
- How much do you know about how, where, and under what conditions this news medium was produced?
- Who profits from this news medium: economically, politically or in other ways?
- How might the answers to these questions shape the content of this news medium?

(Pike 2008), urban landscapes (Hastings 1999) and medical practices (Evans 2006), for example.

Said's notion of 'orientalism' (see Box 2.3) has fostered discussion on the potentially political, oppressive and violent nature of certain discourses, and the ways in which cultural production can be directly linked to broader geographical processes (such as the exercise of colonialist power). While Said's work centred on specific historical and geopolitical contexts, his illustration of the power of socially constructed discourses has provided a vocabulary with which to critically evaluate the role of representations and discourses in diverse contexts. Among cultural geographers, for example, his work has led many geographers to investigate representations of people and places in sources such as travel writing, film and television, news reporting, and art/literature.

A third objection to the myth of cultural creativity is that it overlooks the ways in which certain forms of creativity, and certain cultural products, come to have a high status. We have already noted that certain sorts of cultural practice – such as the composition of classical music, literature, poetry, art – are often valued as special, emerging from the exceptional activities of talented and highly esteemed creators. A key achievement of cultural studies was to critically question the entire, taken-for-granted value system upon which these assumptions are based. The work of the French sociologist Pierre Bourdieu on notions of 'distinction' and 'taste' was influential here, and has been so for many subsequent cultural geographers. In *Distinction* (1984: 466), Bourdieu defines 'taste' in the following way.

> Taste is an acquired disposition to 'differentiate' and 'appreciate' . . . – in other words, to establish and mark differences by a process of distinction which is not (or not necessarily) a distinct knowledge . . . since it ensures recognition (in the ordinary sense) of the object without implying knowledge of the distinctive features which define it. The . . . primary forms of classification owe their specific efficacy to the fact that they function below the level of consciousness and language, beyond the reach of introspective scrutiny or control by the will.

We shall return to this definition in a moment; but before reading on, you may find the following questions helpful in getting to grips with Bourdieu's (initially rather difficult) definition.

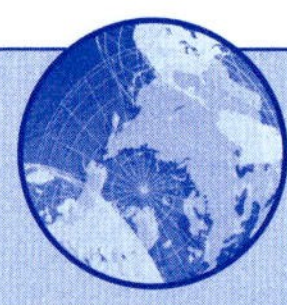

Box 2.2
Michel Foucault on discourse

The work of the French poststructuralist philosopher Michel Foucault (1926–84) has been influential for many researchers working on issues of cultural production. Foucault uses the term 'discourse' to consider how the making of texts, images and meanings is always – at some level – a political act, and how cultural production is specifically used by those exercising political power. Foucault (1972: 80) defines discourse by suggesting that the term is used

> sometimes as the general domain of all statements, sometimes as an individualizable group of statements, and sometimes as a regulated practice that accounts for a number of statements.

Consider each of these meanings in turn (see also Kraftl *et al.* 2012, and Chapter 6 on **texts**). First, Foucault suggests that 'discourse' is often used, particularly in everyday language, as a kind of general, catch-all term for 'all statements'. Second, Foucault argues that we can identify particular, interconnected 'groups of statements' in any given time and place: of particular importance are influential norms and knowledges that are (re)produced and circulated by powerful contemporary institutions and organisations, and which can effectively govern the way in which particular topics, issues or questions are addressed within contemporary society. For Foucault (1991[1975]: 22), discourses are always already central to the pursuit and maintenance of power, because

> power and knowledge directly imply one another . . . There is no power relation without the correlative constitution of a field of knowledge, nor any knowledge that does not presuppose and constitute . . . power relations.

Third, Foucault directs our attention to the practices that produce discourses, challenging us to explore how discourses – and forms of power predicated upon them – are produced and actually done. Social scientists have taken diverse approaches to this challenge, developing diverse methods of discourse analysis that variously focus on:

- verbal and textual techniques of discourse production (e.g. rhetoric, metaphors, grammar, imagery);
- non-verbal and performative tactics that produce powerful discourses (e.g. gestures, staging, presentational skills);
- the materials, tools, texts and technologies that constitute discourses in practice;
- the organisations, institutions and networks that work to produce, disseminate and control discourses; and the emotional/affective components of, and responses to, discourses.

- Think of something that is 'tasteful', or someone who has 'good taste' in some way. Try to pin down exactly what these words mean.
- Conversely, think of something that is rather 'tasteless', or someone who has 'bad taste' in some way. Again, what do these words mean?
- Where do these ideas of 'good taste' and 'bad taste' come from? Where and how do we learn them?

We may find it pretty easy to identify examples of good or bad taste. We find ourselves making value judgements – what we like and do not like, what we value and do not value, what impresses us, what dismays us – every day, in all kinds of ways and contexts. In Bourdieu's terms, we seem to have an 'acquired disposition to differentiate and appreciate': making distinctions between those things we like/love/value/care about and those things we dislike/hate/do not value/couldn't care less about comes easily to us, and seems like a natural characteristic of human social lives. However, Bourdieu questions this taken-for-granted 'disposition to differentiate'. Drawing examples from a study of consumption practices in France during the twentieth century, he demonstrates that 'commonsense' divisions between good/bad taste typically map onto inequalities and divisions within contemporary society. It is no coincidence, he argues, that norms, ideals, practices and preferences of powerful and comfortable social groups have overwhelmingly been understood as 'good taste' – and thus legitimate, normal, superior – within their social-historical context. Likewise, he argues that it is no coincidence

Box 2.3
Edward Said on orientalism

The Palestinian-American postcolonial literary theorist Edward Said (1935–2003) has been important for many researchers exploring the production of images and stereotypes of people and places. Said's most famous work, *Orientalism* (1978), is about how places and people are represented, and the politicised effects of those representations. The book examines how the 'East/Orient' has been represented by Europeans since the eighteenth century. Essentially, Said identifies a recurring set of discourses and representations – what he calls 'imaginative geographies' – through which 'the East' has been repeatedly been portrayed in Western art, literature and popular culture as 'other', exotic, strange, dangerous, frightening, unstable, and so forth. Some examples that Said explored in his work were as follows.

- Representations of Arabs as combatants or terrorists in Western news media. Said analyses discourses within US news media during Arab/Israeli wars during the 1960s and 1970s to demonstrate how Arab soldiers were repeatedly represented as fanatical, treacherous, dishonourable, villainous 'baddies', and reflected upon the ease with which Arab soldiers and terrorists were vanquished by lone Western heroes in Hollywood action movies. Towards the end of his life, Said also reflected upon how these kinds of images of Arabs as villainous, fanatical 'foreign devils' threatening 'us' in the West continued in the wake of 9/11 and via media representations of suicide bombers, Saddam Hussein and Osama bin Laden.
- Representations of 'the East' in classical French painting. Said explores the work of celebrated painters such as Eugène Delacroix (1798–1863), Jean-Auguste Ingrès (1780–1867) and Jean-Léon Gerôme (1824–1904), whose scenes of the Middle East and North Africa were popular in the European art market of the early nineteenth century. He demonstrates that these artists drew upon an iconography of exotic, heady, volatile, mysterious, dangerous/exciting scenes, inhabited by sleazy, suspicious Arab men, and exotic, seductive, mysterious, uninhibited (i.e. scantily clad) Arab women.
- Travelogues of visits to the East by European authors. Said explores the writings of nineteenth-century European travel writers such as Benjamin Disraeli, Gerard de Nerval and Gustave Flaubert, whose travelogues demonstrated, and fed, a fascination with religious fanatics, slave trading, volatile situations, opium dens, sleazy situations and the exotic, 'other' inhabitants of the Middle East and North Africa.

Said argues that this repertoire of 'orientalist' images constitute a 'distorted lens' through which Europeans habitually imagine 'the East': a whole set of ways in which the East, and people living there, are imagined as 'other' and opposite to 'us'. In short, 'they' are represented as 'exotic', primitive, unstable, fanatical, potentially villainous (compared with 'us', who are 'normal', civilised, rational, decent people).

that norms and activities of excluded social groups and (in Bourdieu's terms) 'dominated' social classes have repeatedly been represented as tasteless, vulgar or inferior. Rather, he suggests that notions of 'good taste' are actively (re)produced by those with economic and cultural capital, in order to legitimise and celebrate their own norms and interests. In the process, those who are unable or disinclined to engage in 'tasteful' activities come to be treated with contempt, and positioned as unrefined or inferior. For Bourdieu, then, the production and maintenance of norms about good taste are directly linked to the reproduction of social inequalities and divisions. In *Distinction* (1984), Bourdieu identifies some of the processes and institutions through which the production of taste happens in practice. He highlights the importance of:

- critics, journalists, scholars and other 'cultural intermediaries' within the **cultural industries** (see Section 2.4), who effectively form contemporary tastes by reviewing cultural texts, classifying them (as good or bad, classic or turkey, must-see or miss-able), developing discourses of "what is worthy of being seen and the right way to see it" (Bourdieu 1984: 25);

- family practices, which foster habitual behaviours, shape cultural norms and preferences, and provide spaces/opportunities for certain cultural activities (all depending, of course, on the existing norms and economic and social capital of the family in question);
- educational institutions, which teach particular ways of reading a particular canon of texts, and formally reward particular forms of behaviour (e.g. through the award of qualifications – one form of cultural capital) to those who demonstrate these behaviours and ways of reading (implicitly rewarding those who are prepared and supported for education by family practices).

Through these processes, norms about 'good taste' are deeply embedded and taken for granted. Indeed, returning to the questions at the beginning of this paragraph, it is likely that we shall find it difficult to put our finger on exactly what makes something good/bad taste, or how we acquire our norms and assumptions about taste. This is what Bourdieu means, in a previous quotation, where he describes our ideas about taste as 'not (or not necessarily) a distinct knowledge': we may know 'good/bad' taste when we encounter it, but we are not necessarily able to know or explain exactly why we think in this way. Indeed, Bourdieu goes on to describe taste as functioning 'below the level of consciousness and language, beyond the reach of introspective scrutiny or control by the will'. So, for example, a cultural object, text or practice might affect us powerfully – we may be enraptured by a great artwork, disgusted by a tasteless film, dismayed at the bad taste of a colleague – and we may not be able to fully explain or disguise these reactions.

The concepts of hegemony, discourse, taste and cultural capital explored in this section were developed by disparate thinkers, writing about diverse (and often historical) contexts. However, many cultural studies scholars pieced them together, and applied them to current contexts, to develop new understandings and critiques of the politics of cultural production. Through their engagements with these concepts, cultural studies researchers demanded that all cultural texts, objects and practices should be open to critical examination, to explore how contemporary cultural tastes relate to inequalities of economic and cultural capital (as per Bourdieu), and to unveil the presence of powerful and tactical discourses (as per Foucault) supporting the interests of hegemonic social groups (as per Gramsci). This work provided a set of concepts, methods and vocabularies that many geographers applied, adapted and extended in the new cultural geography: see, for example, our later chapters on **landscapes, buildings** and **texts**. Moreover, there was a realisation that academic researchers had often been complicit in the inequitable processes outlined in this section. In particular, it was noted that scholars in the arts, humanities and social sciences had, more or less systematically, tended to focus on what could be called 'high culture': the kinds of cultural texts, objects and activities that have highly valued status within contemporary societies, which are exclusively the preserve of the socially and economically powerful, and which were valorised by the myth of cultural production outlined at the start of this section. Against the grain of most previous work, many cultural studies researchers turned to investigate all manner of more diverse and popular (previously dismissed as 'low') cultural texts, objects and activities. Again, this turn proved hugely influential for many cultural geographers: see, for example, our later chapters on **consumption** (Chapter 3), **identities** (Chapter 8) and **material objects** (Chapter 10).

2.4 Geographies of cultural production: commodity chains and the cultural industries

As we noted in Section 2.2, cultural geographers were by no means the only or the first geographers to explore cultural production. In this section we introduce two key, related lines of geographical research that were ongoing during the early stages of the new cultural geography. As we shall show, these lines of work were chiefly instigated by researchers who did not define themselves as cultural geographers, but they did explore issues of cultural production, and indeed they did effectively position cultural production as a foundational concern – one way or another – for cultural geographers. First, we introduce economic and industrial geographers' work on commodity chains, and the complex geographical networks through which commodities and cultural texts, objects and media get made in practice. Second, we highlight the work of urban, economic and industrial geographers on the 'cultural industries', and the industrialisation and commodification of processes of cultural production. In each case, we introduce some early, key concepts that proved influential for cultural geographers, before outlining some ways in which cultural geographers extended and developed these concepts.

To understand why a number of economic, industrial and urban geographers had begun to explore processes of cultural production, you may find it useful to know a little about the cultural turn in economic geography. In Chapter 1, we introduced the **cultural turn** in human geography (see Box 1.3), and mentioned that this 'turn' in research interests, methods and conceptual orientation had been evident in practically every area of human geography. Box 2.4 includes some details about the cultural turn in economic geography. You may find these details useful in contextualising the following discussions of commodity chains, cultural industries, and the production of cultural spaces. Essentially, the short version of Box 2.4 is that during the 1980s many economic, industrial and urban geographers began to explore:

- how and where, and under what conditions, commodities are made in practice via increasingly complex and global networks of production;
- the importance of so-called 'cultural industries' within contemporary local, regional, national and global economies.

In this context, geographers used a variety of methods and concepts to explore the processes through which commodities – including cultural objects, texts and media – are produced. Three key, related concepts were as follows (see Leslie and Reimer 1999).

- *Commodity chains* – tracing "the trajectory of a product from its conception and design, through production, retailing and final consumption" (Leslie and Reimer 1999: 404), sequentially joining the dots between the "network of labour and production processes whose end result is a finished commodity" (Hopkins and Wallerstein 1986: 159).
- *Commodity systems* – rather than focusing on isolated case studies of individual commodity types, this concept attempts to map the diverse, interrelated industries, factors and actors that constitute broader systems of, and trends in, commodity production (see Fine and Leopold 1993). That is, individual chains of commodity production are part of much broader, much more complex, systems of cultural production.
- *Commodity circuits* – here, it is argued that commodity chains/systems are not neatly bounded, linear nor unidirectional; rather, they are constantly ongoing and shifting, and there is always potential for the system, commodity and associated meanings to be transformed at any point in the circuit. That is, cultural production processes are part and parcel of much broader, much more complex, and always on going circuits of cultural practices.

These concepts originated largely in analyses of world food systems (Jackson *et al.* 2004), but have subsequently been used to map production processes in every conceivable industrial sector, including systems of cultural production. As we shall detail in Chapter 10, research by geographers – and kindred research by NGOs and charities – has begun to reveal 'hidden' geographies of production behind common commodities, including popular cultural objects, texts and media.

Research on cultural production processes has revealed the agency and economic importance of the 'culture industries': those sectors of the economy that specialise in the creation, distribution and marketing of cultural texts, objects and media. The term 'culture industries' was popularised by the German sociologists Theodor Adorno and Max Horkheimer in a series of publications during the 1940s. Adorno and Horkheimer (1979[1944]) argue that the way in which cultural objects, texts and media were produced changed fundamentally during the twentieth century in Europe and North America. They suggest that cultural objects, texts and media were increasingly

Box 2.4
The cultural turn in economic geography

During the **cultural turn** (see Box 1.3), a number of economic geographers began to critically question two major approaches that underpinned much work in their subdiscipline. These critiques related largely to the limited attention paid to culture within these approaches.

- *Neoclassical approaches in economic geography* - i.e. drawing upon the long-standing traditions, methods, laws and theories developed by economists since the eighteenth century. This approach essentially understands economies as consisting of rational individual actors responding to economic stimuli such as supply, demand, prices, profit margins and market conditions. As Barnes (2005) argues, culture is typically overlooked in this approach, because: (i) theorising economies as reducible to individual actions and responses gives little space for discussion of collective actions and cultural factors; and (ii) the approach's traditional focus on complex mathematical modelling and forecasting, and its aversion to 'soft', 'unscientific' qualitative research methods, preclude rich and effective understandings of cultural processes and factors. We might note, for example, that a number of prominent economic geographers have been quite vehemently dismissive of work by cultural geographers.
- *Marxian approaches in economic geography* - i.e. critiques of neoclassical economics that draw upon Karl Marx's writings on political economy. Marxian approaches often understand economic factors and relations as *determining* social-cultural patterns and inequalities. Here, there is some acknowledgement that cultural forms and processes are worthy of investigation, but typically only as outcomes and indicators of underlying economic issues. Consequently, culture tends to be relatively absent and under-theorised within Marxian approaches: treated as somehow separate from, and of secondary importance to, the really important business of economics. For example, Barnes (2005) notes that culture went almost entirely unmentioned in many large tomes written by leading Marxian economic geographers.

These critiques were sometimes based on caricatures of these approaches, which did a disservice to rich, important work by neoclassical and Marxian economic geographers. However, these critiques prompted many geographers to consider more carefully how culture and economy are complexly interconnected in practice. Massey's (1984) *Spatial Divisions of Labour* is recognised as a key text here. Massey considers the complex relationships between cultural and economic factors in relation to issues such as the location of industries and production processes, and generally calls for a broadening of interests of economic, industrial and urban geographers. For Massey (1984: 7) - and in distinction to neoclassical and Marxian approaches - "the study of industry and production is not just a matter of 'the economic', and economic relations and phenomena are themselves constructed within a wider field of social, political and ideological relations." This concern to understand how economic processes are always already part and parcel of a 'wider field' of contemporary social-cultural geographies prompted diverse, exciting research by what might be called 'cultural economic geographers', as well as many geographers working in related fields. Much key work in this context focused on production processes in relation to two issues:

- how, where and under what conditions commodities - including cultural objects, texts and media - are conceived, produced, circulated and marketed via increasingly complex, globalised production networks, and how these production processes interact with local, regional, national and institutional geographies (Gertler 2003);
- the importance of the cultural, creative and leisure industries within contemporary local, regional, national and global economies, and in relation to urban and regional development or regeneration policies (O'Connor and Wynne 1992).

See Section 2.4 for discussion of these lines of work, and some ways in which they sparked thinking and research within the new cultural geography. See Chapter 3 for discussion of how many cultural geographers worried that these antecedents in economics had led to an overemphasis on cultural production in human geography.

mass-produced, on a vast, industrial scale, in this context, via processes that were increasingly rationalised, upscaled and governed by the imperative to maximise profits. It is fair to say that Adorno and Horkheimer were dismayed at the effects of this shift to mass cultural production: they bemoan the "assembly line character" of many contemporary cultural texts, objects and media, and the "synthetic, planned method of turning out . . . products" (1979[1944]: 163). They argue that cheap, shoddy, unchallenging, populist, 'lowest common denominator' cultural products had increasingly proliferated, and that new technologies of marketing and media 'propaganda' were making cultural consumers increasingly passive, obedient and uniformly uncritical in their 'false desires' for mass-produced cultural products. Just one quotation gives a flavour of the angry pessimism that pervades their work: pause for a moment to consider whether any parts of the following continue to ring true.

> Culture now impresses the same stamp on everything . . . Films, radio and magazines make up a system which is uniform as a whole and in every part . . . As soon as the film begins, it is quite clear how it will end, and who will be rewarded, punished or forgotten. In light music once the trained ear has heard the first few notes of the hit song, it can guess what is coming and feel flattered when it does come. The average length of the short story has to be adhered to. Even gags, effects and jokes are calculated like the setting in which they are placed.
>
> (Adorno and Horkheimer 1979[1944]: 124–125)

Adorno and Horkheimer thus identified:

- a series of historical and technological shifts through which cultural production had become industrialised (and see Negus 1997 on the historical context for these shifts, and for Adorno and Horkheimer's deeply pessimistic outlook in this context);
- a range of new, emerging and increasingly lucrative industries that specialised in the mass production of cultural texts, objects and media.

Later work by economic, industrial and urban geographers has focused on this latter nexus of industrial and economic activities. Whereas Adorno and Horkheimer tended to discuss the 'cultural industries' as a somewhat monolithic and thoroughly problematic entity, more recent work by geographers has explored the complexly interconnected constituents of the cultural industries, and their multiple geographical impacts for local, regional and national economies. Indeed, a key contribution of this recent geographical research has been to highlight the vast, complex range of cultural production processes that are subsumed under the umbrella term 'the cultural industries'.

There have been countless attempts to classify and subdivide these processes. Four methods of classification tend to be widely used in discussions of the cultural industries. First, many classifications subdivide the cultural industries according to the nature of their outputs: so, for example, they can be broken down into sectors such as the television industry, the movie industry, the music industry, the video game industry, the tourism and leisure industries, publishing, the arts, and so on. While this approach can provide a fairly neat and intuitive breakdown of different sectors of the cultural industries, it is also a little problematic, given the increasing prevalence of corporations, activities and cultural products that cut across multiple sectors (see previous discussion of commodity chains).

Second, a number of researchers have attempted to develop classifications of individual occupations: drawing, for example, on census data to count the number, characteristics and whereabouts of individuals working in professions that are classified as constituents of the cultural industries. Note that this kind of classification entails making a judgement about which professions are 'cultural', and includes only those who are in formal paid employment in roles with 'cultural' job titles (perhaps overlooking the role of unpaid practitioners, interns, or those with less obviously 'cultural' job titles working within the cultural industries). Often, there is also an attempt to classify occupations into 'core' cultural activities (typically those directly involved in the making of cultural objects) and 'supporting', 'associated' professions (typically those involved in the distribution, marketing or technical support of production practices).

Third, there are a range of classifications that attempt to identify different processes or moments ongoing within the cultural industries at any given moment. For example, Pratt (2004) recognises that the cultural industries consist of a diverse range of activities, relating to six sets of processes:

- *content origination* – the work of authors, designers, composers or others involved in, or commissioning or supporting, the generation of new ideas, products and intellectual property;
- *exchange* – the work involved in bringing cultural texts, objects and media to audiences and markets,

for example via physical and virtual retail, wholesalers and distributors, and in spaces such as theatres, museums, libraries, galleries;

- *reproduction* – the processes involved in mass production through activities and technologies such as printing, music, broadcasting, production of designed materials and products;
- *manufacturing inputs* – the production and supply of tools and materials used in cultural production (e.g. musical instruments, film equipment, paint, to give just three examples);
- *education and critique* – the work involved in training, fostering and disseminating critical ideas in relation to cultural products;
- *archiving* – including the work of librarians, curators and others engaged in preserving the 'memory of cultural forms'.

Again, in classifications of this kind, there is often an attempt to identify 'core' and 'associated' cultural production activities (see the distinction between cultural goods and cultural services in Box 2.5). Predictably, processes and groupings within the cultural industries are frequently more complex than this kind of neat classification suggests: whenever any classification of this kind is proposed, disputes over which categories are most appropriate, or which occupations fit in which category, tend to follow.

A fourth, and related, way in which cultural industries are often classified involves understanding cultural production processes within the wider context of a 'circuit of culture' whereby

> [c]ultural actors that participate in the cultural production system can be grouped into three categories . . . The first category consists of producers of cultural products and includes designers, artists, architects, and the like, who are involved in the production of the cultural product, which can be a physical good or something more intangible in nature . . . The second category of participants consists of cultural intermediaries – individuals and organizations that are concerned with the communication and distribution (i.e. meaning transfer) of the cultural product to consumers. The final category of cultural actors is the consumers themselves, who transform the cultural products into objects of meaningful consumption experiences . . . The consumer takes the meanings that are transferred from the cultural product as produced and circulated and uses, or transforms, those meanings in the pursuit of consumption and identity construction.
>
> (Venkatesh and Meamber 2006: 13)

Two important characteristics of a 'circuit of culture' approach are: (i) the acknowledgement that cultural production processes are interrelated, and part of a continuously ongoing process; and (ii) the acknowledgement that **consumers** of cultural texts, objects and media have some measure of agency in producing meanings and values related to cultural products (see Chapter 3). As you read round the subject of cultural production, you are likely to see examples of all four of these approaches to classifying cultural industries. All have their merits and drawbacks; however, as we note in Section 3.5, it is always important to acknowledge that cultural production and consumption are complex, and complexly interconnected.

Economic, industrial and urban geographers have explored the importance of cultural industries at different geographical scales. The next three paragraphs present some data from research that has mapped the significance of cultural industries at global, regional and city scales in different contexts.

Since 2008 the United Nations *Creative Economy Report* (UN 2008, 2011) has attempted to map global trends in the economic significance of cultural products. Box 2.5 includes some export data from the 2011 report, providing an indicator of the economic value of cultural goods and services in developed, developing and transitional economies.

Three key geographical trends have been evident in each United Nations *Creative Economy Report,* and in related work by human geographers:

- The cultural industries constitute a significant proportion of most nations' economic productivity, as evident, for example, in import/export and GDP data.
- In most countries, the economic significance of most parts of the cultural industries increased year on year in absolute and proportional terms, even in countries experiencing economic decline or crisis. In developed nations, stimulating growth in cultural industries has often been a key strategy in national policies for urban regeneration and economic growth; it has also been a key component

Box 2.5
Global exports of cultural products and services

Tables 2.1 and 2.2 show the net economic value of cultural products and services in different parts of the world. Note that the data subdivide cultural products into:

- *cultural goods* – relating to the import and export of cultural products including texts, media, artworks, music, design, film, photographs;
- *cultural services* – relating to the import and export of services and expertise that support cultural production, including finance and expertise in intellectual property, telecommunications, information services, business services, advertising, licensing, and technical and audio-visual support.

Take some time to explore some of the key patterns and trends in the data. Notice: (i) how exports of cultural goods and services have changed over time; (ii) any differences between the figures for developed, developing and transitional economies.

Table 2.1 Exports of cultural goods, 2002-08 (in millions of $US)

	2002	*2003*	*2004*	*2005*	*2006*	*2007*	*2008*
Developed economies	127,903	140,884	158,144	171,023	185,895	211,515	227,103
Developing economies	75,835	91,124	109,267	125,321	136,100	156,043	176,211
Transition economies	1,210	1,392	1,920	2,206	2,413	2,741	3,678

Source: from *Creative Economy Report* (UN 2008, 2011), United Nations, pp. 302–303.

Table 2.2 Exports of cultural services, 2002-08 (in millions of $US)

	2002	*2003*	*2004*	*2005*	*2006*	*2007*	*2008*
Developed economies	52,457	61,320	73,185	81,998	125,218	138,045	153,414
Developing economies	7,860	8,185	9,363	12,771	17,133	18,835	21,182
Transition economies	1,910	2,803	3,483	4,467	5,385	7,279	10,491

Source: from *Creative Economy Report* (UN 2008, 2011), United Nations, pp. 319–320.

of many national governments' responses to global economic recessions. In developing nations, investment in cultural and creative industries has increasingly been adopted as a strategy to drive economic development (as evident in some of the rapid sectoral growth rates for developing transitional nations in Box 2.5).

- There are substantial global disparities in the ownership and control of cultural production processes. Cultural goods can be produced anywhere, but profits arising from cultural products overwhelmingly flow towards developed nations, where corporations providing finance, infrastructure and business services for cultural production are typically based. This is reflected in the major disparity in the value of cultural services exports in Box 2.5. As we noted in Section 2.3, the tendency for cultural production processes to ultimately benefit, and be controlled by, a geographically and socially narrow and exclusionary elite was widely critiqued within cultural studies. Research by human geographers has evidenced how, in diverse global contexts, the economic value of cultural creativity is ultimately realised by corporations located in the global North.

Research on the importance of cultural industries to particular national economies has typically revealed the uneven geographical spread of cultural industries, and significant regional disparities in participation in cultural production. Box 2.6, for example, reproduces some data about employment in cultural industries in different regions of the UK: again, spend a few moments exploring some of the key patterns in the data.

Box 2.6
Clustering of creative industries in Great Britain

Table 2.3 shows the degree of clustering evident in different sectors of the cultural and creative industries in regions of Great Britain. The data are presented as location quotients: a calculation of the degree of clustering of each industry, in each region, compared with the national average. A value greater than 1 indicates that clustering in this sector is greater than the national average; a value less than 1 indicates that clustering in this sector is less than the national average. Take some time to explore the geographical patterns evident in the data. Notice where cultural and creative industries tend to be clustered.

Table 2.3 Location quotients of some cultural industries in regions of Great Britain

	North-East	*North-West*	*Yorks/ Humber*	*East Midlands*	*West Midlands*	*East*	*London*	*South-East*	*South-West*	*Wales*	*Scotland*
Advertising	0.69	1.18	0.74	0.72	0.76	0.91	1.77	1.06	0.80	0.42	0.55
Architecture	1.39	1.07	0.86	0.93	0.97	1.04	0.81	1.06	0.96	0.75	1.42
Arts/ antiques	1.09	1.05	1.09	0.98	1.03	0.97	0.82	0.95	1.15	1.10	1.08
Designer fashion	0.64	1.15	0.77	2.73	0.98	0.55	1.73	0.39	0.55	0.48	0.76
Video/film/ photogra-phy	0.55	0.57	0.56	0.49	0.50	0.71	2.68	0.94	0.77	0.55	0.69
Music/visu-al/perform-ing arts	0.55	0.62	0.59	0.59	0.55	0.82	2.36	1.00	0.88	0.73	0.60
Publishing	0.51	0.62	0.65	0.70	0.66	1.06	1.82	1.13	1.07	0.64	0.75
Software/ computer games/ electronic publishing	0.71	0.97	0.64	0.73	0.81	1.09	1.31	1.41	0.87	0.52	0.75
Radio and TV	0.38	0.53	0.36	0.30	0.43	0.56	3.05	0.90	0.74	0.96	0.56

Source: from de Propris, L., Chapain, C., Cooke, P., MacNeill, S. and Malteos-Garcia, J. (2009) The Geography of Creativity, NESTA, London, p. 19., JCIS/ABI and ONS (2007).

In Box 2.6, as in most national surveys of cultural production, it is evident that the cultural industries are unevenly distributed, and disproportionately clustered in a small number of (typically metropolitan) locations. In the UK, for example, it is clear that practically every element of the cultural industries has a major cluster in London and the South-East of England. This disproportionate London-centric clustering of employment in cultural industries reflects the following factors.

- For historical reasons, many major media corporations, and organisations providing finance, infrastructure and business services for cultural production, have headquarters in the city.
- A series of national policy interventions have sought to incentivise location and expansion of cultural industries in this region; simultaneously, the region has been actively promoted as a hub for multiple forms of cultural production.
- Clustering is encouraged by a trend towards agglomeration within cultural industries, whereby increasingly diverse aspects of cultural production are bought out, and come under the control of,

globalised organisations with headquarters in the capital.

- Within many sectors of the cultural industries, this region has come to have a special status as 'the place to be' to get ahead; consequently there are dense localised networks of cultural producers, decision-makers, intermediaries and critics located in particular districts or cultural quarters.
- Clustering and co-location of cultural industries afford economic and practical benefits for organisations and individuals: for example, in the UK, particular groupings of cultural industries (e.g. advertising/design/fashion/electronic publishing, and music/performing arts/publishing/broadcasting) tend to co-locate, producing all manner of mutual benefits, efficiencies, collaborations and spin-offs.

It is also evident from Box 2.6 that, away from the South-East of England, there are a number of regional clusters of more specialised activity: for example, note the clusters of television, photography and film professionals in the South-West, advertising and media in the North-West, and computer games and technological support in the East of England. Geographical research has revealed how these clusters can reflect:

- specific locational and historical factors, such as the location of BBC studios in Bristol in the SW region, and the clustering of technology firms around Cambridge University (in the East region);
- regional economic, regeneration and urban renewal policies that have actively sought to construct spaces for, and regional specialisms in, cultural production;
- place promotion and urban renewal strategies that have sought to market a 'buzz' around the cultural industries in particular locations in order to attract investment and inward migration (O'Connor and Wynne 1992);
- the activities and networks of a relatively small number of cultural producers in the region (see below).

Geographical research focusing on individual cities has revealed how, even at very local levels, there is a significant degree of clustering and disparity in the location and importance of cultural industries. Analysis of these geographies reveals how such clusters exist and function in practice via networks of relationships between individuals – encountering one another as professionals, practitioners and/or friends – within a small number of events and everyday spaces located in particular urban districts. It is through the existence and snowballing of such networks that clustering in the cultural industries proceeds in practice, and regional 'hot spots' of cultural production can come into being (Granger and Hamilton 2011). These kinds of interpersonal networks can, of course, also be rather exclusionary and inaccessible to those who do not have knowledge of, access to, or the inclination to participate in particular networks, spaces and events that are key to the cultural industries in a particular locale.

2.5 Producing and regulating cultural spaces

As we shall explore in Chapter 13, all **places and spaces** are *produced* through, for example: processes of planning, design and **architecture** (see also Chapter 4); regulations and rules (both formal and unspoken); infrastructural and material construction and maintenance; practices, routines and systems; social interactions and relationships; meanings, norms, representations and habits. In this section, we focus on the extensive work of urban geographers on the production and control of cultural spaces. In this context, cultural spaces can be defined (after Cloke *et al.* 2004: 139–141) as spaces in and through which:

- there is some purposeful attempt to plan for, control, direct or educate people towards certain (shared) ways of life (which may vary from global consumer cultures to nationalistic rituals and traditions);
- there is some attempt to cultivate human minds – this can occur via different media whose meaning is meant to guide certain ways of life (these meanings can be represented in diverse ways, including maps, landscape design, spectacular public gatherings, paintings, advertisements and more);
- seemingly non-productive activities, outside the realm of work, may be co-opted by experts (in general terms, 'non-productive' means the kinds of activity

that centre around 'leisure', 'fun', 'relaxation' and learning outside formal, directed educational curricula);

- there is some attempt to plan or direct people's identities. By this, we mean not only the names or types that allow us to classify different groups. In addition, identities may refer to the broader realm of bodily performances and emotions that tend to make cultural spaces somehow 'special' or 'meaningful' to the ways in which people understand themselves (see also Chapters 7, 8, 11 and 12 on **performances, identities, emotions** and **bodies**).

However, as you read this chapter it is important to recognise that cultural spaces are always *contested* – both between experts and 'non-expert' publics, and between the diverse groups that seek to lay claim to spaces both within and beyond the direct control of experts. We look specifically at the issues of contestation, **subcultures** and **resistance** in Chapters 3 and 8.

To explore the planning and regulation of cultural spaces, we shall focus on two distinct forms of spatial production from particular historical moments: early twentieth-century urban planning in the UK, and the making of 'public' spaces in twenty-first-century cities. We then explore the ways in which cultural identities are also regulated and controlled in and through such spatial processes.

When we talk of 'planning' cultural spaces, it is hard to ignore professional disciplines such as urban planning and urban design. The work of urban planners has provoked a diverse field of study for geographers and planning historians alike (Pinder 2005a). We begin with a number of scholars – whom we call 'urban experts' as shorthand – who were becoming increasingly concerned with the condition of Britain's cities after the aggressive industrialisation of the Victorian era. Urban experts fought for greater and more orderly planning of urban spaces on two fronts:

- first, in order to overcome the squalor, poverty, dirt and disorder of the slum-like communities inhabited by poor factory labourers;
- second, in order to counter the increasingly unregulated growth of Britain's suburbs along first rail and then later road arteries.

Consider the first, highly judgemental statement above about the 'slum-like' conditions prevailing in Victorian and post-Victorian cities. It is precisely such assumptions about what had to be planned *out* of such cities that cultural geographers have sought to unpick. In an excellent study, Stallybrass and White (1986: 125) show how the impoverished spaces of the Victorian city were cast by contemporary parliamentarians and social reformers as "the locus of fear, disgust and fascination". Rather than produce objective, rational, 'scientific' knowledge about those places, they created loaded representations of urban landscapes, replete with sewers and street urchins. Once these landscape images had been produced, elite experts could justify their intervention on a number of fronts: from the planning of housing built by philanthropists (Dennis 1989) to the introduction of sanitary measures designed to clean up city sewers, rivers and open spaces (Allen 2008). Hence cultural geographers have shown how moral and subjective images of **landscape** (Chapter 5) are central to the planned interventions made by experts in 'problematic' urban places.

This work has been furthered by a number of historical and cultural geographers who have explored the ways in which experts made moral judgements about the fabric and social lives of cities in the past. In other words, they have explored what we often term the 'moral geographies' of life in those cities (Smith 2000). Those moral geographies often led to overt forms of social and spatial *control* as well as planning. Let us take one example. Howell (2009) has explored the moral geographies surrounding the policing and regulation of prostitution in nineteenth-century cities. He argues that the depth and extent of British regulation of commercial sex practices were much greater than is traditionally imagined. Significantly, Howell also explores congruencies between Victorian Britain and contemporary British laws regarding prostitution (Howell *et al.* 2008), and the transferral of racialised and sexualised ideas about prostitution to Britain's colonies in Gibraltar and Hong Kong (Howell 2004a, 2004b). However, he emphasises that modes of control and regulation were not simply 'exported' from Britain to its colonies. Rather, precise forms of control were formulated at each location, and often informed the regulation of sex work 'at home'.

Look back at the second front on which British urban experts fought their early 'battles': the increasingly

Photograph 2.2 The 'octopus' of urban development? Ribbon development in Llywngwril, Wales.
Source: Gwynedd Archaelogical Trust.

'unregulated' growth of Britain's suburbs. The most famous attack on this unfettered development came in the form of Clough Williams-Ellis's (1928) *England and the Octopus*. Williams-Ellis likened the ribbon developments extending alongside major traffic arteries to the flailing, uncontrolled tentacles of an octopus (Photograph 2.2). At the time, the octopus was an attractive and popular image among urban experts: it represented the rapid colonisation of previously rural landscapes by a monstrous beast that was blurring the boundaries between city and country. For Williams-Ellis and others, these hybrid developments – neither city nor country – were vulgar, tasteless and, above all else, *disordered*.

Although we have considered them separately, it is important to note that early urban experts tended in practice to combine their criticisms of nineteenth-century urban squalor and twentieth-century disorder. As Pinder (2005a: 31–32) puts it: "numerous critics . . . opposed the disorder associated with urban expansion and industrialisation." Hence, as Matless (1998) argues, it is the distinction between spatial order and disorder – in the contexts of expansion and industrialisation – that is key to understanding the geographies of early urban planning. Cultural geographers have therefore sought to understand how planners developed languages, tools and designs that would create more orderly landscapes and, crucially, preserve the distinction between 'urban' and 'rural' places. Matless (1993) shows how early planners in the UK and Belgium drew on the image of the octopus to argue for a vigorous, authoritative antidote. They called "on planning to be itself energetic and visionary, not proddingly piecemeal or cloggingly bureaucratic" (Matless 1993: 172). This was nothing less than a call to arms – to create a unified, systematic and integrated approach to planning at a scale – a national scale – that had never been seen before. Matless (1993) also shows how planners drew upon new techniques for visualising the landscape, such as aerial photography, to express both the expert *mapping* of the problem (the octopus) and the objective, detached manner in which solutions could be planned. Hence the importance of the top-down 'masterplan' to contemporary planning!

Debates about urban planning were not confined to Britain, however. As Pinder (2005a) shows, a turn to modernism in both philosophy and design affected urban planning in many contexts: from the Italian Futurists, who designed cities around light and speed, to the French architect Le Corbusier, who was similarly critical of the squalor and disorder of industrial cities. In many contexts, new cities were meant to both represent and literally embody (in concrete form) the aspirations of a new age based around factory production, machines, and a rational approach to the organisation of society. Where better to start than with orderly, well-laid-out cities connected by efficient public transport and providing healthy high-rise living for the masses – all planned by a new generation of planners?

In its early days, urban planning held at its core a strong sense of hope (for a modern future) and of social-moral duty – utopian ideals to do no less than reconstruct the city in a more modern image. The utopian ideals of the early planners were certainly contested and problematic – not least if we consider the reputation of high-rise social housing in many countries today (Jacobs 2006). But if we move forward to the more recent past, we see that certain aspects of contemporary urban planning bring with them a set of different debates. We focus here on the diverse buildings, streets and urban forms that constitute public spaces. Geographers and other urban scholars have had a long-standing interest in what we call 'public spaces' (Madanipour 2003; Iveson 2007). Iveson (2007: 3) defines public spaces as:

> Spaces where people can (re)present themselves before an audience of strangers . . . The term 'public space' is often used to denote a particular kind of place in the city, such that one could colour public spaces on a map – this is a *topographical approach*. By contrast, however, the term public space is [also] sometimes used to refer to any space which is put to use at a given time for collective action and debate – this is a *procedural* approach.

Public space, then, refers to those streets, parks and other open spaces that – in both commonsense and ideal terms – any member of the public may use. Public open spaces have been a key part of the city throughout history – from the Ancient Greek public square or *agora*, to the medieval English market square, and from the boulevards of nineteenth-century Paris to the vast open spaces and efficient thoroughfares of modernist planning (Madanipour 2003). Arguably, many of these types of public space entailed a different form of *planning*, because each expressed the ways in which dominant members of society sought to define and control social interactions between members of the public.

In the last decade or so, geographers have become increasingly concerned with public spaces. Their interest has been spawned by two related processes that are taking place in many different global contexts. On the one hand, they have suggested that public spaces increasingly reflect the imperatives of consumer capitalism, neoliberalism and postmodern notions of 'heritage' (Zukin 1995). On the other hand – and as a result – geographers have criticised how contemporary urban spaces are becoming increasingly mono-functional, and hence subject to greater forms of control and exclusion than ever before (Sorkin 1992).

So how *do* planners reflect the imperatives of consumer capitalism, neoliberalism and 'heritage' in public spaces? A central argument sits at the confluence of economic and cultural geography (see also Boxes 2.4 and 2.7). It is argued that previously inclusive, multi-functional public spaces in cities are being increasingly financed and planned – at least in part – by private businesses (Harvey 2000). To generalise, this means that many public spaces are now no longer designed, owned or controlled by democratically elected governments who should cater for all members of the public. Rather, public spaces are becoming privatised – their use by ordinary members of the public feeds into the profit-making activities of the corporations that own them. This means that public spaces are increasingly being designed with a narrow, 'ideal' public in mind – the middle-class, often white, adult who will literally buy in to the activities offered and commoditised in these new public spaces. Let us look at some examples.

To give one example of the above trends, many global cities have witnessed the erection of high-profile buildings, often built by 'superstar' architects such as Richard Rogers, Norman Foster or Daniel Libeskind. Indeed, as we wrote this chapter, the tallest building in the world – the 800-metre-tall Burj Khalifa – had just been opened in Dubai, accompanied by massive press attention (Photograph 2.3). Its opening came at a time when Dubai was beleaguered by problems associated with the world financial crisis that began in 2008, and was figured in many press reports as a new sign of hope for Dubai.

Buildings like the Burj Khalifa have been subject to several debates in cultural geography. For some commentators, they represent the pinnacle of competition between cities. As 'entrepreneurial cities' (Hall and Hubbard 1998) seek to press home their status as loci for investment, tourist visits or cultural importance, they increasingly turn to high-profile building

Photograph 2.3 Burj Khalifa building, Dubai.
Source: Shutterstock.com/Kjersti Joergensen.

projects that act almost as advertisements for them. Other commentators show how in some cities – most notably Las Vegas – buildings obtain a high profile because they overtly symbolise what Hannigan (1998) calls the 'fantasy city' (Photograph 2.4). In Las Vegas and elsewhere, urban environments are being planned and sold as playgrounds for leisure, dining, shopping and gambling, where consumption buys temporary escapism and a fantasy lifestyle.

At a perhaps more mundane (and certainly less spectacular) scale, notions of escapism, fantasy and play are built into the shopping malls that have appeared both in city centres and in suburban locations around the world (Goss 1993; see Photograph 2.5). In Britain, the traditional 'high street' – a public thoroughfare in the centre of a town containing the principal shops and public buildings – has in many towns been replaced or accompanied by a covered, climate-controlled shopping centre containing major, globalised retail chains. Such environments are deemed to be safer and more conducive to the main (cultural) business of a town centre – shopping and, increasingly, associated 'café cultures' (see discussion of **consumption spaces** in Chapter 3).

Certainly, in the era of the 'entrepreneurial city', these shopping malls are supported and often planned by public authorities and city planners, as a key way to revitalise urban centres. Indeed, Zukin (2003) shows how in the United States a breed of waterfront shopping centres were, in the 1990s, supported by federal tax laws that attracted investors and paved the way for the commercial redevelopment of 'tired' downtown areas in the guise of 'historic preservation'. Built on disused piers or port areas, projects such as Boston's Faneuil Hall and New York's South Street Seaport sought to retain elements of the architectural past – such as cobbled pavements, old boats and original dockside buildings. Yet these are places based on "visual consumption", which "present shopping as a means of enjoying urban culture" (Zukin 2003: 183). The 'past' is arguably stripped of all but simplistic, visual cues that can be reinserted into a landscape fit for consumption by the (middle-class) public (see Box 2.7).

For cultural geographers, the most worrying feature of all of the above examples is the *façade* of publicness that operates in the service of private profit – in other words, the increased blurring between public and private in spaces designated for 'public' use. Zukin (2003: 183) suggests that "[s]hopping centres have replaced political meetings and civic gatherings as arenas of public life". Similar arguments have been made for a range of other urban forms planned by experts. For instance, Davis's (1990) rather apocalyptic depiction of shopping malls, hotels and gated communities in Los Angeles sees them as preserves of the rich, built and controlled by private development corporations. While presenting often ideal images of 'community' and urban life (Till 1993), they exclude both any notion of 'real' public space *and* anyone who does not conform to the desired 'norm' for clientele in those places. Hence processes of exclusion – which rely on surveillance via closed-circuit television and increasingly militarised forms of control – are involved in the production of cosy, safe spaces where middle-class

Photograph 2.4 The Strip, Las Vegas, Nevada.
Source: Peter Kraftl.

Photograph 2.5 Shopping mall inside the Sands Marina Bay hotel complex, Singapore.
Source: Peter Kraftl.

Box 2.7
Revitalisation of public space in Perth, Australia

Using the example of Perth, Australia, Iveson (2007) shows how a diverse coalition of experts - planners, academics and architects - came together in the early 1990s to revitalise Perth's Central Area. They had three aims: to promote Perth as a 'capital city' for the State of Western Australia; to increase desire for inner city living; and to improve law and order in the city centre. Each of these linked agendas called for the reinvention of public space in the city - ostensibly making the central core more diverse, more attractive to live in and visit, safer, accessible and truer to the city's heritage. In physical terms, a variety of 'public space enhancement' projects took place. These included flagship refurbishment of major public squares, widening of footpaths, new seating and shading in shopping malls, conversion of older buildings into housing, and investment in transport infrastructures. Yet the imperatives of these projects echo many of the processes listed above - from a focus on middle- and high-income housing to marketing campaigns that were "primarily designed to attract shoppers and potential residents from the suburbs" (Iveson 2007: 162).

consumers are cocooned. For further discussion of exclusion, see Chapters 7 and 8 on **performances** and **identities**. Similar arguments have been made regarding the 'end' of public space in fantasy spaces such as theme parks and debates over the intended audiences for public art, statues and memorials.

In sum, then, the planning and control of cultural spaces have *always* entailed a measure of controversy. While experts may be in charge of these processes (at least officially), their work is rarely objective or value-free, often entailing moral judgement. Indeed, if we look back at the history of urban planning from the nineteenth century to the present day, we see a continuing trend whereby processes of planning and control quickly become processes of surveillance, marginalisation and exclusion. Very often, cultural spaces are planned and controlled in the service of both dominant social norms and monetary profit. Hence they fulfil the relatively narrow demands of hegemonic (male, white, middle-class) social groups. In the remainder of the chapter – and with this in mind – we begin to explore how cultural experts have attempted to appeal to wider segments of society by planning and controlling cultural **identities** (see also Chapters 3 and 8).

Cultural geographers have had a long-standing interest in identity. A particular, but key aspect of this study – and perhaps the most overtly geographical – has been an interest in national identity. In the following paragraphs, we shall look at two examples of the production of identity by 'experts' – in many cases ostensibly national and even international identities – before making a couple of concluding points. We focus first on public 'events', and then on education and schooling.

As already discussed, the planning of public spaces involves diverse cultural processes. But the story does not end there. Planned public events, such as festivals and parades, are also crucial cultural elements in the making of public space. Simply put, they provide some of the vitality that makes those spaces 'public' (Watson 2006). In this way, the performative element of planning culture comes to the fore (see also Chapters 7 and 12 on **performance** and **bodies**). Experts do not simply plan public spaces; they also plan – to a certain extent – what goes on in them too. You can probably think of numerous examples – from street entertainers outside shopping malls to bus parades by successful national sports teams. Importantly, those kinds of public event often serve to define, produce and reinforce cultural identities. Although public spectacles are temporary – and often viewed as apolitical and inclusive – cultural geographers have shown how they "convey and perform . . . statements of identity, power and authority" (Hagen 2008: 350). Cultural geographers have focused on a variety of public events, linking these in particular with particular national identities. Hagen (2008) shows how public celebrations and parades in German cities such as Munich were hugely symbolic and powerful expressions of national identity in Nazi Germany before the Second World War. Whereas the Nazi party tried to communicate nationalist conceptions of art, culture and history through exhibitions and built form (for instance in Nuremberg), Hagen (2008: 349) argues that "the parades offered a more participatory, populist format that projected the regime's message on to Munich's public spaces" (see also Hagen and Ostergren 2006). The parades drew on various symbolic resources – from their sheer scale, encompassing tens of thousands of parading participants, to the route through Munich itself, which "intended to depict Nazism as the climatic finale and culmination of German history" (Hagen 2008: 361).

Elsewhere, a number of cultural geographers have assessed the narratives of national and imperial identity involved in exhibitions and expositions aimed at an international audience (Pred 1991; Strohmayer 1996). Domosh (2002) examines the Chicago World's Columbian Exposition of 1893. She suggests that the Expo displayed two types of exhibit: first, depictions of Native Americans that emphasised the superior civilisation of European-Americans and their domination over both 'nature' and the Native Americans who were deemed part of nature; second, commercial displays of American products sold overseas, which were meant to be a showcase for American progress and global economic dominance – a showcase available to the masses. While the Expo was actually designed to market and sell American agricultural technologies, the imagery used there showed farm equipment as the highpoint of American civilisation, even depicting (strong, white) men "as Roman warriors, using harvesting machines to conquer the world" (Domosh 2002: 182). Although the economic and political bases could not be more different in both these examples, we see how public spectacles are used not only to promote a particular kind of national identity but, more forcefully, to symbolise the 'pinnacle' of a civilisation.

The previous examples all have in common a spectacular occupation of public, urban space. They often took place in spectacular settings as a way to reinforce their importance. But cultural geographers have also looked at a range of festivals that – although taking place in less spectacular settings – are no less important to the production of local, regional or national identity. For instance, Mathewson (2000) explores some of the festivals and celebrations that are associated with modern farming practices. These festivals include harvest festivals, local food celebrations, and events that mark, for instance, the maturation of a new season's wine. In common with some of the trends to consumerism noted earlier in this chapter, though, Mathewson (2000: 466) argues that "increasingly, these events are staged promotions of local, regional or ethnic identity" – and not directed solely to a celebration of food or community (Photograph 2.6).

Other work on local festivals has drawn out further fields of contestation along different axes of identity. Waterman (1998) explores the Kfar Blum festival in Israel, an annual chamber music festival held at a kibbutz and organised by three public authorities. The festival was initially attended by elite groups interested largely in the music itself. That elite group saw clear links between the festival and their sense of Israeli national identity. But, following demands to increase participation to a wider range of people, the festival has changed, and become more commercialised. This change is a controversial one in terms of the relationship between the festival and Israeli national identity. On the one hand, it questions the role of public authorities in supporting music festivals and using them to educate the populace about 'appropriate' forms of national identity. On the other hand, these changes are indicative of wider struggles in Israel about who has cultural 'power' and what is the perceived value of 'authentic' (elite) versus commodified versions of culture.

Elsewhere, Greene (2005) shows how, between 1840 and 1940, civic leaders in Milwaukee, USA, held 'multiethnic' festivals. Like other Midwest cities, Milwaukee was – according to the statements of civic leaders – appreciative of its diverse immigrant population and differences in dress, music and dance. Yet Greene argues that they did so in a paradoxical way that allowed immigrants to celebrate and express their ethnic cultures while 'teaching' them English language and American democratic principles in school social centres.

The final two examples cited above highlight the importance of *education* to the planning of cultural spaces. In both cases – indeed, perhaps, in all of the

Photograph 2.6 Union Square farmers' market, New York City.
Source: Peter Kraftl.

events mentioned above – cultural festivals do not only symbolise national or regional identities but also have some kind of educative function. Historical work in cultural geography has made this point most fully. Much of this work makes a direct connection between planned educational spaces and curricula, and forms of national identity. Indeed, one of the key ways in which national cultures *can* be planned is through education – and especially the schooling of young people.

Gagen (2004) argues that these kinds of educational process were explicit in the design and use of playgrounds in early twentieth-century New York. There, new advances in educational theory suggested that physical education (and especially of immigrant populations) would allow children to directly assimilate the more 'desirable' character traits – and especially a sense of national identity and belonging to the American nation. Indeed, echoing examples above, a key part of their education was a series of exhibitions – field days and parades that encouraged children to learn nationalistic rituals. Ploszajska (1994) makes a similar point about embodiment. She shows how in English schools between 1870 and 1944 children were encouraged to learn about the world by making models of volcanoes, mountains and other geographical phenomena. But these models were not simply meant to improve children's learning; rather, they encouraged English children to think about "their individual and collective place as local, national, Imperial or global subjects in the world" (Ploszajska 1994: 389). In particular, the model-making asked children to be more sympathetic to geographical and ethnic differences while simultaneously using this awareness to gain a better appreciation of the value of their own locality and sense of local belonging.

A number of geographers have also made connections between education, national identity and landscape (see also Chapter 5 on **landscape**). Gruffudd (1996) shows, for instance, how the Welsh educational system reflected debates about the 'rebirth' of the Welsh nation between the First and Second World Wars. The system educated children about a new kind of 'patriotic pleasure' – it called upon them to celebrate their Welsh heritage. But it also fostered a sense of national revival – of a nation bouncing back. Gruffudd (1996) shows that a key part of this education involved an appreciation of the natural beauty of Welsh landscapes. Activities included processions around the local area, speeches on 'scenery and why all good citizens should protect it', and the recital of patriotic poems about the Welsh landscape.

This section has shown that the planning of cultural spaces does not simply involve the design of physical spaces. Rather, such processes may mobilise different kinds of cultural practice and place in order to produce specific versions of national identity. Those spaces may range from the classroom and the playground to the spectacular sites of public parades and international exhibitions. But they all have two things in common. First, events and performances are used to enliven cultural spaces and lend them a sense of liveliness, fun, and public togetherness that can rarely be fostered by design or town planning alone. And second, there have been – and continue to be – a range of creative techniques for communicating, educating and in some cases forcing senses of national identity onto citizens. Crucially, these techniques rely on diverse *cultural* practices and spaces – art, dance, sport, poetry and more. These are cultural practices and spaces that lie at the very heart of the study of cultural geography.

Summary

- Cultural production has been a foundational concern for many cultural geographers, one way or another. Much key work in cultural geography has tracked the ways in which cultural texts, media, objects and spaces are made. Conversely, many cultural geographers have begun from a conviction that *too much* geographical work has focused on cultural production (see Chapter 3).
- Some key concepts from cultural studies have been useful to geographers studying cultural production. Notions of discourse, hegemony and distinction recur in the work of many cultural geographers.
- Cultural geographers are by no means the only geographers interested in questions of cultural production. For example, there is a great deal of work by economic, urban and industrial geographers about commodity chains and the cultural industries. Elsewhere, urban geographers and planners have made important contributions to our understanding of the creation, regulation and maintenance of different kinds of cultural spaces and identities in diverse contexts.
- Although many researchers have focused on cultural production, it is important to understand that production is not separate from other processes: see Section 3.5 for a discussion of this point.

Some key readings

Crane, D. (1992) *The Production of Culture: Media and the urban arts*, Sage, London.

A useful resource for anyone interested in how cultural production processes, creative industries and urban spaces are interrelated.

du Gay, P. (ed.) (1997) *Production of Culture/Cultures of Production*, Sage, London.

Many students find this a really accessible and helpful introduction to questions of cultural production. The book is particularly useful in the way it grounds concepts of discourse, hegemony, distinction and circuits of production in real-life case studies.

Eldridge, J. (ed.) (1993) *Getting the Message: News, truth and power*, Routledge, London.

A fascinating collection of essays considering the processes through which news media are produced. It provides numerous examples illustrating how hegemonic discourses are produced in practice, and circulated in everyday news media.

Gertler, M. (2003) A cultural economic geography of production. In Anderson, K., Domosh, M., Pile, S. and Thrift, N. (eds) *Handbook of Cultural Geography*, Sage, London, 131–146.

A thoughtful consideration of the cultural turn in economic geography, which outlines how questions of cultural production should be important for human geographers.

Granger, R. and Hamilton, C. (2011) *Breaking New Ground: Spatial mapping of the creative economy*, Institute for Creative Enterprise, Coventry.

An interesting study of local and regional clustering in the cultural and creative industries, and the causes and consequences of this clustering. Think about how the patterns observed in this case study might apply to your home town.

Massey, D. (1984) *Spatial Divisions of Labour: Social structures and the geography of production*, Macmillan, London.

A classic text, which argues that geographers should consider connections between economic factors, urban spaces and cultural processes – for example, as evident in the location of production processes and cultural industries.

UN (2008, 2011) *Creative Economy Report*, United Nations, New York.

A fascinating overview of recent data about cultural and creative economies, and geographies of cultural goods and services. Check the most recent edition for global background information to the issues discussed in this chapter.

The key readings suggested at the end of Chapter 10, on the **production of material things**, are also relevant to this chapter. In particular, we strongly recommend the work of the geographer Ian Cook to anyone interested in exploring the complex practices through which commodities and cultural products are produced in practice.

Cultural consumption

Before you read this chapter . . .

- Make a list of all the cultural objects, texts and commodities you have used so far today.
- What have you *done* with these objects in practice?
- How do these practices matter to you and your sense of identity?

Chapter Map

- Introducing consumption
- Consumption: *doing* culture
- Geographies of cultural consumption
- Consumer agency: subcultures and resistance
- Connecting cultural production and consumption

3.1 Introducing consumption

In Chapter 2, we noted that a great deal of classic work by cultural geographers and cultural studies researchers focused on processes of **cultural production**: that is, how cultural objects and spaces are made, regulated and come to have particular associated meanings and norms. We reiterate that this body of work was very significant in the development of cultural geography, and in unveiling previously hidden cultural processes and politics. However, if you have read Chapter 2, a few nagging questions may already have occurred to you:

- What do people actually *do* with all these cultural objects?
- What *actually* goes on in cultural spaces once they have been produced?
- How *do* people deal with hegemonic forms of cultural power in their everyday lives?

Questions like these have motivated a second major body of work within cultural geography and cultural studies that has called for a focus on 'consumption', exploring consumers' practices *with* cultural objects and *within* cultural spaces. In this chapter, we introduce notions of consumption and consumer culture, and explore how geographers have contributed to understandings of them. Section 3.2 describes a 'turn' towards the study of consumption within cultural studies and human geography during the 1990s. The chapter's next two sections outline two important lines of work that resulted from this 'turn'. Section 3.3 focuses on research that addresses the complex geographies of consumption, encompassing how consumption matters in everyday lives, the production of spaces *for* consumption, and practices that involve the consumption *of* particular landscapes or spaces. Section 3.4 foregrounds research about links between consumption and identities, which has been important in exploring how people deal with, and often resist, hegemonic cultural politics within their everyday lives. Finally, Section 3.5 is a kind of concluding reflection for Part 1 of the book: here, again, we emphasise the importance of thinking about connections between the processes described in Chapters 2 and 3.

3.2 Consumption: *doing* culture

During the early 1990s many cultural theorists and human geographers called for a turn away from the kinds of production-focused research featured in Chapter 2, towards a focus on 'consumption'. So what are cultural geographers talking about when they talk about 'consumption' in this context? In Box 3.1 we offer both a short answer and a long answer to this question. For the moment, the short answer will suffice: consumption refers to the purchase and use of commodities, texts and services. To study consumption is therefore to explore what people actually *do* with cultural objects and spaces in practice, once they have been produced. However, as you read this chapter, you should begin to notice that consumption is no simple process: this is where the 'long answer' in Box 3.1 serves as a reference point. For, ultimately, it is important to recognise that buying and using any commodity is a deceptively complex process; after all, it involves all manner of actions, feelings, meanings, decisions, desires and social interactions, all contextualised by the norms, discourses and cultural politics detailed in Chapter 2. Moreover, it will become clear that the term 'consumption' has been used to refer to a vast array of practices: "everything from art to

Box 3.1
What do cultural geographers mean by 'consumption'?

The short answer . . .

> [T]he complex sphere of social relations and discourses which centre on the sale, purchase and use of commodities.
> (Mansvelt 2005: 6)

The long answer . . .

As the cultural theorist Raymond Williams (1976: 78–79) outlines, the word 'consumption' has had a complex history.

1. In its earliest uses, from the fourteenth century onwards, to 'consume' (from the Latin *consumere* – 'to completely use up') meant to physically devour or ravage: "to destroy, to use up, to waste, to exhaust" (Williams 1976: 78). This meaning survives in terms such as 'consumed by fire' or 'consuming a meal', or in metaphors such as 'consumed by grief' or 'consumed by envy'. Similarly, until the nineteenth century, the disease now known as tuberculosis was commonly called 'consumption', to evoke the rapid, devastating and often terminal way in which the disease 'laid waste' to human bodies.
2. A closely related word, 'consummate' (from the Latin *consumare* – 'to bring to completion') has been used since the sixteenth century to describe a state of fulfilment attained through the completion of a particular task or transaction. The term has a particular risqué meaning too: traditionally, consummation refers to 'completing' a marriage via the first sexual relations between newlyweds. In each of these senses, consummation refers to both the act of completion and the sense of satisfaction arising from doing so.
3. In eighteenth-century Europe a new notion of 'the consumer' emerged to describe someone making a purchase or making use of some kind of good or service. Here, 'consuming' meant the act of buying and using some kind of commodity. Williams (1976) suggests that the emergence of this new notion of 'consumption' was closely linked to the emergence of new, increasingly large-scale and organised forms of capitalist political economy. He suggests that traditional forms of buying and selling had involved close, loyal relationships between suppliers and their customers. New, large-scale capitalist commodity markets led to the breakdown of these kinds of relationship, and a new vocabulary ('the consumer', 'consuming') was needed to describe and legitimate the more detached, alienated relationships that existed between buyers and

Box 3.1 continued

sellers in this context. This notion of 'the consumer' has become increasingly popularised over the last century, in terms such as 'consumer rights', 'consumer culture', 'consumer goods' and so on.

Clarke *et al.* (2003: 1) note that these different meanings are kind of "rolled into one" when cultural geographers talk about 'consumption' now. So consumption is about:

- the process of selecting, purchasing and using some kind of commodity or service (as in point 3 above);
- *and* the literal, physical, bodily act of using up, wearing out or making use of that commodity or service (point 1);
- *and* the ways in which commodities or services are put to use; the kinds of things that consumers do with them in practice, and what they get out of them, and the resulting meanings, feelings and accomplishments (point 2).

All of this, and more, is going on in "the complex sphere of social relations and discourses which centre on the sale, purchase and use of commodities" (Mansvelt 2005: 6).

shopping, to watching television, receiving welfare and visiting a zoo" (Edgell and Hetherington 1996: 1). Or – as in this chapter – everything from strolling in a shopping mall, to photographing a tourist attraction, to pursuing a hobby, to having a particular hairstyle, to engaging in forms of protest and resistance.

The turn to consumption emerged from a set of critiques of classic **production**-focused research (see Chapter 2) in cultural studies and cultural geography. Some key critiques from this context were as follows.

- It was argued that it was necessary to correct a "productionist bias" in the methods and concepts used to study cultural processes (Urry 1990: 277). Classic research had produced numerous rich accounts of cultural production, but it was increasingly evident that this told only part of the story. Many researchers called for new methods and debates to explore how cultural texts and media are experienced, used, debated and enjoyed (or not) by consumers and audiences.
- It was argued that classic accounts of cultural politics substantially underestimated the role and agency of consumers. Consumers were, arguably, too often positioned as blank figures or passive, powerless dupes at the inevitable mercy of systems of cultural production. Therefore many researchers called for acknowledgement of consumers' significant and potentially subversive creativity in their engagements with popular cultural texts and objects (Seiter *et al.* 1989; Miles and Paddison 1998).
- Many critics felt that the 'productionist bias' in classic approaches betrayed a kind of cultural elitism, closely linked to hegemonic notions of 'high culture' (see Chapter 2). That is, it was increasingly felt that many forms of popular consumption were implicitly treated as perhaps frothy, insubstantial and irrelevant: certainly unworthy of academic concern, compared with the weighty analysis of cultural politics of production. Treating popular cultural consumption as a legitimate topic of academic research was seen as an effective way of unsettling this elitism. It was also seen as a way of recognising the fundamental importance of consumption practices for individuals' everyday lives and identities.
- Perhaps most significantly, it was argued that a 'productionist bias' had led to many researchers and theorists overlooking very substantial social and geographical transformations arising from the rise of 'consumer culture'. 'Consumer culture' is widely used as a shorthand for complex cultural-historical shifts variously ascribed to: the development of European capitalist commodity markets during the eighteenth century; the proliferation of affordable mass-produced consumer goods afforded by the nineteenth-century European industrial revolution; the expansion of leisure time and disposable income for many people in Europe and North America in the aftermath of the Second World War; and ongoing processes of globalisation and connectivity via communications technologies. In this context, a vast proliferation of consumer goods and popular

cultural forms has become available for a mass audience. It is argued that the acquisition and use of consumer goods has thus become increasingly central to many people's everyday lives and identities. It is also noted that in many parts of the world there has been a decline in the importance of manufacturing industries and an increase in the economic importance of cultural industries devoted to the production of objects and spaces for consumption. As we outline in Section 3.3, it is also argued that there has been a vast proliferation of specialised spaces of and for consumption in this context.

3.3 Geographies of cultural consumption

The debates and critiques outlined in Section 3.2 sparked a substantial enthusiasm, across many disciplines of the social sciences, for consumption-focused research: essentially exploring the diverse things that people do with cultural objects, and within cultural spaces. Many cultural geographers were very much caught up in this turn to consumption, and in the following subsections we highlight three key ways in which geographers have contributed to understandings of consumption. The following points can also be read as answers to the question: why have many geographers cared so much about consumption?

Consumption matters in everyday geographies

It is often argued that the rise of 'consumer culture' has had two important consequences: a vast proliferation of consumer goods relating to practically every aspect of human life, and a shift in the way in which many consumers perceive and care about all these goods. In short, it is suggested that there exists an ever-increasing array of commodities, and that consumers perhaps value, debate, worry about and attach value to commoditised consumption as never before. Certainly, consumption should be understood as fundamentally important for everyday geographies in four ways.

First, most simply, people spend so much time in the act of consumption in the spaces we occupy. Think back to the question at the beginning of this chapter: just how many consumer goods or services have you been using today? Much of this consumption may pass unnoticed or be taken for granted, but consumption practices are central to *so many* aspects of everyday lives: eating, drinking, shopping, wearing clothes, using products and technologies, watching, reading or listening to cultural products and media, discussing or thinking about any of the above. It is actually difficult to think of moments when, as inhabitants of consumer culture, we are *not* engaged in some form of consumption: as we go about our everyday geographies, commodities really do surround us and prove to be central to all manner of everyday activities. Even as we sleep, we are typically surrounded by, and actually in the process of using, consumer goods: think of pillows, alarm clocks, pyjamas, teddy bears (Kraftl and Horton 2008; Horton and Kraftl 2009). So, as a basic point, it should be clear that consumption practices – the things that people actually do with cultural products – should be important to anyone seeking to understand everyday geographies. To explore this point further, see the later chapters on **performances** (Chapter 7) **identities** (Chapter 8), **everydayness** (Chapter 9), **material objects** (Chapter 10) and **bodily practices** (Chapter 12).

Second, consumption practices can significantly affect how people experience spaces. As one example, Box 3.2 presents some findings from Anderson's (2002, 2004) research about listening to popular music. Note how, in these moments, listening to even a short burst of a particular tune has turned these individuals' moods, and affected how they feel about themselves and the space they occupy. These examples reveal the potential of particular consumption practices to transform one's mood and experience: such that a loud blast of ska can make one forget about worries and responsibilities, or a favourite tune can make time pass a little more swiftly when doing chores or homework. We might think about other kinds of example too: listening to music on headphones can transform the experience of a long journey; a shopping trip might cheer us up (or make us despair); a book or film might play on our mind for days, or change the way we see the world; a television programme might form the basis for an ice-breaking conversation with a stranger; a shared cultural interest

Box 3.2
Moments of cultural consumption: listening to popular music

Photograph 3.1 Listening to a favourite tune can transform how we feel, and the space we are in.
Source: Alamy Images/Fancy.

Anderson's (2002, 2004) research on practices of listening to popular music reveals how cultural consumption can be profoundly important for individuals' everyday lives. Anderson's research participants – from the North-West of England – describe moments where they were aware of how music mattered to them within everyday spaces. Consider three examples. Note how the act of listening to a particular tune changes these individuals' mood, and their experience of everyday spaces and tasks. Can you think of instances when you have used music in this kind of way, or when hearing particular music has affected you in a similar way?

That evening I had the house to myself . . . in a temper, put [my son] to bed and I came downstairs and I would normally do a bit of housework, might watch TV or I might read my book and I just thought . . . I want to put some music on LOUD. So I put ska music on and I danced and danced and it were BRILLIANT, absolutely brilliant. I managed to switch off . . . I had a great, great time . . . It's funny – after that I stopped and thought 'GOD, you're 35, stop this . . . put the brakes on, grow up, turn it down . . . get the washing up done' and I screeched to a halt with it and felt a bit silly for a bit . . . but in that moment I . . . just danced, danced me heart out and it all lifted, just for a bit. It's a cliché but I WAS somewhere else.
(Anderson 2002: 211)

Things like cleaning the bath . . . I just hate it, there are certain things I don't like that have to be done . . . I was really, really bored so then I went downstairs and got some music. David Gray, and it made me forget what I was doing, I was on automatic pilot now, it was just my hands doing it . . . It makes it more bearable and I almost actually enjoyed it, it definitely made me feel better about . . . having to do it.
(Anderson 2002: 214–215)

I'd just got home from school. I was like hurggh can't be bothered to do any work . . . I started doing it but I was just SO bored . . . I just couldn't concentrate so I put on some music while I was doing my work . . . It was S Club 7, quite happy, you can just go along to it . . . It did make me more happy whilst I was doing the homework, changed the way I felt about it, I sung along to it sometimes and then got on with my work . . . sort of forgot that I was having to do homework . . . You can't be bored to that song anyway.
(Anderson 2004: 744)

might form the basis for a friendship or love affair. As these examples suggest, consumption practices can sometimes transform everyday geographies, even though consumption is often a taken-for-granted part of everyday life (see also Horton 2010, 2012).

Third, relatedly, consumption can really matter to people: enthusiasm for particular consumption practices can occupy our thoughts and, again, transform our experiences of everyday geographies. Sometimes, this kind of enthusiasm can be short-lived and intense: perhaps as we get caught up in a particular tune (see Box 3.2), or a cultural phenomenon, fad or fashion. Other forms of enthusiasm can be much longer-enduring and – as with many hobbies, for example – can entail a lifetime of patient work, collection and expenditure (see Box 3.3). However, both of these sorts of enthusiasm are significant for geographers because of their capacity to transform everyday spaces and routines, because of the numerous spaces and practices that have developed around these activities, and because people *care* so deeply about them within their everyday geographies.

Box 3.3
Hobbies and craft-based consumption: 'a hedgehog there, a stoat there'

Photograph 3.2 A model railway: an example of craft-based consumption.
Source: Alamy Images/Picturesbyrob.

Yarwood and Shaw's (2009) research with model railway enthusiasts explores forms of hobby- and craft-based consumption that occupy many people for "tens of years, if not longer" (p. 428). Consider one of their interviews with a model railway enthusiast at a convention in South-West England. In the following extract, the creator of a model railway layout gives a 'guided tour' of his layout, and reveals some of the patient, detailed work involved in its creation.

> I had to go to Shepton Mallet, the newspaper records, and they wrote what was happening there. Then I went to Bath to find out what was happening up that end . . . It took me a year to get it right, and two years to get it sort of all worked out and a year to build it . . .
>
> This is all made of polystyrene . . . And these are made from what you get your bacon or your fish in . . . So much is thrown out. This you see, you cut this near enough to shape and you use a soldering iron just to get your little curves in it. Just put it here and it will fizzle back so that you get all these marks from a hot soldering iron . . .
>
> Everything has got to have a story; it's got to make sense. [Pointing out individual elements of the railway layout] You have got a hedgehog there, a stoat there, a badger . . ., here, they have had a load of sleepers and ballast that they are tidying up. He is just trimming the weeds off. That's the gang for this side. You have got the quarry workers . . . There is a horse and cart, the old coal delivery, you are in the fifties, and granny watching the coal delivery . . . Granddad is talking to her and he is feeding the sheep with hay; a shepherd and his dog.
>
> (Yarwood and Shaw 2009: 428)

Yarwood and Shaw (2009) note that this kind of hobby- or craft-based consumption has several key characteristics:

- It involves significant time commitment, expertise and attention to detail. This kind of activity often involves decades of careful work, craft or collection (somewhat in contrast to the fleeting moments of pleasure and excitement in consumption related in Box 3.2).
- It can involve significant expenditure, over a long time period, on specialist materials and resources. Many of these materials are commercially mass-produced (e.g. Hornby trains, in the example of model railways). However, consumers often also exercise a great deal of creativity to make their own components (as in the enthusiast above, making scenery from bacon packets), or to incorporate remarkable elements of humour and realism (as in the stoats, hedgehogs and archival information from Shepton Mallet above), to 'bring to life' mass-produced commodities.
- Hobby- and craft-based consumption practices are served and maintained by networks of enthusiasts and experts. These networks gather and share, compare and contrast expertise at particular spaces and events such as conventions, clubs, trade shows, exhibitions, specialist retailers and online discussion boards. Within these networks, individuals are noted as particularly 'good at', 'serious about' or 'devoted to' their hobby or craft. There also exist numerous subdisciplines and interest groups within these networks: for example, at the model railway convention discussed above, Yarwood and Shaw note that enthusiasts tend to focus on either models of later diesel engines or nineteenth-century Great Western Railway steam trains.

Fourth, consumption practices are absolutely central to identities: how people see and construct themselves and 'others'. Consumption is therefore recognised as important for understandings of social and cultural geographies, and this realisation has been key for the work of many human geographers. As we explore in detail in Chapter 8, people's decisions, tastes and usage of consumer goods and services can be instrumental in the way in which they present themselves, connect with fellow consumers, or construct some people as 'other'. Consumption practices can effectively cut across or compound other lines of social difference, such as age, gender, ethnicity, disability or social class: sometimes serving as a point of connection between previously disparate social groups; sometimes deepening the perceived 'otherness' of already excluded groups; sometimes creating divisions and tensions within apparently homogeneous groups of people. These points are pursued in Section 3.4, where we focus on the role of consumption in the constitution of 'subcultures' and acts of resistance in relation to contemporary constellations of cultural politics.

There are countless spaces of/for consumption

A great deal of consumption-focused research within the social sciences deals with the proliferation of spaces that are expressly designed and used as venues for specific cultural practices. Research in this context has identified two broad spatial trends that have characterised the development of contemporary consumer culture. First, cultural theorists, geographers and historians have charted the inception and early development of distinctive consumption spaces such as coffee shops, pubs, department stores and shopping malls. The *Arcades Project* by the German cultural theorist Walter Benjamin (1892–1940) is widely hailed as a classic work in this context. This text compiles Benjamin's profuse, detailed notes and observations made during strolls through glass-roofed shopping arcades in Paris during the 1920s and 1930s. These arcades were among the first of their kind: they were constructed in the 1820s, at a time when similar spaces for shopping, strolling and 'window-shopping' were being constructed, and were considered fashionable and tasteful (see Chapter 2), in many European cities. Benjamin's notes are usually interpreted as an attempt to understand the emergence of modern consumer culture by investigating details of these particular consumption spaces, and observing traces of their history (see Box 3.4). Here and now, spaces such as shopping malls, department stores and coffee shops can seem totally commonplace and taken for granted. However, historical research by a number of geographers – taking a somewhat similar approach to Benjamin – has revealed how such spaces emerged at particular social-historical moments, and how they embody particular norms, rules, tastes and cultural politics (see Chapter 2) of their context. As noted in Chapter 2, the continued, present-day production and regulation of consumption spaces such as shopping malls remains a key concern for many geographers. Crucially, as part of the turn to consumption, many geographers have begun to explore the activities, mobilities, subcultures and social interactions of people within consumption spaces. For example, see Box 3.5 for examples of how people may bend the rules, make their own entertainment, and do their own thing within consumption spaces.

Second, cultural theorists have attempted to describe a series of ways in which consumption spaces have undergone restructuring and transformation over the last century: how certain spaces have, perhaps increasingly, been reshaped as centres *for* consumption, specifically designed to afford and encourage consumption. These trends – which Paterson (2006: 58) wittily summarises as 'McDisneyfication' – are thought to have resulted in the homogenisation of consumption opportunities, the ubiquity of global brands, and the prevalence of particular kinds of highly 'stage-managed' consumption spaces.

This notion of 'McDisneyfication' gives a nod to several key sociological theorisations of these trends, which are summarised in Box 3.6. These ideas have been widely cited in academic and popular literature, and certainly in many ways they ring true: witness the homogenisation of many high streets, supermarkets and shopping malls, the global reach of many brands, and the global spread of so many consumption fads, fashions and phenomena. In the following subsection, too, we note that claims in

Box 3.4
A shopping arcade, 1822: Benjamin's *Arcades Project*

Photograph 3.3 Galerie Vivienne: a Parisian shopping arcade, which opened in 1826.
Source: Alamy Images/Alex Segre.

Walter Benjamin's *Arcades Project* is a brilliant, enigmatic and famously large piece of work: most editions are more than 1000 pages long. The book was unfinished at the time of Benjamin's death in 1940, so we are left with a vast, fragmented assemblage of notes, observations and reflections on glass-roofed Parisian shopping arcades. The arcades were built during the 1820s (Photograph 3.3) and encountered by Benjamin about a century later. It is an in-joke among social scientists that no-one has actually read the book from start to finish, but it is well worth exploring: practically every page contains some kind of interesting observation or thought about the nature and development of modern urban consumption spaces. Benjamin's reflections on the arcades are characterised by numerous thought-provoking points going on simultaneously. Let us consider just one paragraph:

> Most of the Paris arcades are built in the fifteen years following 1822. The first condition for their development is the boom in the textile trade. *Magasins de nouveautés* [fashion boutiques], the first establishments to keep large stocks of merchandise on the premises, make their appearance . . . The arcades are centres of commerce in luxury items. In fitting them out, art enters the service of the merchant. Contemporaries never tire of admiring them. For a long time, they remain an attraction for tourists. An *Illustrated Guide to Paris* says: "These arcades, a recent invention of industrial luxury, are glass-roofed, marble-panelled corridors extending through whole blocks of buildings, whose owners have joined together for such enterprises. Lining both sides of the arcade, which gets its light from above, are the most elegant shops, so that the passage is a city, a world in miniature. The arcades are the scene of the first attempts at gas lighting. The second condition for the emergence of the arcades is the beginning of iron construction."
>
> (Benjamin 1999: 15)

It is fair to say that there is a lot going on here. In just one passage, Benjamin directs us towards four key characteristics of the arcades, and modern consumption spaces in general.

- The arcades were constructed in a particular social-historical context, particularly: a booming textile trade, new practices of merchandising and stock control, new technologies of iron production and construction. (Elsewhere in the *Arcades Project*, Benjamin discusses at length how these contexts of cultural production were afforded by the contemporary economic, political and industrial context of France and the French empire.)
- In the construction of the arcades, great attention was paid to aesthetics, mood and design. Lighting, textures, construction materials and techniques of display were deliberately employed to create pleasant spaces conducive to strolling and window-shopping, as well as making commodities appear as attractive as possible. Hence Benjamin's description of the arcades as exemplifying a shift towards 'art in the service of the merchant'.
- By bringing together different fashionable shops in one place, the arcades created specialist spaces that were destinations *for* consumption. New ideals of luxury, elegance, fashion and taste (see Chapter 2) were closely attached to these spaces: the arcades were at once spaces where one could purchase fashionable luxury goods *and* luxurious spaces where it was fashionable to be seen (and unfashionable *not* to be seen). They

Box 3.4 continued

were also spaces where new national and international fashions could be encountered and purchased. In this context, many consumers changed their shopping habits, choosing to travel to shop in particular arcades rather than local shops.

- The development of the arcades afforded new consumption practices. Benjamin particularly notes the way in which absent-minded strolling and window-shopping in arcades became fashionable in Paris (and many European cities) during the nineteenth century. The arcades also created new routes through the city, and afforded new habits and experiences: for example, Benjamin notes that it became commonplace for passers-by to shelter in the arcades during sudden rain showers, and spend time window-shopping until the weather improved. The arcades also became destinations for tourists from all over France, and were thus integral to the development of contemporary travel and tourism industries. (As such, the arcades themselves became the focus for consumption practices of tourists – see the subsection 'Many consumption practices involve the consumption of spaces themselves' below.)

Box 3.5
A shopping mall, 2000: teenagers' geographies of a consumption space

Photograph 3.4 The Grosvenor shopping centre, Northampton.

Source: Alamy Images/Colin Palmer Photography.

A number of geographers have investigated how different people inhabit, and *do* consumption, within consumption spaces such as shopping malls (Shields 1989), coffee shops (Laurier and Philo 2006a, 2006b) and bars (Jayne *et al.* 2006). For example, Matthews *et al.* (2000) conducted research with 400 young people aged 9–16 in five shopping malls in the English East Midlands. Consider three extracts from Matthews *et al.*'s (2000) interviews with teenagers in the Grosvenor shopping centre in Northampton (Photograph 3.4). In particular, note what these young people are actually *doing* in this space designed for consumption:

> I like it here [outside HMV music shop]. We can listen to the music, chat with each other, see our friends, two or three of us will turn up here and soon there'll be five maybe more of us. We know that we will see someone we know.
>
> (p. 286)

> We meet here and then move on. What's good is you don't get bored. There's always someone to talk to, someone to see. We can check out the clothes and that too. See what's new, see the latest fashion. Can't afford none of it: just look and that.
>
> (p. 287)

> See that geezer [a security guard], he picks on us. What a prat! We don't go though. If he says "get going" or something, we then mooch around a bit until we find somewhere else where he can't see us. We won't leave until we want to go!
>
> (p. 290)

A number of key observations are evident in these quotations.

- All kinds of social interaction and practice were going on within the space of this shopping mall. At any given moment, young people were meeting, hanging out, listening to music, chatting, moving on, getting moved on, window-shopping, or 'mooching around'. Indeed, the interviews suggest that young people spent most of their time in the shopping mall doing *anything but* shopping. (Although, of course, practices such as window-shopping and listening to music are themselves consumption practices and may lead to purchases.) The shopping mall was a key place in these young people's lives even though, in many cases, they were unable to buy anything; indeed, they often related the mall's importance to its usefulness as a place to meet and shelter as much as to its shopping opportunities.
- Shopping malls are complex social spaces where multiple social groups interact, and can sometimes come into conflict. For example, in this shopping mall, Matthews *et al.* (2000) recorded the existence of at least 20 different local gangs of young people who inhabited the mall, and interacted in different ways, at the time of the research. Moreover, the research describes tensions between adults and these gangs, manifest in the actions of security guards and police, who were frequently called upon to 'move on' young people, following (or anticipating) complaints from store managers or members of the public.
- Many young people actively attempted to bend the rules, evade the attentions of security guards, and make their own entertainment within the regulated, consumption-focused space of the shopping mall. This could involve a range of activities, from window-shopping, chatting and strolling to running, skateboarding, shoplifting, smoking or alcohol consumption.

Box 3.6
McDonaldisation–Disneyfication–Coca-colonisation

A number of social scientists have attempted to define how consumption spaces have undergone transformation over the last century. Three widely cited definitions – which identify the tactics of major global brands as key case studies – are outlined below. Try to identify: (i) examples of spaces in your everyday life that have some of these characteristics; (ii) instances where these features of consumption spaces are unsuccessful or resisted by consumers.

McDonaldisation

This term was coined by sociologist George Ritzer (1993) to describe a set of processes through which consumption spaces have been increasingly rationalised and regulated over the last century. He argues that these processes can be observed in any fast food restaurant, and should be understood as exemplifying four broader trends within global cultural, service and leisure industries.

- Towards greater efficiency and cost-effectiveness of functioning. Consumption spaces are increasingly designed to facilitate rapid throughput of customers, to maximise expenditure, and to minimise staffing costs (e.g. the proliferation and normalisation of drive-thru, self-service and self-clearing fast food restaurants).
- Towards greater consistency and predictability of products. Chains of consumption spaces are designed to offer consistent, predictable experiences and products, irrespective of location (e.g. one can buy a Big Mac, *un Big Mac*, and *eine Big Mac* in Los Angeles, Paris and Berlin respectively, and expect to receive essentially the same product).
- Towards greater calculability of outputs, stock control and profit margins. Consumption spaces are specifically designed and regulated to ensure this calculability (e.g. through technologies and processes that strictly and precisely regulate portion sizes in fast food restaurants).
- Towards heightened control of production and consumption practices. Employees and customers are subject to intense regulation in many consumption spaces, to the extent that their agency and decision-making are limited (e.g. in fast food chains burgers are pre-prepared, pre-measured, cooked in automated fryers, and prepared to predetermined instructions; improvisation, creativity and imagination on the part of staff are limited and discouraged).

Box 3.6 continued

Disneyfication

Bryman (1999) suggests that consumption spaces are increasingly 'stage-managed' to direct consumers to purchasing opportunities, and create atmospheres and scenarios where expenditure is encouraged. He suggests that this stage management involves four main strategies, which have been perhaps most spectacularly applied in Disney theme parks.

- Theming and branding - the creation of spaces with particular themes, concepts, atmospheres and brand associations (e.g. Disneyland is not just a theme park, but a Disney theme park: it is marketed and staged as a magical, special place; it contains multiple themed zones associated with different Disney characters and brands).
- De-differentiation - the removal of barriers between different kinds of space (e.g. in theme parks, attractions, gift shops, retail opportunities and food outlets are closely integrated; parks are designed so that visitors pass numerous consumption spaces while they are in the process of visiting the park).
- Merchandising - the proliferation of branded goods, souvenirs and memorabilia, and spaces in which they can be purchased.
- Staff training in emotional and performative labour - the work involved in staging and maintaining the theme and atmosphere of a consumption space.

Coca-colonisation

This term was coined by Wagnleitner (1994) to describe the spread of global brands of consumer goods. He identifies two geographical processes, as exemplified by the global ubiquity of the Coca-Cola brand.

- Global colonisation - the way in which particular products become available and desired worldwide; such products are produced, marketed and consumed via increasingly complex global commodity networks (see Chapter 2), and these networks have complex geographical consequences.
- Lifestyle colonisation - the way in which consumer capitalism colonises even the most intimate spaces of everyday lifestyles; how branded consumer goods can come to surround us, and matter to us, as we go about our everyday lives.

Box 3.6 about the heightened commodification, 'staging' and merchandising of experiences can be useful when thinking about the consumption *of* spaces via tourism and heritage consumption.

However, although these ideas do provide a useful frame for understanding broad characteristics of many present-day consumption spaces, it is important to recognise that they do not tell the full story. As with the production-focused approaches summarised in Chapter 2, these accounts of the production and restructuring of consumption spaces can have the effect of overemphasising ways in which consumers are regulated, duped and constrained by market forces and cultural producers. For this reason, many geographers have increasingly researched the ways in which consumers actually behave, and often develop creative, resistive and 'alternative' practices through their consumption behaviours: see Section 3.4 for more information about this line of work. It is also the case that social scientists writing about consumption spaces have often tended to focus on rather spectacular case studies (such as mega-malls, or multinational corporations noted for aggressive and ethically dubious business practices). See Chapter 9 for more discussion about why this could be considered problematic, and for an introduction to the work of geographers who have sought to acknowledge other, more **everyday geographies**, not least through a focus upon more small-scale, low-key consumption activities and spaces.

Many consumption *practices* involve the consumption of spaces themselves

The last subsection described how certain kinds of space (such as arcades, malls or coffee shops) were created as venues for particular consumption practices. We also noted how many consumption spaces have arguably been restructured to facilitate increasingly intense, homogenised and 'stage-managed' forms of consumption. Research of this kind has been important in highlighting the centrality of **spaces** (see Chapter 13) in many consumption practices. Extending this point, many cultural geographers have noted that many consumption practices involve the *consumption of spaces themselves*. In other words, they suggest that spaces are not just where consumption happens; rather, it is often the case that spaces are the 'thing' being consumed, and the whole point of much consumption is to spend time somewhere or to relate, somehow, to a particular space.

The work of the sociologist John Urry (1990, 1995) has been really important for geographers thinking along these lines. Urry (1995: 28–30) points to a whole range of examples of consumption *of* particular spaces. These include: visiting a tourist attraction or landscape; paying an admission charge to access a particular space or viewpoint; choosing a place to go on holiday; buying gifts and souvenirs related to a tourist destination; purchasing products because of their association with a place (perhaps buying a particular bottle of wine because it reminds you of holidaying in the Loire); or visiting a place to do a specific, renowned activity (perhaps seeing a musical on Broadway). He also suggests that, as people go about their daily lives and consumption practices, they often 'buy in' to more abstract feelings, ideas or myths about particular landscapes or localities. So, for example, if you love the English countryside, it is quite likely that all kinds of consumption practice will relate to that broad sense of the countryside's loveliness. You might: visit country parks, spend holidays in rural locations, buy souvenirs of your trips, enjoy books or television programmes that evoke England's green and pleasant land, spend loads of money on walking gear, maps and guidebooks, furnish your home in rustic cottage style, or even buy a house in the country. Indeed, a large body of geographical research has investigated how British rural spaces are imagined and represented as 'idyllic', and how a vast array of products, texts and commodified experiences allow consumers to 'buy in' to this notion of 'rural idyll'. We can also consider how this notion of consuming place is relevant to other types of enthusiasm or worldview: for example, what kinds of spaces might someone be drawn to consume if they love heritage, adventure, nature, nightlife or the arts? Or if they belong to a particular subculture (see Section 3.4 and Chapter 8), or if they are a fan of a popular cultural phenomenon or have a hobby (see Box 3.3)?

Notably, Urry also considers how spaces are actually consumed in practice. Urry's most famous work *The Tourist Gaze* (1990) considers how spaces, landscapes and attractions are consumed by tourists via particular practices of looking. He argues that 'gazing' is quite central to being a tourist: fundamentally, it is what tourism is about, and what the consumption of space, in this context, *is*. For:

> the . . . characteristic of tourist activity is the fact that we look at, or gaze upon, particular objects, such as piers, towers, old buildings, artistic objects, food, countryside and so on. The actual purchases in tourism (the hotel bed, the meal, the ticket, etc.) are often incidental to the gaze, which may be no more than a momentary view. Central to tourist consumption, then, is to look individually or collectively upon aspects of landscape or townscape which are distinctive, which signify an experience which contrasts with everyday experience.
>
> (Urry 1995: 131–132)

Urry (1990) identifies some key characteristics of this 'tourist gaze' (see Photograph 3.5). In so doing,

Photograph 3.5 Tourist photography: an example of the tourist gaze.
Source: Getty Images/Tim Graham.

he begins to highlight what people actually do when they consume a space as tourists – although we could note here that Urry's notion of gazing can be criticised for over-privileging visual consumption (see Chapter 12 for a sense of why this may be problematic). Some characteristics of tourist gazing are as follows.

> The tourist gaze arises from movement of people to, and their stay in, . . . sites which are outside the normal places of residence and work . . .
>
> The places gazed upon are for purposes which are not directly connected with paid work and normally offer some distinctive contrasts with work (both paid and unpaid) . . .
>
> Places are chosen to be gazed upon because there is an anticipation . . . of intense pleasures, either on a different scale or involving different senses from those customarily encountered . . .
>
> The viewing of . . . tourist sights often involves . . . a much greater sensitivity to visual elements of landscape or townscape than is normally found in everyday life. People linger over such a gaze which is then visually objectified or captured through photographs, postcards, films . . .
>
> [T]he gaze is constructed through signs and tourism involves the collection of signs. When tourists see two people kissing in Paris, what they capture in the gaze is 'timeless, romantic Paris'; when a small village in England is seen, what they gaze upon is the 'real olde England'.

Urry (1995) also notes how different kinds of tourism can involve subtly different kinds of gazing: seeking different 'signs' through different activities. For example, we might consider how gazing practices might differ between tourists who are interested primarily in activities as diverse as backpacking, wining and dining, sunbathing, boozing, extreme sports, heritage, wilderness or volunteering. Any such gazing practices are invariably framed by contemporary systems of **cultural production** (see Chapter 2), and served by an ever-increasing array of cultural and tourism industries. Thus Urry (1990: 2–3) suggests that the tourist gaze also has the following properties.

> A substantial proportion of the population of modern societies engages in such tourist practices and new socialised forms of provision are developed in order to cope with the mass character of the tourist gaze . . .
>
> Such anticipation is constructed and sustained through . . . film, newspapers, TV, magazines, records and videos which construct that gaze . . . [and] provide the signs in terms of which the holiday experiences are understood . . .
>
> An array of tourism professionals develop, who attempt to reproduce ever-new objects of the tourist gaze."

We have quoted at some length from Urry's work here because his work has been so significant for geographers studying tourism, leisure and many forms of consumption. A substantial body of geographical research has followed Urry's lead by exploring how spaces of leisure and tourism have increasingly been restructured, commodified and 'stage-managed' in order to regulate, direct, and maximise profitability of the tourist gaze at particular sites.

3.4 Consumer agency: subcultures and resistance

In Section 3.1 we noted that, during the early 1990s, many geographers, sociologists and cultural theorists advocated a 'turn' to consumption, to correct a 'productionist bias' in much previous research. These critics called for research exploring the practices, experiences and agency of consumers: in other words, what consumers do, what they think and feel, and how they make decisions about consumption. Two decades on, there exists a vast and diverse body of consumption-focused literature dealing with these issues, and exploring all manner of consumption practices and spaces. Type 'consumption AND' into a university library catalogue, or any database of academic publications, and you will encounter an extraordinary array of resources. In this subsection we highlight three key, recurring lessons from this body of work. We recommend that you familiarise yourself with these points and then spend some time exploring academic literature that relates to a form of consumption you care about or find interesting. Reflect on the second and third questions at the start of this chapter: what forms of consumption have you been doing today, and how are these practices important to you? As you read this section, consider how each of the following points relates to your own experiences.

Point 1: Consumers are active and creative

Underlying all consumption-focused research in the social sciences is a realisation that consumption always involves *doing* something. Even sedentary activities such as reading, watching television or Internet browsing, or taken-for-granted habits such as shopping and snacking, entail some degree of activity, engagement and decision-making. In this respect, all consumers can be thought of as active, to a greater or lesser degree, as they undertake consumption practices. Moreover, a great deal of consumption-focused research has explored how consumers are often very deliberate, knowing and purposeful: they actively decide to do (or not to do) certain consumption practices for particular reasons and to achieve certain outcomes. As the cultural theorist Paul Willis notes, consumption practices are often done to make a point or to express something about oneself. For consumption is central to

> [the] vibrant symbolic life and symbolic creativity in everyday life, activity and expression . . . through which individuals and groups seek creatively to establish their presence, identity and meaning . . . all the time expressing, or attempting to express, something about their actual or potential cultural significance.
>
> (Willis 1990: 1)

As this quotation suggests, consumption practices also involve some degree of creativity on the part of consumers. The activities through which cultural objects and spaces are interpreted, invested with meanings, and appropriated into one's everyday life are central to consumption, and can all be understood as creative acts: each can spark new understandings, experiences and ways of being. Much consumption-focused research can be read as celebrating the capacity of individuals and groups to engage creatively with cultural objects and spaces, and thus make their own meanings, styles and statements. As Willis (1990) notes, creative acts of consumption are extraordinarily diverse, and can be profoundly important and inspirational for individuals and groups. The examples in the following quotation relate specifically to young people's consumption practices, but the points about 'symbolic creativity' can be applied to any social group.

> We are thinking about the extraordinary symbolic creativity of the multitude of ways in which young people use, humanise, decorate and invest with meanings their common and immediate life spaces and social practices . . . [through] personal styles and choice of clothes; selective and active use of music, television, magazines; decoration of bedrooms; the rituals of romance and subcultural styles; the style, banter and drama of friendship groups; music-making and dance.
>
> (Willis 1990: 2)

Activities such as these often involve a great deal of work, knowledge and collaboration among consumers (for an example, see Box 3.9) Crucially, such practices can result in activities, meanings, spaces and styles that are novel and unexpected. These practices can be subversive and critical towards cultural norms and **hegemonic** forms of cultural power (see Chapter 2). As we outline in points 2 and 3 below, consumer goods can be used in ways that differ markedly from the intentions of cultural producers. Consequently, the active and creative nature of consumption can be strongly politicised as consumers seek to refuse and resist contemporary cultural norms through practices that draw on available cultural resources. The work of the cultural theorist John Fiske (1989) was particularly important in theorising this potentially politicised nature of consumption. Fiske recognises that

> the resources [used by consumers] – television, records, clothes, video games, language – carry the interests of the economically and ideologically dominant; they have lines of force within them that are hegemonic and that work in favour of the status quo. But . . . these resources also . . . carry contradictory lines of force that are taken up and activated differently by people situated differently within the social system.
>
> (Fiske 1989: 2)

Points 2 and 3 below provide some examples of how such 'contradictory lines of force' have been activated in practice through the lifestyles of particular subcultures, and through resistive and alternative consumption practices.

Point 2: Consumption is central to identities

Theorists of consumer culture often argue that consumption practices have become increasingly central to individuals' identities over the last century. It is argued that consumption is widely approached (and promoted) as a kind of "life project" where consumers "display their individuality and sense of style in the particularity of the assemblage of goods, clothes, practices, experiences, appearance and bodily dispositions they design together into a lifestyle" (Featherstone 1991: 86). In other words, the things that we do with cultural objects and spaces are fundamentally important to our everyday lives, our sense of who we are, and how we present ourselves to others.

The work of the University of Birmingham's Centre for Contemporary Cultural Studies (CCCS) (1968–2002) has been hugely influential for social scientists seeking to understand the importance of consumption for lifestyles and everyday geographies. A succession of classic studies by CCCS researchers investigated how 'subcultural' groups resist, and do their own thing within, contemporary hegemonic norms. The term 'subculture' is a popular one as well as an academic one: broadly, it refers to "groups of people that are in some way represented as non-normative and/or marginal through their particular interests and practices, through what they are, what they do, and where they do it" (Gelder 2005: 1). The classic CCCS studies focused on postwar UK youth subcultures, particularly mods, skinheads and punks. It was noted that these diverse groups shared some key characteristics.

- Consumption practices were central to becoming a member of, and belonging in, these groups. Listening to particular music, wearing particular clothes, consuming particular drinks and substances, and sporting particular hairstyles or bodily adornments were central to the practice of being a mod, skinhead or punk.
- These subcultural groups were created and maintained through particular "expressive forms and rituals": using consumer goods, clothes and everyday objects in the context of lifestyle practices that made a particular point and/or symbolised group membership and "self-imposed exile" from the "natural order" of society (Hebdige 1979: 2). These forms and rituals were often described as a process of 'bricolage'. This was originally an anthropological term describing how people have historically attempted to comprehend the world by using resources and material objects that happened to be at hand. Thus it was suggested that youth subcultures used diverse available cultural resources, drawing on objects and influences from old, new, popular and obscure contexts. In the process, new, often subversive, meanings were attached to familiar objects, and new statements, styles and emotions were created.
- This subcultural bricolage was politicised and resistive on three levels: (i) as a way of upsetting, and distancing oneself from, parents' norms and values; (ii) as a statement of dissatisfaction with the music, stars and styles of contemporary popular culture, and the limited possibilities offered by contemporary cultural industries; and (iii) as a broader symbolic rejection of the repressive conformity, small-mindedness, conservatism, austerity, social injustice and lack of opportunities that were felt to characterise postwar Britain (Hall and Jefferson 1976).
- Youth subcultures were thus understood as a way in which young people created *their own* spaces and practices, resisting rules, norms and expectations imposed by parents, the cultural industries, and contemporary society in general.
- Youth subcultures have often been figured as problematic, deviant, antisocial and immoral by contemporary mainstream media. Members of subcultural groups have often been treated as 'folk devils' by many adults. They have often been the subject of 'moral panics', and represented as symptomatic of contemporary societal ills.
- Although initially treated as deviant and alarming, the styles of youth subcultures are typically very quickly appropriated and repackaged as 'cool' by the mainstream cultural industries.

The CCCS approach to subcultures remains profoundly influential for social scientists exploring the role of consumption in identity formation. This body of work presents cultural geographers with a rich seam of research and theory. Studies such as Hebdige's writings on the 'semiotic guerrilla warfare' of punk (see Box 3.7) provide vivid, powerful and affecting accounts of the practices and spaces of particular subcultural groups.

However, a certain degree of caution is required when approaching this body of work. Classic work on youth subcultures has been criticised for:

- focusing almost exclusively on the subcultural practices of white, British, male protagonists, and ignoring the exclusionary whiteness, maleness and Anglo-centrism of these groups;
- focusing on *youth* subcultures, and ignoring the cultural activities of other age groups (and also being complicit in labelling young people as deviant and antisocial);

Box 3.7
Hebdige (1979) on the 'semiotic guerrilla warfare' of punk

Photograph 3.6 Punk style: semiotic guerrilla warfare?
Source: Shutterstock.com/Lakov Filimonov.

The following extracts are taken from Dick Hebdige's (1979) *Subculture: The meaning of style*, a classic CCCS text that explores the meanings and politics of subcultural groups and consumption practices. The book includes a well-known analysis of 1970s punk as a form of 'semiotic guerrilla warfare', and the following quotations outline how this cultural warfare was waged. As you read, think about the following questions. (i) How would you describe Hebdige's style of writing? (ii) How does this writing make you feel? (iii) What does Hebdige's account tell us about the lives and experiences of punks themselves?

> Punk reproduced the entire sartorial history of British working class youth cultures in 'cut up' form, combining elements which had originally belonged to completely different epochs. There was a chaos of quiffs and leather jackets, brothel creepers and winkle-pickers, plimsolls and pakamacs, moddy crops and skinhead strides, drainpipes and vivid socks, bum-freezers and bovver boots – all kept 'in place' and 'out of time' by the spectacular adhesives: the safety pins and plastic clothes pegs, the bondage straps and bits of string which attracted so much horrified and fascinated attention.
>
> (Hebdige 1979: 26)

> The names of the groups (the Unwanted, the Rejects, the Sex Pistols, the Clash, the Worst, etc.) and the titles of the songs ('Belsen was a Gas', 'If You Don't Want to Fuck Me, Fuck Off', 'I Wanna Be Sick on You') reflected the tendency towards wilful desecration and the voluntary assumption of outcast status which characterised the whole punk movement. Such tactics were, to adapt [famous French anthropologist] Lévi-Strauss's famous phrase 'things to whiten mother's hair with'.
>
> (Hebdige 1979: 109–110).

- glamorising a small number of iconic, spectacular, resistive examples of subcultures (and thus promoting a somewhat partial and caricatured view of the role of consumption in resistance and identity formation);
- treating subcultures as fairly coherent, singular, stable entities (e.g. talking about 'Punk', but failing to acknowledge differences between punks in different communities or spaces);
- overlooking some social-cultural changes that render the notion of 'subculture' obsolete – it no longer makes sense to speak of a single mainstream culture (because of globalisation, simultaneous multimedia, and the multiple geographies of cultural production, as outlined in Chapter 2, there are many coexisting cultures and no straightforward relationship between 'culture' and 'subculture');
- suggesting that subcultural groups are always in the process of resisting hegemonic power, and shoehorning all subcultural practices into this narrative of resistance;
- over-theorising subcultural practices, and writing in a style that is inaccessible and irrelevant for members of subcultures themselves;
- methodologically, failing to engage with the opinions, experiences and everyday geographies of the members of subcultures themselves.

This last point is perhaps especially problematic for cultural geographers. As exemplified in Box 3.7, much classic work on subcultures was essentially a theorist's *reading of* subcultural activity. The extract in Box 3.8 features a punk's reaction to reading Hebdige's account of the 'semiotic warfare' of punk: contrast Hebdige's exciting, influential, beautifully written rhetoric with Muggleton's somewhat less glamorous, but more nuanced, reflections on his own life as a punk. Recognising this gap between theorisations of subcultural consumption and consumers' actual everyday lives has latterly led many social scientists to undertake more careful, in-depth research *with* (rather than merely *about*) members of diverse subcultural groups. Consequently, it is now possible to access a wide range of detailed qualitative studies based on research with all manner of popular cultural consumers. Box 3.9 presents an extract from just one example of research of this kind: drawing on a sociologist's research with extreme metal fans from Scandinavia. Again, contrast this with Hebdige's writing in Box 3.7. At first glance, the lives of Kahn-Harris's (2004) research participants may seem rather banal and undramatic, but note how this kind of research reveals complexities and subtleties of subcultural practices: for example, the sense of care and community between these fans, the careful work involved in developing a record collection, and the listening

Box 3.8
A punk's perspective on CCCS studies of subcultures

In the following extract, Muggleton (2000), who had been a teenage punk before training as a sociologist, recalls his first experience of reading Hebdige's (1979) *Subculture: The meaning of style* (see Box 3.7). As you read, note the contrasts between Muggleton's experiences and the vivid claims made about punk in Box 3.7.

> From late 1976 onwards I became increasingly involved in the emerging provincial punk scene. I took on certain aspects of the subculture's dress code, jettisoning my flares for straights, platforms for trainers, my bright orange, wing-collared shirt for a white button-down bri-nylon . . . I listened less and less to Pink Floyd and more to the Ramones . . . I never sported safety pins through my lip, nor stopped wearing my beloved denim jacket . . . I wasn't aware of Britain's economic decline or the fracturing of the political consensus, nor was I angry about unemployment . . . Some years after, I was browsing in a . . . bookshop and my eye caught the lurid cover of Dick Hebdige's . . . *Subculture: The meaning of style* . . . [T]he cover image and the title prompted me to buy, hoping perhaps that it might help me to recapture the feeling and spirit of those heady times. I took it home and began to read and could hardly understand a word . . . I fought my way through . . . and was left with the feeling that it had absolutely nothing to say about my life as I had once experienced it!
>
> (Muggleton 2000: 1–2)

Box 3.9

Show No Mercy: extreme music but mundane consumption practices

Photograph 3.7 The thrash metal band Slayer.
Source: Getty Images.

> First time I heard Slayer, I couldn't handle it . . . the first album *Show No Mercy* . . . I'd heard about it, I knew it was fairly, totally fast, totally manic, hardcore metal and I couldn't handle it – man it was like, all this satanic stuff, it's like 'whoa, what's going on here?' (Extreme metal fan, quoted in Kahn-Harris 2004: 111)

The album *Show No Mercy* (1983), by Californian thrash metal band Slayer, is widely hailed as a classic of the genre. The album cover features a painting of Satan holding a sword next to a pentagram, and photographs of scowling band members, clad in black leather and metal spikes, posing with inverted crosses. The music is characterised by ultra-fast guitar riffs and drumming, and screamed lyrics about torture, the Blitzkreig, Satan's impending rise, violent hatred, serial murder, black magic, mutilation, souls tormented in hell, the Third Reich, battlefield slaughter, and a world where there is no future and, indeed, 'no fucking world to be saved'. Give it a listen: maybe you too will be 'like whoa, what's going on here?' Or maybe not.

In its style, content and iconography, *Show No Mercy* could be read as a fairly spectacular example of a subcultural phenomenon. Indeed, fans of Slayer (and countless other extreme metal bands and sub-genres) have been analysed as subcultural bricoleurs, appropriating violent, shocking, misanthropic imagery to rebel against parental conservatism and suburban alienation (Harrell 1994; Petrov 1995). However, Kahn-Harris's (2004) research with Scandinavian extreme metal fans suggests that this interpretation misrepresents the actual lives and experiences of these fans in at least three senses.

- Interpreting the extreme metal scene as antisocial and shocking overlooks strong bonds of friendship, care, mutual support and a kind of collegiality that constitute the scene. Kahn-Harris's (2004) work describes the more 'mundane' everyday pleasures of being an extreme metal fan: listening to music, hanging out, debating and bantering about new releases, anticipating and attending gigs, collecting merchandise and

Box 3.9 continued

so on. He also notes the enduring friendships and romances that can blossom within the scene. Also - and rather contrary to the assumption that extreme metal is purposefully repulsive to parents - he notes how extreme metal fandom can sometimes bring families together via discussions, common musical reference points, or forms of parental support (providing lifts to gigs, lending money for recording sessions, buying tickets, giving CDs as gifts, etc.).

- Most academic readings of subcultures such as extreme metal scenes overlook the significant care, commitment, knowledge and work involved in 'scene membership'. Kahn-Harris draws attention to the mundane everyday practices that are central to the consumption of extreme metal: cataloguing and sharing CDs and mp3s; writing emails and letters to other fans; contributing to blogs, zines, mailing lists and message boards. He also notes the listening practices through which the ostensibly spectacularly violent sounds of extreme metal are incorporated into individuals' everyday lives and spaces. Although many fans are - as in the quotation above - initially overwhelmed by the music, we might say that they 'domesticate' it by listening with headphones, with friends, and in everyday contexts (while driving or doing housework or homework). These practices can be rather bookish and scholarly, too: Kahn-Harris notes that most extreme metal fans have a working and opinionated knowledge of several thousand different pieces of music, and countless sub-genres and sub-sub-genres of extreme music.
- Kahn-Harris (2004) also problematises the assumption that extreme metal is an aggressively and purposefully shocking subculture, by exploring how extreme metal fans interact with 'non-scene' friends and family members. He notes that considerable care is often taken to avoid causing offence or creating tension. This can involve concealing CDs and posters, not wearing lurid or offensive band merchandise, turning music off or down, or choosing to play more 'accessible' or 'popular' tunes while in the company of 'non-scene' friends.

Incidentally, one of the authors of this book once saw the singer from Slayer in the audience at a gig. He appeared to be clutching a hotdog, a broadsheet newspaper, and a packet of throat lozenges: more evidence that 'extreme' subcultural identities invariably involve mundane consumption practices!

practices through which 'extreme' sounds are domesticated. For many cultural geographers it is precisely this kind of detail and subtlety that is crucial for understandings of everyday geographies of consumption.

Point 3: Consumers can actively resist and subvert cultural politics

In point 2, above, we noted the importance of classic research about subcultures in recognising the role of consumption practices in subverting or resisting contemporary cultural norms. However, we also suggested that this work often foregrounded just a few glamorous, dramatic, iconic examples, and thus tended to overlook other more subtle, complex and everyday forms of resistance and lifestyle. Consequently, we should be aware that classic work on subcultures presents us with a particular, and often slightly caricatured, understanding of the roles of cultural consumption in forms of resistance. By contrast, geographers such as Pile (1999: 14) draw attention to the complexity and plurality of everyday practices of resistance, highlighting the vast repertoire of actions through which people can refuse, resist, subvert or protest particular spaces and circumstances "through giving their own (resistant) meanings to things, through finding their own tactics for avoiding, taunting, attacking, undermining, enduring, hindering, mocking the everyday exercise of power". As Pile (1999: 14) notes:

> [p]otentially, the list of acts of resistance is endless - everything from foot-dragging to walking, from sit-ins to outings, from chaining oneself up in tree-tops to dancing the night away, from parody to passing, from bombs to hoaxes, from graffiti tags . . . to stealing pens from employers, from not voting to releasing lab animals, from mugging yuppies to buying shares, from cheating to dropping out, from tattoos to body piercing, from pink hair to pink triangles, from loud music to loud t-shirts, from memories to dreams.

Acts of resistance are discussed in more detail in later chapters on **performances** (Chapter 7) and

identities (Chapter 8). In the quotation above, note how forms of consumption (listening to loud music, getting a tattoo or haircut, buying a loud t-shirt) are explicitly or implicitly central to some of Pile's examples. Not all consumption is resistive, and not all resistance involves an act of consumption; but, in many cases, practices of consumption and resistance are closely interrelated. This relationship can take a range of forms: some examples are illustrated below.

Resistance using icons and objects of consumption

There is a rich, diverse heritage of forms of activism and protest that deploy subversive humour, slogans, iconography and public performances or demonstrations to broad politicised ends. In this context, icons of contemporary consumer and popular culture have often been spoofed, defaced or subverted in order to make a particular point, to highlight some absurdity or injustice, or to critique the activities and interests of the economically and politically powerful.

The tactics of 'culture-jamming' activists provide one example of this form of protest. The term 'culture jamming' refers to a range of politicised "communication strategies that play with the branded images and icons of consumer culture to make consumers aware of surrounding problems and diverse cultural experiences that warrant their attention" (CCCE 2011: unpaginated). Culture jamming can involve the production and circulation of posters, pamphlets, graffiti, internet memes, emails or blogs, and/or the staging of campaigns, protests, stunts, parodies, hoaxes, or website hacks. These practices are often humorous and satirical in nature (Woodside 2001), and explicitly inspired by theorists' critiques of cultural production (see Chapter 2; also Dery 1999). Culture jams seek to "communicate with citizen-consumers who often avoid political problems and topics in favour of building an isolated reality around consumer comforts" (Pickerel *et al.* 2002: 3). Many culture jams are designed to puncture this sense of consumer comfort, by directing attention to problematic media representations, inequitable cultural politics and unethical corporate practices and interests:

> [m]any culture jams are simply aimed at exposing questionable political assumptions behind commercial culture so that people can momentarily consider the branded environment in which they live. Culture jams refigure logos, fashion statements, and product images to challenge the idea of 'what's cool', along with assumptions about the personal freedoms of consumption. Some of these communiqués create a sense of transparency about a product or company by revealing environmental damages or the social experiences of workers that are left out of the advertising fantasies. The logic of culture jamming is to convert easily identifiable images into larger questions about such matters as corporate responsibility, the 'true' environmental and human costs of consumption, or the private corporate uses of the 'public' airwaves.
>
> (CCCE 2011: unpaginated)

Subvertisement – a contraction of 'subversion' and 'advertisement' – is one culture-jamming tactic, as popularised by groups such as Adbusters in Vancouver (www.adbusters.org), the Billboard Liberation Front in San Fransisco (http://www.billboardliberation.com/), and countless local groups and campaigners. See Box 3.10 for an introduction to this practice. Subvertisement is just one example of the way in which particular icons and objects of consumption have been purposefully transformed (e.g. through graffiti, defacement, or vandalism) or recontextualised (e.g. through performance or parody) by activists to make their point.

Resisting particular consumption spaces and practices

Some acts of protest and resistance target specific sites of/for consumption. For example, the opening of corporate retail outlets, or spread of branded consumption opportunities, frequently prompts localised opposition, often with an underlying concern to preserve traditional or local consumption spaces and practices. In some contexts, this kind of protest intersects with other forms of activism. See Box 3.11 for one example: note how the manifesto links local concerns (e.g. disruption to a bus stop) with much broader concerns about consumerist lifestyles and corporate practices and consumption spaces.

Adopting alternative or ethical consumption practices

Lifestyles based upon purposeful ethical decisions about everyday consumption practices have proliferated, in many parts of the world, over the last century. Consumption practices such as vegetarianism and environmentally friendly living – or the boycotting

Box 3.10
A subvertisement from Bristol, UK

Photograph 3.8 Subvertisement.
Source: Rex Features/Invicta Kent Media.

Subvertisement involves the manipulation or defacement of advertisements for politicised ends. Frequently, this practice is intended to problematise contemporary consumerist norms and products: using the medium and iconography of commercials to disseminate critical and politicised messages about the product or corporation being advertised. The manipulation of advertising billboards is perhaps the most common form of subvertisement. Subvertisements can be witty and intellectual, or crude and direct; they can entail activities ranging from one-off, impulsive defacement to organised national or international campaigns (see LibCom 2009). See Photograph 3.8 for a recent example. The billboard originally advertised the British Legion's Poppy campaign and read 'for their sake . . . wear a poppy', but has been transformed into a critique of former Prime Minister Blair's foreign policy interventions in Iraq and Afghanistan. Consider:

- How effective is this subvertisement in subverting the original advert?
- What tactics are used in this subvertisement?
- What is your reaction to this image?

of commodities that originate in unethical political contexts, or have substantial carbon footprints, or which are not fairly traded – are increasingly commonplace. Nevertheless, as anyone who has been quizzed about being vegetarian knows, such practices go against some powerful cultural norms about consumption, and continue to be considered weird and 'alternative' by many people.

The extent to which these kinds of consumption practice are resistive and politicised varies considerably between individuals. It can be instructive to think about such consumption practices in terms of a broad spectrum of decisions and behaviours: from personal, private, mild or occasional engagement with ethical consumption to deep, politically activist commitment to wholesale transformation of lifestyles and norms. There are numerous typologies of consumers' attitudes and behaviour (see Tallontire *et al.* 2001; Connolly and Prothero 2008). Typically, these typologies classify consumers depending on how critical and aware they are of impacts of consumption decisions, and the extent to which they acknowledge the need to change their habitual consumption practices. See Box 3.12 for two examples.

Producing alternative consumption spaces

Dissatisfaction with spaces of/for consumption, and their underlying cultural politics, can prompt the

Box 3.11
The 'No Tesco in Stokes Croft' campaign

Photograph 3.9 Anti-Tesco graffiti in Stokes Croft, Bristol, UK.
Source: Alamy Images/Adrian Arbib.

On 1 December 2010, a group of 40 protesters dressed as a Tesco delivery lorry walked up Stokes Croft Road in Bristol, UK, and stopped to make a 'delivery' – a banner reading 'NO TESCO' – to the site of a proposed Tesco supermarket. The protest aimed to demonstrate how traffic, pedestrians, a cycle lane and a bus stop on a busy arterial road would be disrupted by delivery lorries in the event of the store opening.

The 'human delivery lorry' was just one event in a year-long campaign by local protesters opposing the opening of this consumption space. In the 12 months following the announcement that Tesco intended to open a store on the road, local residents orchestrated a campaign of emails to TESCO HQ, 2,500 postcards sent to local councillors and planning officers, attendance and impassioned contributions at public meetings, letters to Members of Parliament, petitions and contributions to planning and judicial review processes, a massed rally, and picketing and placards outside the proposed store. Graffiti (Photograph 3.9) and subvertisements appeared along Stokes Croft Road: 'NO TESCO' was stencilled on every other paving slab, and a banner reading 'Tesco: every little hurts' (spoofing the supermarket's advertising jingle 'every little helps') was raised above the building. The building was squatted and temporarily used as a community space, hosting gigs, films, language classes and a soup kitchen.

When the store opened in February 2011 – despite these forms of local protest and opposition – it became a focal point for more violent forms of protest: windows were smashed, paint was thrown, the store was looted, and riot police and 24-hour security guards were required to disperse a number of riotous community members. The celebrated Bristol graffiti artist Banksy created images, across the UK, depicting Tesco flags being hoisted and saluted. Banksy also produced a limited edition print, with all proceeds donated to fund ongoing contributions to planning and judicial review processes, and to support community members arrested while protesting.

A number of local groups, campaigners and bloggers coordinated or publicised these diverse forms of activism. The following manifesto was circulated within this network (http://notesco.wordpress.com/thecampaign/why-don't-we-want-a-tesco/). As you read, note the reasons given for opposing the opening of this Tesco store, and the links made between local concerns and broader politicised concerns with corporate practices and consumption spaces.

Why Don't We Want a Tesco?

Tesco and big supermarkets generally have adopted a model which is good for short-term gain, making money for distant shareholders. Unfortunately, this is bad for sustainable economic growth, drives out local business, undermines workers' rights and damages the environment.

- **Local Economy:** A thriving local economy feeds off itself with money being re-circulated between different trades and businesses. This sustains a diverse range of jobs and enterprises. The **money spent in a Tesco**

Box 3.11 continued

would go one way, out of Stokes Croft and out of Bristol. To grow a resilient local economy Stokes Croft needs independent local businesses that buy their products and services in Bristol . . .

- **A Fair Deal for Producers:** Supermarkets have a history of slashing the price they pay suppliers under the banner of serving us cheap food. Often our food isn't actually that much cheaper and this behaviour is mainly about increasing their margins. The end result is **farmers and suppliers struggling to survive**. Often they get bought out by bigger businesses willing to disregard issues like fair trade, deforestation, and workers' rights. Tesco are now strongly fighting the introduction of a supermarket watchdog that will stop unfair practices.
- **Playing Dirty:** The big supermarkets have become **increasingly devious** with their marketing giving us the impression food is cheaper than it really is. A couple of tactics that are hard to spot: price-cut campaigns on high-profile products that coincide with a net increase of the price of all their goods; price flexing, changing prices up and down over a number of months so customers don't know whether they're getting a bargain or being ripped off - generally it's the latter (No Tesco 2011: unpaginated).

Box 3.12
Two typologies of consumer behaviour

Two typologies of consumer behaviour are outlined below. As you read, think about the following questions:

- How would you classify yourself using these two typologies?
- Why would your consumption habits be classified in this way?
- How useful are typologies like these for understanding individual consumption practices?

A typology of green consumption

Steffen (2009) classifies consumers into four categories, depending on their commitment to environmental-friendliness:

- *Dark green environmentalists* - advocates of substantial changes to consumer behaviours; often deeply critical of contemporary consumerist norms and practices; they may call for withdrawal from contemporary globalised systems of cultural production, adopt activist lifestyles, and advocate community-level transformations as solutions to environmental problems; they may also seek to persuade others of the need to adopt these actions and changes.
- *Bright green environmentalists* - consumers and activists who believe that environmental friendliness must be accompanied and incentivised by economic prosperity; a belief that entrepreneurial spirit, technological advancement, and new consumer products are the best path to environmental sustainability.
- *Light green environmentalists* - consumers who may adopt modest individual lifestyle and behavioural changes in order to be more environmentally friendly; they may be ultimately content and comfortable within a consumerist context; they may be concerned about, and interested in, environmental issues, but may not desire, or recognise the need for, more fundamental social and environmental change.
- *Greys* - those who deny the need for individual or societal change in the interests of the environment; they may be 'dishonest and self-interested' (e.g. lobbyists for fossil fuel companies), or 'contrarians' (who may essentially enjoy opposing the pro-environmental opinions of individuals or social/political groups with whom they disagree on other grounds), or conservative or cautious in outlook, or may lack information to accept more environmentally friendly lifestyles.

A typology of ethical consumption

Tallontire *et al.* (2001) classify consumers into three categories depending on their commitment to ethical consumption:

- *Activists* - committed consumers of fairtrade and ethical products; consumers who wish to know the actions that they can undertake, and the changes they can make, to be more ethical in their consumption practices. 'Persuader activists' also seek to persuade others of the need to adopt these actions and changes.

- *Ethicals* – occasional to regular purchasers of fairtrade and ethical products; consumers who are concerned about, and wish to know more about, ethical issues associated with their consumption habits; however, these consumers often do not change their consumption habits in the light of this information.
- *Semi-ethicals* – infrequent purchasers of fairtrade and ethical products; consumers who seem to be rarely concerned about ethical issues associated with their consumption habits. These consumers may or may not become more regular consumers of ethical products if these products are cheaper or more readily available.

production of alternative spaces and opportunities for consumption. Consider the example of illicit raves and squat parties in Box 3.13. Events such as these require considerable skill and commitment on the part of individuals and groups typically motivated both by disappointment with the forms and cost of nightlife within their local context and a broader critique of the commercialisation of urban consumption spaces. These activities often originate, and attempt to proceed, in out-of-sight or derelict spaces, beyond the norms and regulations of contemporary geographies of cultural production. As such they are often viewed and valued as 'underground' or 'alternative' spaces for consumption. The production of 'alternative' spaces

Box 3.13
'Pound for the sound': creating alternative consumption spaces

> "Jay, where's the party?" . . .
>
> "On the other side of the river – there's a geezer meeting people on the bridge once an hour to take people over."
>
> "What's the place like?"
>
> "It's a knackered old pub, man . . . , derelict."
>
> The . . . pub had been derelict for several years after the docks had closed. Whoever had broken in and taken the metal grille off the windows had cleaned it up and got the electricity on and fixed the water in the toilets. A bloke on the door repeated to everyone "pound for the sound".
>
> (Chatterton 2003: 217–218)

Chatterton's (2003) research on geographies of urban nightlife in a British city reveals numerous case studies of consumers working to create alternative consumption spaces. He explores the motivations of individuals and groups who – as in the preceding quotation – put on parties, raves, gigs, club nights and events in squatted buildings, derelict spaces, houses or warehouses. These events are often organised on an unlicensed, not-for-profit or cooperative basis. Organisers are often motivated by dissatisfaction with what are seen as limited, homogeneous, inauthentic nightlife opportunities in their hometown, and are often actively politicised and linked to broader anti-consumerist campaigns, and critiques of the state of contemporary urban nightlife. The staging of these events involves considerable work on the part of organisers: consider the work involved in making a derelict pub suitable for a party in the example above. This work often entails networks of trust, fundraising, and collaborative work linking diverse committed local individuals:

> drawing upon their own energies and resources, creative communities of designers, promoters, musicians, fanzine makers, pirate radio producers, DJs, film-makers, writers, squatters, punks and artists have inhabited the margins, signalling their opposition to corporate activity.
>
> (Chatterton 2003: 216)

Chatterton (2003) explores how the production of such alternative spaces is often linked to subcultural lifestyles and identities, and can be part of a broader repertoire of activities through which consumers can resist and avoid mainstream consumption spaces. He also notes the tactics through which organisers seek to conceal alternative consumption spaces from police and local authorities, or at least regulate the behaviour within the space to the extent that events are tolerated despite their illegality. The 'pound for the sound' night passed off without intervention by the police.

> The police drove past a few times . . . but did nothing to stop the party. They knew it was contained and out of the way and that it wasn't the type of crowd to cause trouble. And anyway, it would be more hassle than it was worth to kick everybody out.
>
> (Chatterton 2003: 218)

of/for consumption is practically a characteristic of urban spaces: there are countless examples of this kind of alternative cultural production, spanning the work of disgruntled concert promoters, to the practices of young people hanging out beyond the gaze of adults, to opportunities for the consumption of criminalised substances, to the committed and passionate work of anarchist and activist collectives.

3.5 Connecting cultural production and consumption

Chapter 2 introduced the notion of **cultural production,** and the present chapter has explored some issues relating to **cultural consumption**. As we have discussed these cultural processes, we have explained that much research by social scientists, including human geographers, has tended to be *either* production-focused *or* consumption-focused. That is, much classic (and especially early) work on cultural processes and politics tended to focus on *either* cultural production (how, where, and under what conditions cultural stuff is made) *or* cultural consumption (what is done with cultural stuff in practice, and how these practices are important). There were usually reasonable and logical reasons for this either/or focus: simply, different groups of researchers – with different motivations, interests, expertise, methods and concepts – tended to gravitate towards research on one topic or the other. However, as a consequence (and as we have reflected in the foci of Chapters 2 and 3), we are left with two rather separate bodies of work, dealing with production and consumption respectively. This can give a sense that cultural production and consumption are quite different, neatly separable, processes. However – and this is *a big, important point* – it is important to recognise that processes of cultural production and consumption are actually always interconnected in practice.

The interconnectedness of cultural production and consumption has been increasingly recognised by social scientists. Researchers in media and cultural studies have attempted to describe this interconnectedness in different ways. In the remainder of this section we outline six key ways in which relationships between production and consumption have been theorised. You may find the diagrams in Figure 3.1 helpful in making sense of these concepts. As you read on, think about whether these concepts resonate with your own experience. For each of the concepts shown in Figures 3.1a–f try to identify an example from your own experience as a consumer in contemporary society: do some of the concepts ring truer than others?

a) Production-led understandings of culture

In Chapter 2 we noted the importance of concepts of **hegemony, ideology** and **discourse** in many classic studies of cultural production. We introduced a line of influential research in cultural studies that sought to unmask hegemonic, ideological meanings – implicitly reflecting, serving and promoting the interests of those who are most often involved in cultural production – in many popular cultural texts and media. Accounts of this kind tend (see Figure 3.1a) to understand cultural texts and media – and their producers – as overbearingly powerful (also consider the power ascribed to corporate cultural producers in ideas of McDonaldisation, Disneyfication and Coca-colonisation in Box 3.6). Conversely, in this kind of account consumers can seem to be rather passive, being influenced, conditioned or duped by hegemonic discourses. A similar logic is commonly found in anxieties about the effects of popular media: for example in media scares about the effects of the latest scary movie, the latest violent computer game, the latest controversial television programme, and so on. Here, again, consumers are positioned as vulnerable recipients of powerful media messages created via contemporary processes of cultural production. Perhaps, as you read this, you can think of a current example, wherever you are, of anxiety about the effects of a cultural text or medium: to what extent do you think this anxiety is justified?

b) Consumption-led understandings of culture

In Chapter 3 we have introduced research exploring the importance of **consumption practices** and the **creative agency** of consumers. We noted how, in distinction to earlier production-led understandings, many social scientists sought to acknowledge, and give voice to, the audiences of popular cultural texts and media (see Figure 3.1b). The recognition of the importance of creative cultural bricolage – using texts, media and anything else to hand as resources for practices of identity formation – was a major component of work in cultural studies and, subsequently, the **new cultural geography** (see Chapter 1). As we suggested earlier, in the discussion of punk culture, researchers have often been drawn to instances where consumers can be seen to be powerful, autonomous and creative: making their own lifestyles and cultures from resources available to them in contemporary cultural contexts. Sometimes, this has produced accounts in which cultural production processes seem rather incidental to the more important, exciting, and autonomous agency of consumers. However, it is more often the case that foregrounding consumption requires researchers to puzzle through the nature of the relationship between production and consumption: concepts c, d and e below are most typical of work of this kind.

c) Culture as dialectic

It is now the case that a great deal of work in cultural studies has tried to reconcile the hegemonic power of cultural producers (see concept a) *vis-à-vis* the agency

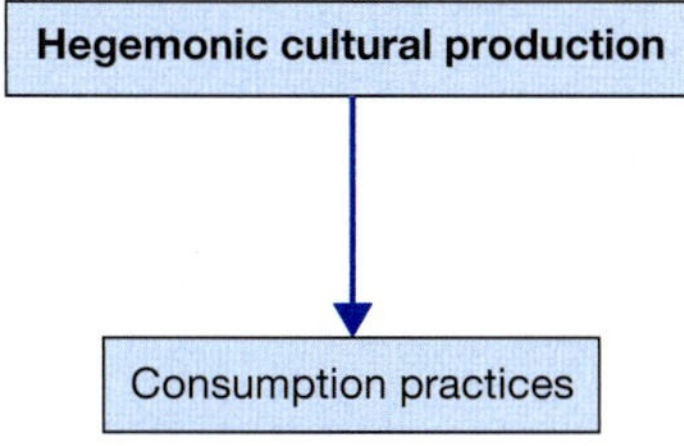

(a) Production-led understandings of culture

Key point: powerful, hegemonic cultural products affect consumers' lives

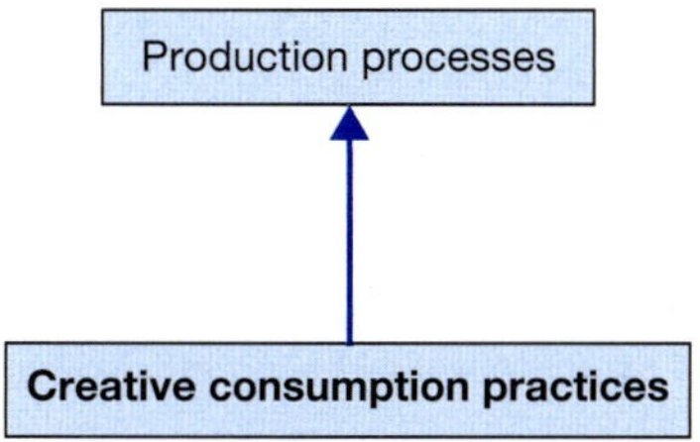

(b) Consumption-led understandings of culture

Key point: consumers do autonomous and creative activities with cultural products

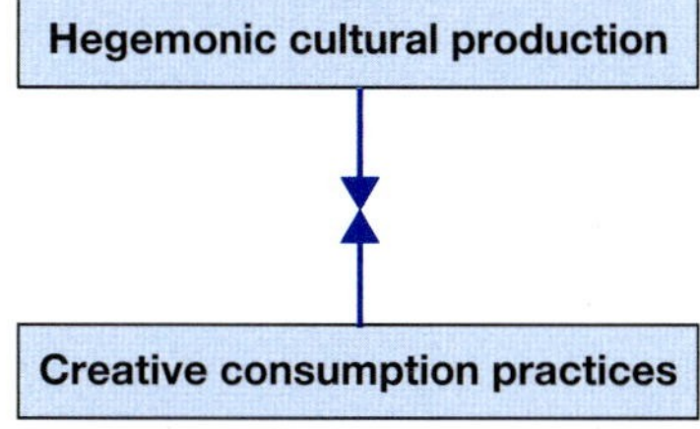

(c) Culture as dialectic

Key point: there is tension and conflict between the 'top-down' power of cultural producers and the 'bottom-up' agency of consumers

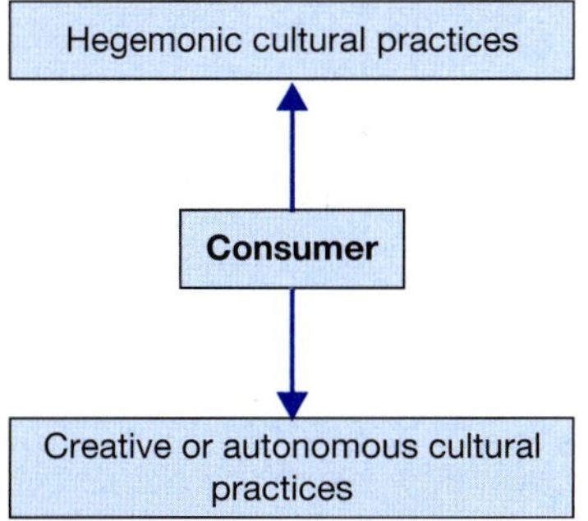

(d) Culture as ambivalent and paradoxical

Key point: Individual consumers may, on the one hand, do some cultural practices that are creative and autonomous but, on the other hand, do some practices that are uncritically aligned with hegemonic cultural power

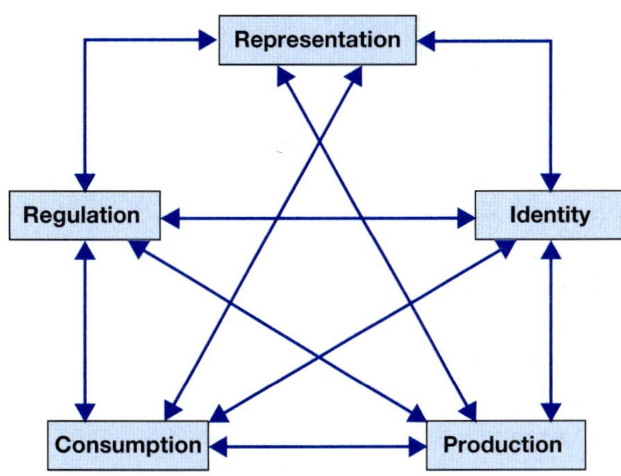

(e) Culture as a circuit (after du Gay 1997)

Key point: culture is an ongoing, continuous process, in which production and consumption are two key, interconnected elements

Figure 3.1 Six ways of understanding relationships between cultural production and consumption.

Source: Figure 3.1(e) after *Doing Cultural Studies: The Story of the Sony Walkman*, 1 ed., Sage, London (P. du Gay 1997) introduction, Circuit of Culture p. 3.

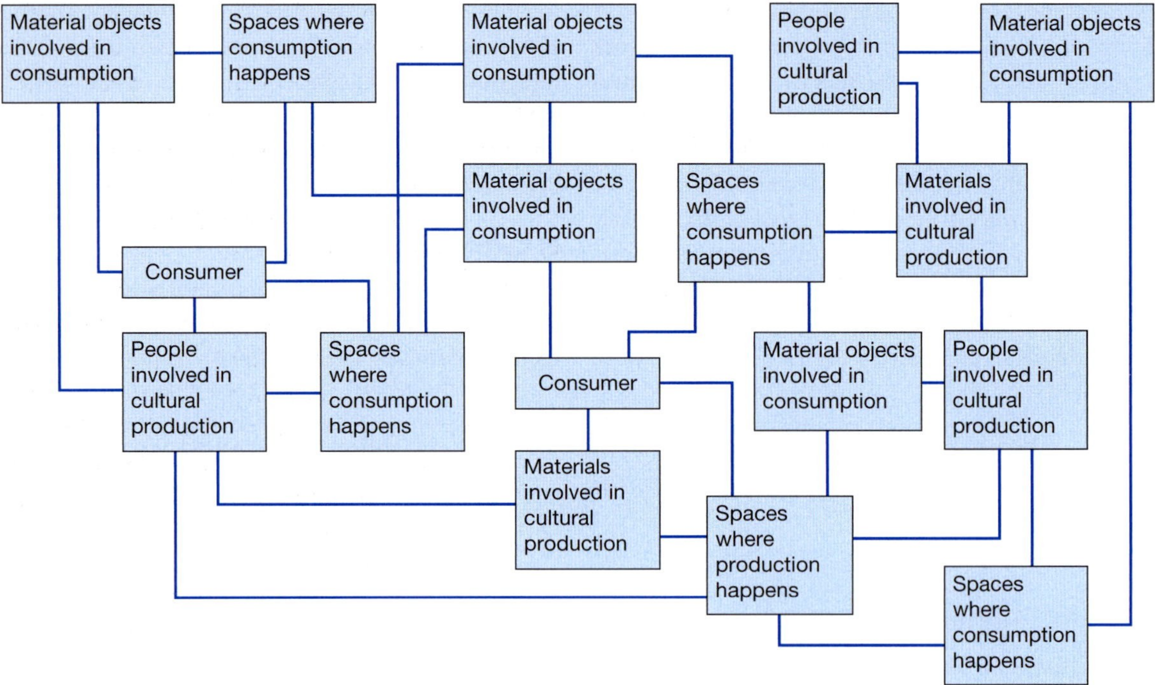

(f) Culture as an actor network

Key point: cultural phenomena are very complex processes in which diverse humans, non-humans, communities, spaces and processes are interconnected

Figure 3.1 continued

of consumers (see concept b). We have already noted in this chapter how the work of the CCCS was especially important in this respect, in exploring consumers' lifestyles and their relationships to hegemonic cultural formations. Frequently, this relationship was imagined as a dialectical tension (indeed, often a conflict or struggle) *between* the 'top-down' power and cultural forms imposed by cultural producers *and* the 'bottom-up' agency of groups of consumers (see Figure 3.1c). The examples of **subcultural** and **resistive** consumption practices highlighted earlier in this chapter can be seen as neatly exemplifying this kind of agency, and this kind of dialectical cultural clash, between the interests and norms of powerful, elite cultural producers versus the creative and subversive activities of contemporary consumers, who use and refashion hegemonic cultural products to their own ends, in their own ways, to create their own lifestyles and identities. Try to think of an example of this kind of lifestyle activity, wherever you are. Do cultural producers or consumers ultimately have most power in this instance? To what extent does this kind of lifestyle resist, or to what extent is it compromised by, hegemonic cultural power?

d) Culture as ambivalent and paradoxical

As some of the examples in this chapter have suggested (see Boxes 3.3, 3.8 and 3.9), the relationship between cultural production and consumption is not always a straightforward struggle between hegemonic cultural production and resistive consumer agency (although one could easily get this impression from classic work on subcultures – see Box 3.7). Many cultural theorists now recognise that things are often simply more complex, subtle and puzzling than this: not all cultural producers can be easily painted as hegemonic, ideological bad guys (e.g. in Box 3.3, consider the company that manufactures miniature stoats and hedgehogs for model railway enthusiasts), and not all consumers are necessarily engaged in powerfully purposeful projects of subversive identity formation (consider Boxes 3.8 and 3.9). Moreover, 'consumption' involves acts of creative cultural productivity (and besides, cultural producers are, themselves, consumers of cultural products too).

This kind of realisation has prompted many cultural theorists to consider the subtleties of cultural processes in practices (see Figure 3.1d). For example, Hall (1980) recognises that many consumers can have an ambivalent relationship to hegemonic culture: on the one hand (and at certain moments) we may be reasonably or unquestioningly satisfied with dominant-hegemonic cultural products; on the other hand (or in certain situations) we may find ourselves in distinct opposition to the ideological meanings inherent in contemporary popular culture. More commonly, he argues, consumers engage in a process of cultural 'negotiation', appropriating some elements of hegemonic culture, while ignoring or learning to live with others.

Others have reflected upon some of the quirks and paradoxes inherent in this kind of negotiation (Miles 1998a, 1998b): for example, the way in which consumers can utterly reject some hegemonic discourses and yet apparently unquestioningly live with others, or the way in which apparently 'top-down' and mass-produced hegemonic cultural forms can enable individual experiences and personal meanings (see also Box 3.2). Again, try to think of an example from your own life. How do you 'negotiate' with contemporary popular culture? Which elements do you take on board, which do you reject, and which do you put up with? (And how does this actually happen in practice?)

e) Culture as a circuit

Cultural theorists have increasingly recognised that cultural production and consumption should be understood as interconnected and part of broader global, political and cultural processes. du Gay's (1997) 'circuit of culture' is the most widely cited attempt to model this interconnectedness (see Figure 3.1e). du Gay (1997: 3–4) identifies five major, interconnected processes – production, consumption, as well as representation, identity and regulation – which together

> complete a sort of circuit – what we term the circuit of culture – through which any analysis of a cultural text or artefact must pass if it is to be adequately studied . . . We have separated these parts of the circuit into distinct sections but in the real world they continually overlap and intertwine in complex and contingent ways.

The key point here is the recognition that culture is an ongoing, continuous process, in which production and consumption are just two (albeit key) elements. The circuit of culture also recognises the complex connections between production, consumption and other key processes on the circuit: note that the arrows in Figure 3.1e go in all directions, to evoke the way in which processes can affect one another in practice (often in very complex, iterative ways). Note, too, the point that diverse aspects of any cultural phenomenon must be considered to get any kind of holistic, realistic sense of what is going on. How useful do you find the circuit of culture as a way of making sense of cultural processes? (What would you say are the key pros and cons of this kind of model?)

f) Culture as an actor network

The circuit of culture is widely admired and used as a basis for mapping cultural processes. Certainly, it is a very helpful tool for recognising the interconnected processes that are inherent in all cultural phenomena. However, it is increasingly argued that it is problematic to reduce cultural processes to a neat diagram, with clear, linear boxes and arrows. Here, it is argued that cultural processes are just more complex, messy and contingent than can be represented in this way. In Chapter 10 we shall introduce approaches such as **following the thing, material culture studies,** and **actor-network theory,** which have been important for many geographers and social scientists seeking to articulate the complex ways in which commodities are produced. Likewise, many cultural theorists have begun to interpret cultural phenomena as very complex actor networks: heterogeneous associations of humans, non-humans, communities, spaces and processes (see Figure 3.1f). We recommend that you visit Chapter 10 to get to grips with this idea, and then return to this section, to consider the usefulness of these ideas for thinking about cultural production and consumption.

In this section we have sketched six ways in which relationships between cultural production and consumption can be understood. Each of the concepts in Figure 3.1a–f has its strengths and weaknesses, and it is the case that all of them can ring true in certain contexts (hopefully, as you have gone along, you will have identified how these concepts are relevant, or not, to your own experiences as a cultural consumer). It is not our intention to prescribe which of these models is the best (although, as you read on, it will probably become clear that we favour the recognition of complexity and everdayness in non-representational and actor-network approaches to human geography). Instead, we want to highlight the key point that runs through each of these concepts: the way in which cultural production and consumption are always interconnected, even though they may have been researched and written about by somewhat disparate groups of human geographers in the past. We shall return to this point in Chapter 10, where, in particular, we shall recommend the work of the geographer Ian Cook as important, thought-provoking research that connects processes of commodity production and consumption. Moreover, we hope that our later chapters on **performances** (Chapter 7), **identities** (Chapter 8), **everydayness** (Chapter 9), **emotions** (Chapter 11), **bodies** (Chapter 12) and **space and place** (Chapter 13) will enable you to develop an increasingly complex and sophisticated comprehension of the processes introduced in Chapters 2 and 3.

Summary

- Many classic accounts of cultural processes tended to focus on cultural production (see Chapter 2). To correct this 'productionist bias', many social scientists have explored the lives and geographies of cultural consumers.
- The work of cultural geographers has been important within a turn to explore cultural consumption. There is now a great deal of research that addresses the complex geographies of cultural consumption, exploring, for example, how consumption matters in everyday lives, the production of spaces *for* consumption, and practices that involve the consumption *of* particular landscapes or spaces.
- Consumption practices are often very important in the constitution of individual and group identities. For example, a great deal of work by social scientists has explored how consumption practices are central to the lifestyles and spaces of subcultural groups. This work highlights how consumption practices can enable people to deal with, and often resist, hegemonic cultural politics within their everyday lives. However, as we have discussed, accounts of subcultural resistance and creativity can sometimes be a little over-romanticised, and overlook complex geographies and interconnections between production and consumption processes.
- Much classic work on cultural processes and politics tended to focus on *either* cultural production *or* cultural consumption. This can give a misleading sense that cultural production and consumption are neatly separable processes. However – as we map out in Section 3.5 – it is important to recognise that processes of cultural production and consumption are actually always interconnected in practice.

Some key readings

In addition to the following list, we recommend you also refer to the suggested readings at the end of Chapter 10, many of which deal with the consumption of commodities and material objects.

Bell, D. and Valentine, G. (1996) *Consuming Geographies: We are where we eat,* Routledge, London.

An excellent introduction to ideas of consumption and their relevance for human geographers. The book focuses on one particular form of consumption – eating – and explores geographies of eating practices in different spaces and at different scales.

Crewe, L. (2000) Geographies of retailing and consumption. *Progress in Human Geography* **24,** 275–290.

This engaging paper challenged geographers to take consumption seriously. The author's specific interest is in retailing and shopping: she argues for geographical work on these topics, drawing on some of the concepts we discuss in Chapters 9–13.

Clarke, D., Doel, M. and Housiaux, K. (eds) (2003) *The Consumption Reader,* Routledge, London.

An excellent collection of classic and more recent essays on consumption. The collection is particularly rich in conceptual material that illustrates the complexly spatial nature of consumption practices.

Lury, C. (1996) *Consumer Culture,* Polity Press, Cambridge.
An excellent introduction to concepts of consumption and consumer culture. The book is wide ranging in its influences, and is particularly strong on concepts of material culture (which we discuss in Chapter 10).

Mansvelt, J. (2005) *Geographies of Consumption,* Sage, London.
An outstanding, detailed reflection on the complexly spatial nature of consumption practices. The book is rich in case studies (both historical and recent), and provides detailed introductions to key thinkers and concepts.

Paterson, M. (2006) *Consumption and Everyday Life,* Routledge, London.
A very readable and engaging introductory text that situates consumption practices in the contexts of everyday geographies. We particularly enjoy Paterson's way of selecting thought-provoking and very current examples to illustrate and ground concepts relating to consumption.

Seiter, E., Borchers, H., Kreutzner, G. and Warth, E. (eds) (1989) *Remote Control: Television, audiences and cultural power,* Routledge, London.
A classic collection of essays dealing with issues of cultural consumption and audience agency in the context of television viewing. The chapters by Ien Ang, David Morley and John Fiske are widely cited as important in the development of the idea of consumer agency.

Urry, J. (1995) *Consuming Places,* Routledge, London.
As discussed in the chapter, Urry's work on the consumption of places – and his concept of the 'tourist gaze' – have been widely used by geographers. We like the way in which Urry's work attempts to detail how consumption practices – such as gazing – happen in practice, and some of their consequences.

Several cultural geographies

Part Map

- Chapter 4 – Architectural geographies
- Chapter 5 – Landscapes
- Chapter 6 – Textual geographies
- Chapter 7 – Performed geographies
- Chapter 8 – Identities

In Chapter 1 we identified three major ways in which the work of cultural geographers is important:

- researching cultural processes and their complex geographies and politics;
- exploring the importance of cultural materials, media, texts and representations in all kinds of spaces and geographical contexts;
- engaging with new lines of social and cultural theory, and reflecting upon their importance for human geographers.

Part 2 deals with the second of these themes. The following chapters introduce examples of geographical work focusing on some particular types of cultural phenomena and objects. To provide a sense of the breadth of recent work in cultural geography, we have selected five key themes, and have highlighted a sample of geographical work in each context. These chapters can be read as stand-alone introductions to each particular topic, or the whole section can be read as something of a cross-section of key interests and topics that have preoccupied many cultural geographers over the last two decades. The list of topics is by no means exhaustive, and no doubt different cultural geographers would come up with different lists of key topics, but we hope that these chapters will give you some 'ways in' to the wider subdiscipline of cultural geography.

Chapter 4 introduces geographical work about **architecture,** buildings, design and planning. Drawing on both earlier work by **traditional cultural geographers** and more recent work by **new cultural geographers** (see Chapter 1 for an explanation of these terms), we highlight how buildings and building processes are important for cultural geographers. Chapter 5 focuses on a topic that has long been central to the work of human geographers: **landscape**. We outline how landscape has been conceptualised in different ways by different cultural geographers – ways that reflect the broader, shifting concerns of the subdiscipline. Chapters 6 and 7 explore geographies of **texts** and **performances,** respectively. Here we introduce geographers' engagements with a diverse range of cultural texts, objects, phenomena and media: encompassing cartographic documents, contemporary fictions, policy discourses, autoethnographic diaries, musical and sporting performance, dance and performance art. In the process we outline the importance of everyday texts and performativity, more generally, for the work of cultural geographers. Finally, Chapter 8 extends some of the points made in Chapter 3 to consider the importance of **identities** and identity formation for the work of cultural geographers. We hope that, somewhere in this suite of chapters, you will find something that sparks your interest and prompts you to explore some of these topics in more detail: the lists of suggested readings at the end of each chapter should help you to find an area of cultural geography that interests you.

After reading Part 2, there are a number of activities which could help you to extend and develop your understanding of recent work in cultural geography.

- Return to the long list showing 'Cultural geography, 2009-10: from A to Z' in Box 1.4. Try to identify how the research topics in the table link to the broad themes - architecture, landscapes, texts, performance, identities - we have introduced in this section. You could also take a look at the most recent edition of the academic journals *Social and Cultural Geography* or *Cultural Geographies*: choose the paper that sounds most interesting to you and, again, try to link it to the broad themes from this section of the book.
- Having read Chapters 7 and 8, think about how these discussions of **performativity** and **identity formation** extend and add complexity to the ideas of cultural production and consumption outlined in Chapters 2 and 3.
- Choose one of the broad themes - architecture, landscapes, texts, performance, identities - that we have introduced in Part 2, and bear this topic in mind as you take a look at the chapters in Part 3. In Part 3, we explore ideas of **everydayness** (Chapter 9), **materiality** (Chapter 10), **emotion** and **affect** (Chapter 11), **embodiment** (Chapter 12), and **spatiality** (Chapter 13). Consider how these concepts are relevant to the topic you have chosen. Try to identify some ways in which these concepts extend or complexify your understanding architecture, landscapes, texts, performance or identities.

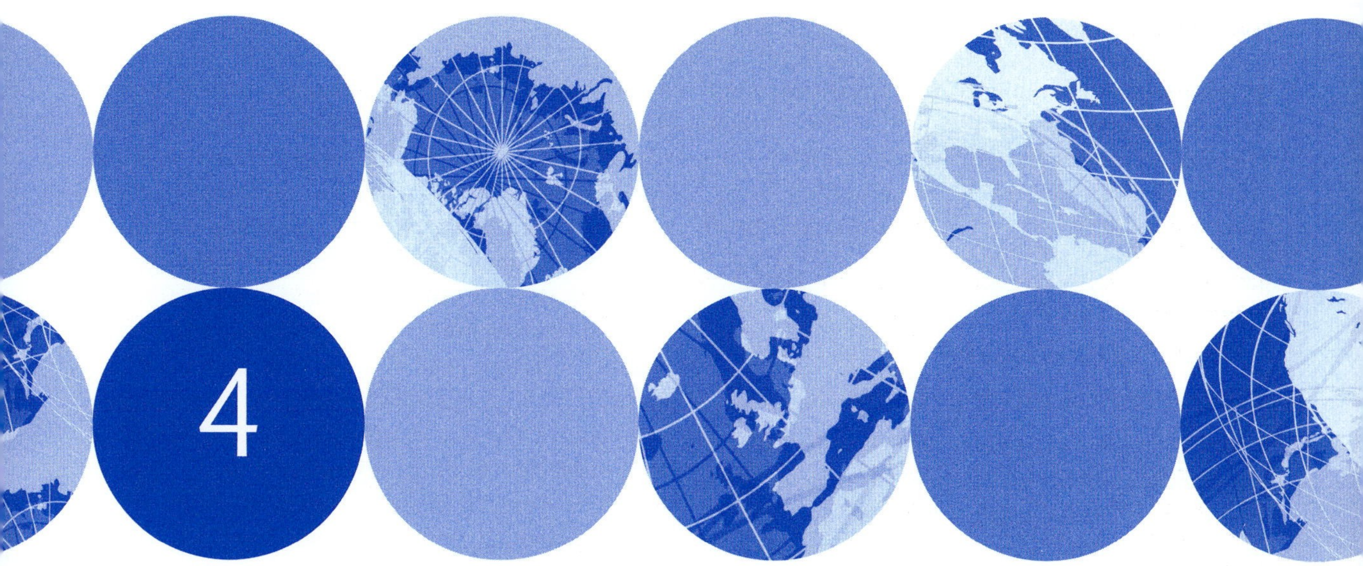

Architectural geographies

Before you read this chapter . . .

- Take a moment to look around you. Think about where you are. We would anticipate that you are either in, or near, a building (or perhaps more than one). Focus upon one of these buildings.
- Now write, or sketch, your answer to the following question: *what does this building mean to you?*

Chapter Map

- Introduction: taking notice of buildings
- Why do cultural geographers study buildings?
- What are buildings – and what do they do?
- What are buildings made of?
- What happens in and around buildings?

4.1 Introduction: taking notice of buildings

In January 2008, one of the authors moved house. As a first-time buyer, he had no real idea of how this worked. With the prospect of parting with well-earned cash looming large, he had to learn about the process. For the first time, ever, he had to *know* a tremendous amount about one of the most commonplace of built forms in the UK: a semi-detached (two-family) house. He had to know detailed things, such as:

- the precise dimensions of a new conservatory built by the previous owners, which might or might not have required planning permission;
- the odd activities prohibited by the vendors of the land on which the house was built (in 1959) – including a ban on keeping chickens;
- whether legal indemnification for missing guarantee certificates was required for the double-glazed windows;
- whether the cracks under the front window meant the house was slowly, inexorably, falling down (and many other seemingly insignificant, mundane details).

His experience should remind us that a building is not *just* a building. There are many different ways of seeing, feeling, being in, and writing about, buildings This is even true for small family houses.

But this chapter is not simply about houses. It is about various types of architecture, both 'big' and 'small' (a distinction we discuss in the chapter). As the introductory exercise suggested, this chapter is about how geographers have constantly sought to question the material spaces around us, and in particular those that we commonly call 'buildings' or 'architecture'. After discussing why geographers study buildings, the chapter introduces a range of approaches that geographers have used to answer three key questions about buildings:

- What are buildings?
- What are buildings made of?
- What happens in and around buildings?

In the process we explore some ways in which a geographical approach has extended understandings of the built environment, and how taking notice of buildings has enriched the work of cultural geographers.

4.2 Why do cultural geographers study buildings?

Buildings have been an object of study for geographers – especially cultural geographers – for nearly a century. Before we discuss how geographers have studied buildings, we want to take a moment to consider why this might be, and how this impulse has reflected broader work in cultural geography.

So why do cultural geographers study buildings? We want to highlight two key reasons, although you will come across many more if you read around the subject. First, and quite simply, the practice of architecture is one of place-making. This is the case from the small scale to the very large: from the family unit designing and building a shelter for themselves to the architect, contractors and builders involved in realising the largest skyscraper. Architecture is fundamentally about turning spaces into places, and/or turning places into other places. Architectural practice involves a massively complex combination of ideal designs and intentions with material technologies (bricks, windows, steel girders) to create what are often the very fundamental spaces in which we live our lives.

Second, following from this latter idea, buildings are the taken-for-granted spaces in which we spend a good proportion of our lives. Wherever we live, a large proportion of our lives is spent in, around, underneath or outside buildings. The context may vary, but the fact is that buildings are fundamental parts of our lives. Yet we very often take them for granted – that is, unless we need to change them (as per our introduction), or mend them, or want to visit unusual examples. But a key part of cultural geography has been unpick, question and even problematise the spaces and places we most take for granted. As we shall see, buildings are massively complex places, and the geographical study of buildings combines work on technology, power relations, aesthetics and consumption. Scratching the surface, buildings are also at the centre of many debates that concern not just cultural geographers, but most of us – debates about the regeneration of city spaces, about the meaning of home, or about the value of public places such as the municipal library.

A final reason for undertaking geographical work on architecture is that buildings provide excellent case studies for the broader debates, issues and concepts that interest cultural geographers (see Chapter 1). For instance, in this chapter we shall outline how the mapping of building types across the United States was a key activity of the very first group of knowingly 'cultural geographers': the **Berkeley School**. Then we shall explore how **new cultural geographers** chose certain kinds of building (often large-scale, imposing, well-known urban landmarks) as key examples of the ways in which places could symbolise the political intentions of the rich and powerful. And we shall explore how, increasingly, cultural geographers have turned to buildings as some of the key sites at and with which we can explore how material spaces (walls, floors, corridors, rooms) interact with and produce bodily movements, feelings and meanings. First, though, we turn to the question of what we mean when we talk about buildings.

4.3 What are buildings – and what do they do?

In the scheme of things, buildings have received relatively little attention from human geographers. Perhaps this is because – like semi-detached houses, or the building you're in right now – it is easy to just take buildings for granted. Yet architecture is a fascinating topic for cultural geographers. A mixture of art and technology, architecture involves the designation of boundaries, volumes, façades, movements – in short, the very process of *making a place*. Buildings are required to do many things – to house, to entertain, to incarcerate, or, like airports, to literally move people along corridors, through security gates, and into departure lounges (Adey 2008).

So the questions that give this section its subtitle are far from straightforward. In fact, the act of asking these questions is an important one. By so doing, we can open up what Jenkins (2002) calls the 'black box' of architecture. By this he means that, too often, a building is simply seen as a static, safe, material thing whose internal workings are opaque. We shall return to this point at the end of this section, because it is directed at earlier work by geographers – and others – on architecture. For now, though, bear in mind that it is this act of *questioning* buildings that has driven a number of cultural geographical engagements with architecture, specifically since the early twentieth century. So, we ask again: what are buildings – and what do they do?

Early geographies of architecture

The earliest geographies of architecture formed part of the work of the **Berkeley School** (Chapter 1). Their work was concerned more broadly with the study of landscapes. Yet in their search for the unique forces that created a landscape, they turned to buildings. For them, a building could be defined quite simply. It was – like the roads, technologies and tools they also studied – an expression of a 'way of life'. As John Goss (1988) argued much later, buildings were 'cultural artefacts'. He states: "although constrained by environmental conditions and available construction materials, form and style in architecture reflect the level of technological development and the values of a culture" (Goss 1988: 393). Once buildings were viewed in this way, they could be mapped – and so could the unique culture that had produced them. In this way, Wilbur Zelinsky called for the study of vernacular architecture, such as agricultural barns, arguing that:

> The many varieties of barns, sheds, wells, and other farmstead structures [merit] close attention. Enough work has been done on barns to indicate that they are perhaps the most conservative of structures in general use and may serve quite as effectively as house types to indicate sources, routes and dates of settlers and the spatial outlines of culture areas. Distinctive barn types are common to the South, others to the Midland and French Canada. Much of older New England is characterised by the readily recognisable connecting barn.
>
> (Zelinsky 1973: 100)

Finding little historical record of either barn types or design, Zelinsky (1958) documented how barns had been transplanted from central Germany to the eastern United States. He described the main characteristics of New England 'connecting barns' (barns connected to the main farmhouse) in terms of size, shape and construction. He then mapped 'frequent occurrences' of these barns, explaining differences in their construction by both climate and the influence of other cultural groups' building styles. Also part of the Berkeley School, Fred Kniffen (1965) undertook similar studies on vernacular or folk housing in the eastern United States. He mapped the distribution of different housing types, suggesting how particular architectural features (building materials, roofs) varied over space. For example, Figure 4.1 shows one map from Kniffen and Glassie's (1966) study of methods for building in wood across the eastern United States: the terms on the map refer to different methods for laying and interlocking horizontal logs in wooden buildings.

The work of the Berkeley School meant that architecture – even the architecture of barns and folk houses – would be taken seriously by geographers. This meant that, for the first time, geographers were paying attention to buildings as appropriate objects for geographical enquiry. Buildings were viewed as especially important, visible traces of different cultural groups, demonstrating historical influences as well as the interplay between culture and the environment as people had created settlements in the United States. This meant that mapping their distribution was a key way to understand the location and diffusion of different cultural groups.

Despite their popularity (which continues today), later cultural geographers criticised the Berkeley School's work on architecture (see Chapters 1 and 5). Chiefly, the Berkeley approach tended to simplify the definition and meaning of architecture such that it

Figure 4.1 Distribution of methods of horizontal log construction in the eastern USA.

Source: from Building in wood in the eastern United States, *Geographical Review*, 56, pp. 40–66 (Kniffen, F. and Glassie, H. 1996).

could only describe historical patterns. It was argued by the critics that buildings were created by – and were part of – much broader fields of social, political and economic relations than recognised by the Berkeley School.

The new cultural geography: reading the symbolism of buildings

During the 1980s, cultural geographers advanced studies that explored the symbolism of buildings.

This research – which was deemed controversial by supporters of the Berkeley School – was part of a broader change in so-called **new cultural geography** (Section 1.4) that tried to 'read' landscapes, places and built forms as if they were representations (see Chapters 1 and 5 for further discussion). It was argued that buildings conveyed particular meanings. Some buildings do this more obviously than others – it is worth stopping for a moment to think about how the building *you* are in (or outside) might be doing this (Box 4.1).

In recognising the symbolism of buildings, geographers did three things (Box 4.2 provides a case study of these three things in action). First, they tried to get behind the description of built forms to explore why and how they looked the way they did. They tried to understand the economic imperatives that guided the size and location of a building. They tried to understand the interests and ideologies guiding the 'personalities' behind a building, such as its owner and its architect (e.g. Gruffudd 2001).

Second, geographers demonstrated how buildings are not neutral containers for human action. Buildings – like other landscapes – can represent particular power relations (Dovey 1999). They can exclude some social groups and include others – either directly or indirectly. Imrie (2003) shows, for instance, how architects hold 'ideal' images about the human bodies that will inhabit their buildings. Rarely do architects think about the size, shape and especially dis/ability of the people using their buildings, and they may (perhaps inadvertently) exclude people who do not fit their ideal. Elsewhere, the gendered assumptions of modern architecture have created particular meanings for particular types of building based around social expectations of what men's and women's roles are in modern societies. For instance, the home was (and often still is) viewed as a private space for women, and therefore architects in the 1930s tried to ensure that the design of modern kitchens was appropriate to the ways in which a modern 'housewife' was understood. In particular, the design of modern kitchens could be seen as representative of the ways in which wider (British) society required middle- and working-class women to engage in domestic chores that would ensure the smooth function of the ideal, nuclear family (Llewellyn 2004).

Third, geographers revised the ways in which we see buildings. Rather than expressions of a cultural group, buildings could be viewed as **texts** (Chapter 6), representing the interests and power relations mentioned above. There were considerable disagreements about how those social, political and economic contexts were represented by buildings (compare Cosgrove 1997 with Duncan and Ley 1993; summarised by Lees 2001). Yet architecture was viewed as a kind of language (Goss 1988, 1993). Hence colours, shapes, symbols (whether religious, sporting, artistic, or historical) could literally be 'read' from a building, just as letters, words and common phrases make up what we usually understand to be a written text. The difference was that the architectural texts referred to different parts of a building, such as statues, columns or roof style. In turn, with a little training, geographers could look for particular meanings – be they political, cultural or historical – just as we learn to understand

Box 4.1
What does 'your' building mean?

Think again about the building you are in or outside:

- Is there any decoration?
- Is it built in a recognisable 'style'?
- Does its size or shape convey anything beyond its use?
- Can you tell who built it (who its architect or owner is/was)?

Compare this building with some others you might have come across – religious buildings, shopping malls, skyscrapers. Perhaps these other buildings have more or less obvious messages to tell people who look at them.

Box 4.2
The New York World Building (from Domosh, 1989)

Photograph 4.1 The New York World Building.
Source: Corbis/Bettmann.

New cultural geographical approaches to architecture enabled geographers to focus on one particular landscape or building. Domosh's study of the New York World Building (Photograph 4.1) was one of the first attempts to "rescue some of the meanings that are normally lost in functional analyses of skyscrapers" (Domosh 1989: 348). This early skyscraper, constructed in 1890s New York, was seen to represent a number of ideals around American literature and business endeavour.

Domosh does two main things in her analysis of the building. First, she places this newspaper building in the political-economic context of late nineteenth- and early twentieth-century New York. She details how New York elites used their buildings as displays of ostentatious consumption, and how skyscrapers expressed economic power and growth through getting taller – much in a way that historically had been reserved for public or religious buildings. Second, Domosh explores the values expressed in the building's symbolism. These were related to the political-economic context of the city, but also to the aspirations of the building's owner, newspaper magnate Joseph Pulitzer. For him, the building should not merely reflect his economic power or the commercial success of the newspaper. Rather, its Renaissance arches and decoration were designed to express his civic pride in the city and his passion for public service.

For Domosh (1989: 350), both of these "layers and types of explanation" were critical to understanding the functional and the symbolic elements of Pulitzer's concerns and prevailing class-based conditions of late nineteenth-century New York. The New York World Building could be understood only in terms of these elements.

the meaning behind political slogans, advertisements or fictional novels. Looking at the façade of a building – at its different symbols like letters and phrases – allowed one to read off the intentions of its architect, and the broader socio-political context of its making (Box 4.2 provides an example of what we mean by this).

An approach that looks at the symbolism of buildings is especially important, because it tells us whose interests are served by a building (for instance those of a multinational corporation or a political party). This approach also helps us to look behind *who* creates a building to explore *why* they have built it in the way they have. That is, in recognising symbolism, we recognise that, often, buildings are not neutral parts of our everyday environments, and that the way they look is often a deliberate attempt to promote this or that message, or to serve a particular person's interests. After all, symbolism is, unquestionably, a big part of how architects and owners make their buildings stand out – take the brash, glitzy, playful casinos of Las Vegas as an extreme example. Those casinos use enormous, spectacular displays in order to entice consumers and gamblers inside. They use (admittedly extraordinary) references to other places as diverse as New York, Egypt and Venice. In this case, casino owners use symbolism to distract consumers from the true purpose of their buildings; but they also do so to display the glitz, glamour and lure of their casino above their competitors.

In most cases (apart from Las Vegas), buildings often seem too everyday to be remarked upon. But this approach teaches us that, if we look behind the façade, we can uncover and explore all kinds of political debates, tensions and problems that go into the very making of places – if you like, the literal material creation of 'human' geographies.

As indicated above, the symbolic approach to architecture has been pigeonholed as part of the new cultural geography of the 1980s. As a result, many of the **criticisms of new cultural geography** can also be applied to the textual or symbolic approach to architecture. These criticisms are outlined in Section 1.4. However, we want to point out one important critique of the symbolic approach to architecture. That is, quite simply, that architectural geographers had missed out a good deal of what architecture is, means, and does (Lees 2001). For Jenkins (2002), then, the 'black box' of architecture had not yet been opened out by these two groups of scholars. For him, "[t]hey all treat the individual building as a blank canvas on which another discourse is illustrated" – the processes of constructing, inhabiting and maintaining a building as *this* building (not another) are simply assumed (Jenkins 2002: 225). In other words, geographers have tended to focus on the meanings of what buildings do – but in so doing, they have missed out both on how buildings are constructed and on the many ways a building is used after it is officially completed.

We began this section by asking what a building was, and what a building could do. Jenkins' (2002) phrase "a blank canvas" neatly summarises how many geographers have understood buildings as something that symbolises something else – either a culture, in the case of the Berkeley School, or a set of political or cultural ideals, in the case of new cultural geographers. But this is *not* the end of the question of 'what a building can do': we shall return to that question later in this chapter. The next two sections explain cultural geographers' attempts to overcome the missing elements in the architectural geographies being written by the Berkeley School and new cultural geographers. The literature covered in the next two sections should be interpreted as *building* on that earlier work – you will recognise elements of both approaches in what follows.

4.4 What are buildings made of?

As our introduction indicated, when one of the authors moved into his new house, he became obsessed with questions about windows, cracks, paving slabs and the other material things that make up a house. His anxieties remind us of the reliance of human beings – and human geographies – on **material things** (Chapter 10). Clearly, buildings also rely on material things. Think for instance of the many properties of bricks that make them such useful blocks from which to make low-rise buildings. Think (perhaps for the first time) about the properties of concrete that mean it can be moulded into sinuous curves – for example in multi-storey car parks. Or, think (again, probably for the first time) about the properties of steel that make it an ideal skeleton for tall buildings. Certainly, after the 9/11 attacks on New York's World Trade Center towers, many commentators were struck by how certain configurations of certain material things (and certain forces) could bring about the destruction of a building. All of this should set you thinking: what are buildings made of, and what actually holds a building together so that it looks like a piece of architecture?

Having criticised the Berkeley School above – and having said that an interest in material things is a recent one for cultural geographers – it is instructive to briefly revisit the work of Zelinsky, Kniffen, Sauer and others. This is because these early cultural geographers were, principally, interested in the material manifestations of 'culture' in the landscape. Their work may have been limited – it was certainly biased towards rural regions of North America, and by present-day measures was theoretically thin (Goss 1988; Lees 2001). Yet they ascribed great value to everyday landscapes, and to simple material artefacts, much in the ways an ethnographer still might. They allowed for the very study of things such as barns – their different façades, their connection to homesteads, their shape, the type of wood or metal from which they were made. They began the process of inscribing them with cultural meaning for geographical study – even if that meaning and its socio-political context were rather bare. Their work should also be read in a North American academic context, where settlement geographers were struggling to assert the validity of non-indigenous North American settlement patterns against the far longer sweep of European history. Thus vernacular architecture and folk artefacts became a key set of resources for the development of North American cultural geographies. Put simply: the material bits and pieces that made up barns and folk housing were made to *matter*.

Globalisation and 'global' cultures of housing

As might be guessed from the introduction to this section, we can say much more about the bits and pieces that make buildings what they are. We shall call this the 'materiality' of architecture. Significantly – like our introduction – much of this work has focused in one way or another on homes and housing.

We begin with the seminal work of Anthony King (1984, 2004). King's enduring interest has been on placing built forms at the centre of contemporary theorising about globalisation, postcolonialism and modernity. As King himself states in the preface to his *Spaces of Global Cultures* (2004: xvi), he is concerned with "architectural and building cultures as they are affected by transnational processes". His is a work of *flows* around the globe – of people, ideas, forms and materials – which then take local forms (compare with Ian Cook's work, explored in Chapter 10). There is therefore a constant interchange of scaled effects. The global makes the local, and vice versa (Massey 2005); crucially, this is all happening through the very process of architectural forms being built in different places. Famously, King illustrated his earlier arguments with the seemingly ubiquitous form of the bungalow (King 1984). He demonstrated that the bungalow was a hybrid form, taking on particular shapes and characteristics as it was internationalised via particular institutions and professional actors (architects and planners). So a bungalow in Kent, England, could look, feel and mean something very different from one in New Delhi, India. Yet both those countries – and therefore both those forms – were linked by the processes of colonialism, the internationalisation of architectural practice and the building technologies proffered in those two contexts.

Jacobs (2006) argues that King's work is significant when compared with that of the Berkeley School because, unlike the Berkeley School, King does not simply use a building as an example of, or proxy for, a cultural group or cultural region, ready to be mapped. Rather, he picks apart the different elements of the forms that were being mapped. It is, effectively, a step back, and a step closer. It is a step back: a critical questioning of how built forms emerge in the first place (which can *then* be mapped, or 'read'). It is a step closer: an in-depth examination of the planning procedures, flows of information, and types of material technology that constitute those built forms and make them identifiable as buildings of particular styles. These are both crucial steps if we are to fully understand how buildings matter in terms of their making at different, interlinked spatial scales. While our own homes might feel unique, they share certain components, certain assumptions and certain laws with many other homes, sometimes in some pretty unlikely places.

New cultural geography: political-economic approaches to architecture

We consider more recent work in this vein in a moment. But first we want to pause to consider again the work of new cultural geographers. If you remember, we suggested earlier that their work involved moving on from a common approach to 'reading' symbols from architectural façades. In particular, a number of these geographers were concerned with what we might term the 'historical-material' conditions that produced a building. You may come across the term 'historical-material' in relation to Marxism – and this is no coincidence. Significantly, Marxism is not simply about class relations or class struggle, but about the value, control and ownership of material things, at particular times, in particular places. This approach to material things can more broadly be labelled a 'political-economic' one.

In political-economic analyses, buildings can, with some care, be considered material things with ascribed values, and with (often emotionally charged) struggles over their ownership. For example, some parts of David Harvey's (1973) seminal *Social Justice and the City* are concerned with the location and ownership of housing in cities with respect to different socio-economic classes. Harvey's arguments about class and social difference have informed the ways in which many cultural geographers have studied how landscapes may include the views of some (privileged, mainstream groups) while excluding others (in Harvey's case, the working classes). Geographies of architecture have been no different, and in fact have provided some excellent examples of political-economic approaches to culture (e.g. Mitchell 2000).

Take the housing market as an example of the above point. It represents the locally variable system of buying and selling homes that involves valuation, real estate agents and mortgages, and which was deemed to be at the centre of a global financial crisis (generally considered to have started in 2008). The housing market could be viewed as a key 'circuit' in the accumulation of capital. But in thinking about the geography of architecture, Goss (1988) develops Harvey's point, arguing that buildings be viewed as commodities because they can be bought and sold. But houses are more than simply parts of a system of buying and selling, because they gain value in other ways. For Goss, the value of a building is not simply determined by the price of the land it sits on or by some relevant planning laws, but also draws on a range of cultural practices typical of globalising consumer cultures (Chapter 3).

The cultural processes that Goss cites – such as advertising – provide houses with cultural meanings additional to their simple economic value. So the designers of exclusive 'gated' housing use architectural design (symbolism), advertising, leisure pursuits (golf, water sports, tennis courts) and the promise of security to persuade their target market to move in (Till 1993; Al-Hindi and Stadder 1997; Phillips 2002). Just like any other consumer product (such as a car or a new gadget), this turns houses into commodified objects surrounded by carefully planned images and stories that aim to persuade consumers to buy them. Sometimes estate agents even try to associate particular kinds of lifestyle (e.g. retirement or 'traditional village life') with a newly planned community (Till 1993). If successful, these images and stories may influence the value of a property – just like the planning laws and land values that also affect the value of a piece of real estate.

In a slightly different way, David Ley (1993) used a political-economic approach to stress the relationship between cooperative housing *designs* and the *historical-material conditions* under which cooperative social formations were possible in 1980s Vancouver (Photograph 4.2). By 'historical-material conditions', we refer to the kinds of economic, social and political context (local and global) that allowed groups of residents to come together against the grain of capitalist processes such as gentrification.

Ley provides a more optimistic reading of postmodern capitalism that allows for alternative social formations, many of which remain able to express their communal (shared) visions through built forms. Ley explores how the kinds of house they built – their design, layout, legal ownership – directly expressed the shared values of these alternative groups. Ley's work is an excellent example of a cultural geography of architecture that *combines* the 'textual' approach to symbols with an in-depth analysis of the political-economic, material conditions through which certain types of building are allowed to flourish.

But what are buildings made of? Actor-network theory and the 'materiality' of architecture

Most recently, geographers have re-engaged with the materiality of architecture through actor-network theory (see Chapter 10). While the approaches above provide a pretty comprehensive framework for studying buildings, there is, it seems, still more we can say about the role of material objects in architecture. There is also more to say about buildings *as* material, technical achievements that are part of the kinds of network and flow that King writes about so eloquently.

Photograph 4.2 The design of cooperative housing in Vancouver. The architect wanted this 60-unit structure to be 'memorable', 'a good neighbour', ' a source of pride'. 'The gables and clapboard (in powder blue) [reflect] . . . nearby older houses, the postmodern [elements reflect] . . . new town houses in this gentrified district; . . . roof gardens and false chimneys [are] part of the iconography of "home"'.
Source: Ley (1993: 140). Alamy Images/Mike Dobel.

This, of course, brings us back to the question of what a building is and what it does. As Jenkins (2002) argues, an actor-network approach to buildings questions the given materiality of a building. It questions the feeling we have that a building has always been there, and always will be. As Dewsbury (2000) notes, as other material things decompose or wear down, even buildings are falling down, however slowly!

Actor-network theory actually requires us to ask two questions about buildings. First – and we have already considered this in some depth – what is it that makes a building a building? Second, how and why do buildings become included in our lives? Another way of framing this second question is to ask how buildings become enmeshed in social relations such that they become acceptable, usable, and a part of our lives that almost fades into the background (for some people this is the hallmark of a 'good' building). The trick is to try to answer both of these questions together, all the time. Rather than consider a house as simply a technical achievement, *or* view a house in the context of the particular political-economic conditions that surround it, we have to try to do both together. Jacobs (2006) labels this a 'socio-technical' approach to buildings, since it combines social processes with technical processes, and does not prioritise one or the other in advance. Notably, geographers have, in the past, tended to prioritise social processes over technical processes when writing about architecture.

Jacobs' (2006) work promises to help us rethink what we mean by buildings, and how we even *start* to write geographies of architecture (see Box 4.3 for a case study of her work). What she terms the 'thingness' of a building (Jacobs 2006: 11) – what makes a building self-evidently *this* building – should not simply be assumed. It is an achievement. Many things, people, laws and practices have had to go right for it

Box 4.3
Buildings as 'socio-technical' achievements

Photograph 4.3 Red Road high rise block, Glasgow.
Source: Alamy Images/John Peter Photography.

Jane Jacobs' (2006) and colleagues' (Jacobs *et al.* 2007) work on tall buildings has been instrumental in helping geographers to think again about architecture. She explores the rise (pardon the pun) of super-tall housing blocks – using the example of Red Road in Glasgow (Photograph 4.3).

Jacobs tries to rethink what makes a building, in a number of ways. For example, she argues that residential towers can be seen as socio-technical achievements. That is, they proffer new combinations of social relationships with technologies. So these combinations provide new ways of removing garbage, dealing with noisy neighbours, or doing the washing. This means that buildings are not just big, amorphous blocks of concrete. Instead, they are held together by all sorts of complex, seemingly mundane relationships between people and bits of technology. Some of these technologies were new when Red Road was built (in the 1960s) – technologies such as lifts, windows and garbage-removal systems. Each technology has its own story to tell – but only in taking part in the story of the building and the people who built and inhabited it. Hence a set of stories about washing, or about windows, that combine personal histories with detailed understandings of the technologies themselves.

Another key contribution of Jacobs' work is to challenge the ways in which we understand the impact of buildings on people's lives. Many of us assume that the shape, form and layout of a building can control our actions. But Jacobs argues for a different way of thinking about the relationship between people and buildings. In particular, buildings are often viewed as being 'just there', in the background. They are seen as an undeniable and unchanging part of our lives – a 'black box'. Instead, every building, Jacobs argues, has a 'turbulent' history. There are disagreements about how it should be built, the legal decisions surrounding it, and the technologies to be used – especially in such innovative buildings as Red Road. Quite simply – and this is the neat thing – we need to open out the 'black box'. This means that we can look again at what makes a building a building – at the controversies,

Box 4.3 continued

personalities and sheer work involved in turning a pile of bricks and steel into what we term (and assume to be) a 'residential high-rise'. In other words, Jacobs argues, doing this allows us to find out what are the 'claims' made about a socio-technical collective that mean it simply becomes a more-or-less stable part of our lives.

to get to the point of even being finished. Similarly, when buying a house, many things, people, laws and practices have to go right for that house to become *labelled* as *my* house, my home, my bit of architecture. To be sure, we can then also involve in our discussions the symbolisms that add to my house *becoming* my house – perhaps by thinking about the colour I have painted my garage door. But we must, in effect, cast the net much wider. Only then we can begin to think (Box 4.3) about how all the messy and complex things that end up (usually deliberately) at one site can be claimed as 'architecture' (Jacobs 2006).

Think again about the building you're in, or outside. If it is a library, what particular *claims* have to be made to make it a library? Or, to put it another way, what changes (to what happens inside, to certain laws, to its layout, to its contents, to its form) would have to occur before the library became a bookshop . . . or a factory . . . or a block of flats? We've come back – again – to the very question with which we started this chapter. So one of the key ways in which cultural geographers approach architecture is *still* to keep on asking: what is architecture, and what can it do? Finding out what buildings are made of is just part of this process.

4.5 What happens in and around buildings?

Take a look at Photograph 4.4. It is a railway station. Using a Berkeley School approach, we could consider whether and how this design of railway station is distributed regionally (in this case, in the English

Photograph 4.4 Leicester railway station, UK.
Source: Dr Sophie Hadfield-Hill.

East Midlands). Using a symbolic/textual approach, we could try to read the station's interior, façades, its layout and its style to uncover the intentions of its architect or the types of bodies that might pass through. Using a material approach, we could unpick which technologies 'made it' in the original construction and the ongoing refurbishment/maintenance of this station in the contexts of changing train usage in the UK.

Using each of the approaches outlined so far in this chapter, then, we could build three quite distinct, but overlapping, geographies of this scene. Let us be clear: we could *never* write a 'complete' geography for this railway station. But we are missing one ingredient: the practices that we call the 'inhabitation' of buildings. Advocates of both the political-economic (e.g. Ley 1993) and materialist (e.g. Jacobs 2006) approaches to architecture promote attention to the daily practices that are also key to giving buildings their meanings. They both highlight that **bodily practices** (Chapter 12) of being in a building matter – using it, walking through it, sitting in it, redesigning it, cleaning it, or pretty much whatever *you* are doing *now* in your chosen building.

The reasons for thinking about 'what happens' in and around buildings – the title of this section – are fourfold. Bear in mind as you navigate these four reasons that although this section appears last in the chapter, it does not mean that an interest in the practices of inhabitation is the most recent area of research in architectural geography. Indeed, some of the reasons we are about to list stretch back to correspond with the relatively early days of the Berkeley School's research.

First of all, geographers have – quite naturally – not been the only ones with an interest in architecture and built spaces more generally. A quite diverse set of philosophers, architects and urban theorists have tried to understand the complex ways in which people use built spaces. Many of these characters – from Walter Benjamin to the Situationists – are considered elsewhere in this book. So let us take a couple of examples that seem to be cited quite often in geographical texts about not only architecture, but also built forms. Notably, each draws out the practices of **everyday geographies** (Chapter 9) as a creative force within society that can disrupt the dominant ways in which powerful groups (big business, politicians, planners, famous architects) intend the public to use buildings. For instance, Michel de Certeau (1984) asks us to contrast the view of life from the top of a Manhattan skyscraper with the hustle and bustle on the street below. He argues that the 'stories' created by thousands of footsteps on the ground are far more meaningful than those of the grand stories told about cities by politicians. More recently, Iain Borden, Jane Rendell and others (Borden *et al.* 2001) have tried to rewrite architectural histories from the viewpoint of the diverse people and practices that move through and reside in buildings, rather than from the 'official' stories that architects and historians tell about buildings. The fascination with the everyday life of buildings – think of it a little like idle gossip – is an enduring one. But it is one that simply allows us to become aware of the diverse ways in which 'ordinary' people use, think about, experience and feel buildings when they are in them. This impulse – which is a political one more than anything else – also allows cultural geographers to listen to a much greater range of stories about buildings, often from marginalised groups such as the homeless, whose voices are not often heard in academic research (Lees 2001).

Second, geographers have reminded us that everyday life means so much more than idle gossip; or, rather, they have demonstrated how *even* such things as idle gossip, or the interpersonal encounters that happen in a coffee shop, are hugely significant (Laurier and Philo 2006a, 2006b). For an interest in everyday life has also meant an interest in what is often termed the **consumption** of places (see Chapter 3). A coffee shop, for instance, is not just a place where we go to consume coffee, but a place that – with its clever design, its smells, its atmosphere – we go *to* consume. The difference is a relatively subtle one. We consume a place itself – just, as we noted in the previous section, as we can market a place or sell it. The most obvious example of the way in which we can consume a place is an enduring one for cultural geographers – how, when on holiday, mass tourists simply seem to 'collect' photographs and images of places rather than spend time experiencing them at length (Urry 1995). So a place – including a building – can be produced, sold, marketed, and now consumed.

A focus on consumption shifts attention from the **production** of a building (the materials, symbols and contexts of its making; see also Chapter 2) to how people use it as part of their everyday life. Perhaps we should not assume a ready distinction between production and consumption – and perhaps we should not read the consumption of a building exactly like that of a tin of soup. The point is, though, that consumption is both a practice that matters, and one that, significantly, produces its own meanings. This allows Lees (2001) to suggest yet another move beyond reading the symbolism of buildings like a text. She says that "if we are to concern ourselves with the inhabitation of architectural space as much as its signification . . . we must [also] engage . . . actively with the situated and everyday practices through which built environments are used" (Lees 2001: 56). Lees describes how a number of cultural geographers interested in consumption have explored the meanings that people make of everyday consumer objects – such as the personal stereo (Bull, 2000). In the same way, cultural geographers can explore how people 'consume' buildings – both by asking (interviewing) inhabitants or users about their experiences of being in buildings, and by observing the kinds of things that people do in them.

Lees uses observations to create a series of notebook vignettes. She recounts some of the many uses of a public library in Vancouver – some unexpected – by newspaper readers, vagrants, and playing children. Her work has inspired others to explore how people make their own meanings from schools (Kraftl 2006a, 2006b) and blocks of houses (Llewellyn 2004). So, Lees urges geographers to look both at the meanings that 'ordinary people' make of buildings, and at the things they do in them. In doing so, they do not just look at how people 'consume' buildings, but at the kinds of bodily practices they perform there – what kinds of movement, activity and bodily sensation a building allows. This approach therefore provides a further example of the many cultural geographies of the body that we discuss elsewhere in this book.

Third, some geographers have drawn on a recent interest in **emotion** and **affect** (Chapter 11) to explore how the meanings that people attach to buildings are not always or easily expressed in words (Rose *et al.* 2010). Certain buildings – like the home you come from – have many sentimental meanings attached to them. This is, partly, because so many memorable things have happened there – so many embodied practices have given rise to so many memories. You may look back and laugh. Or, if you are less fortunate, you may look upon your home as a place of incarceration, trouble or misery. During the 1970s, humanist geographers (Chapter 5) tried to explore what it is about places that makes us love or hate them (Tuan 1977). They used poetry, film, literature and a whole range of other artistic devices to understand why particular places provoked emotional reactions. Very often, they focused on buildings, especially those 'traditional' types that evoke shared meanings in a nation's history (traditional homesteads, for instance). Very recently, architectural geographers have tried to understand how architects *and* users of buildings manipulate built spaces to deliberately create certain emotions among other users. They have focused on the manipulation of certain 'homely' atmospheres in schools for young children, and on the creation of calm places in airports (Kraftl and Adey 2008; also Adey 2008).

Finally, this recent work on affect reminds us that practices of inhabitation do not take place *after* a building has been 'finished'. Some people – including many architects – argue that a building is never finished, and that we should embrace the idea that it is always incomplete (Lerup 1977). Rather, the process of building is a much more complex, related form of negotiation between the 'builders' and the 'users'. Therefore, if we combine Jacobs' (2006) arguments with those of Lees (2001), we see that we need to constantly revise our ideas not only about what makes this or that building what it is, but also about what kinds of inhabitation are made possible by the building in combination with its architect and its users. It is not as simple as saying that people inhabit buildings. Buildings and people produce various forms of inhabitation, *together*. So buildings are an effect of all the things that make them up (including inhabitation). And inhabitation is just one effect of the building!

Take again the building you're in, and consider what you're doing right *now*. What else could you do, at this building? Being honest, and being realistic, what types of thing are likely to happen in the next five minutes in your building? Now look back at all the things we've considered in this chapter and think about *why* your building means that only certain things are likely to happen. Is it design? Is it the people around you? Is it just a feeling you get?

Summary

- Geographies of architecture ask us to question buildings in new ways. Buildings often seem to be 'just there' – but since the early twentieth century, geographers have provided a number of ways to show how they are not simply 'just there'.
- Early approaches to architecture (Berkeley School) tried to describe and map the characteristics of particular types of building – such as barns and folk houses.
- Symbolic approaches to architecture were diverse, but involved 'reading' the façades of buildings like 'texts'. Then the hidden meanings, histories and political contexts of a building could be uncovered. Sometimes those contexts could then be challenged by geographers with those people who lived in (or could not live in) those buildings.
- Materialist approaches to architecture do two things. On the one hand, they remind us of the importance of material things in our everyday lives – and of the political economies that structure our lives. On the other hand, they remind us that buildings combine social and technical factors equally; if we disentangle these factors we can rethink what makes a building a building!
- Approaches stressing everyday practice remind us that people make their own meanings with – and feel differently about – the buildings they live in and with. It is important to find out about these meanings and emotions in order to challenge the view that buildings are 'just there', and that architecture finishes when the builders leave.

Some key readings

Duncan, J. and Ley, D. (1993) *Place/Culture/Representation*, Routledge, London.

Chapters 6, 7 and 8 provide examples of a political-economic 'reading' of buildings – in this case housing. This has become a seminal text, because it not only provides an example of treating buildings as 'texts', but also allows for a consideration of the political contexts surrounding buildings, and the everyday practices of 'ordinary' people who use them.

Lees, L. (2001) Towards a critical geography of architecture: the case of an ersatz colosseum. *Ecumene*, **8**(1), 51–86.

Lees calls for a geography of architecture that pays greater attention to people's embodied practices. She uses vignettes from ethnographic observations to explore the many conflicting meanings made by users of a public library in Vancouver. Note that the journal is now called *Cultural Geographies*.

Jacobs, J. (2006) A geography of big things. *Cultural Geographies*, **13**, 1–27.

This is an example of how a materialist or actor-network theory approach can be applied to buildings. Jacobs advocates looking at the social and the technical equally, so that we can unpick the ways in which a building gets built – and the ways in which people make claims about and for 'architecture'.

Kraftl, P. and Adey P (2008) Architecture/affect/inhabitation: geographies of being-in buildings. *Annals of the Association of American Geographers*, **98**, 213–231.

This paper provides an example of how emotional geographies are built, controlled and managed in buildings. It uses case studies from school architecture and airport architecture, and shows how both designers and users are part of the ongoing ways in which a building can move people.

Zelinsky, W. (1973) *The Cultural Geography of the United States,* Prentice Hall, London.

An example of cultural geography using a Berkeley School approach. Browse the index to find examples of the ways in which housing, barns and other vernacular buildings are treated using this approach.

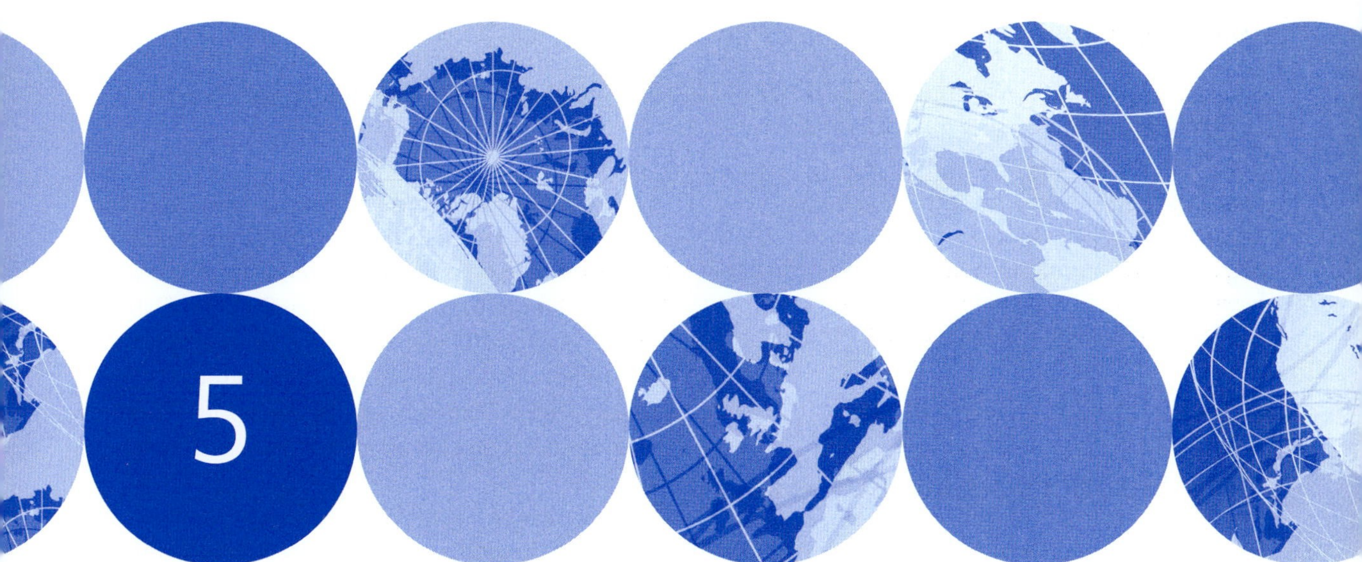

Landscapes

Before you read this chapter . . .

Look at Photograph 5.1.

- What does the image show?
- What does the image mean?
- Do you think the image is 'accurate'?
- Does the image hide anything?
- How does it make you feel?

Can you answer all the above questions just by looking at the image?

Photograph 5.1 The Grand Canyon, Arizona.

Source: Peter Kraftl.

Chapter Map

- Introduction: landscape as . . .
- Defining 'landscape': some (brief) wordplay
- Landscape *as* . . . material
- Landscape *as* . . . text
- Landscape *as* . . . performance/feeling

5.1 Introduction: landscape as . . .

The word 'landscape' is one of those words that is hard to pin down. Often, it is easier to think of an example – or show somebody a picture – than it is to define, in words, what a landscape is. Some landscapes are personal to us as individuals. We have emotional reactions to particular places, scenes, photographs, sounds. Certain landscapes are evocative and have particularly personal meanings. Photograph 5.1 is a particularly important landscape for one of the authors. But other landscapes are more than personal. They hold meanings for local communities – even for whole national populations. You might be able to think of a landscape that is really indicative of the country you are from. Landscapes also bear the marks of history – landscapes collect, collate and layer social interactions with the environment.

But the question is: what makes a *landscape* different from, say, a *place*? Or a *view*? Or a seemingly *natural* thing like a chain of mountains? The answer is, sometimes, not very much. This is why the term is so hard to define – to untangle from the other things that cultural geographers (and other geographers) study.

So, rather than worry too much about what landscape is (or is not), in this chapter we consider landscapes as something *active* (Mitchell 1994). After looking at some definitions of the term, we explore what landscapes do, and what is done to them. We suggest that many cultural geographers do not consider what landscapes are, but study them in more open, unfinished ways – captured by the term 'landscape *as* . . .'.

5.2 Defining 'landscape': some wordplay

We begin by looking at the dictionary definition of the word 'landscape'. First, though, think for a moment about the word 'landscape'. Bear in mind the questions about the picture of the Grand Canyon, shown above. How would *you* define the word 'landscape'? How many *different* definitions can you think of? The *Oxford English Dictionary (OED)* divides its definition into two categories: landscape as a *noun* and landscape as a *verb*.

Landscape as a noun. Here, landscape is considered a material thing. Hence it is a "tract of land" with particular features. These features are usually "considered as a product of modifying or shaping processes and agents (usually natural)". A tract of land might be a named valley, a chain of hills, or a coastline. The *OED* also draws on early work by physical geographers such as Woolridge and Morgan (1937). They claimed that knowing the physical geography of a landscape was 'indispensable' for any cultural geography thereof. So the material bits and pieces that make up landscapes – rocks, vegetation, water courses – matter to our understandings of what a landscape is.

Earlier uses of the term – from the seventeenth century onwards – find landscape being deployed as a term for 'pictures' and 'representations' *of* a tract of land. For this reason, landscape has had an enduring relationship with landscape painting. One of the roots of the word – the Dutch word *landschap* – was used as a technical term by European landscape painters in the sixteenth and seventeenth centuries. Indeed, many have suggested that the practice of landscape painting emerged in Holland in the seventeenth century (Adams 1994). So particular ways of representing landscapes are not just given, but have *developed* over time (see Box 5.4).

The *OED* notes one further use of the term as a noun, which is somewhat related to landscape painting. That is, 'landscape' denotes a "view", a "prospect", a "vista" or "the object of one's gaze". It is a material tract of land that has been turned into a landscape by virtue of being gazed *at*. Often this gazing is done in a particular way (Cosgrove 1985). It might be done by following the conventions of style in painting, or the techniques associated with taking a digital photograph (Box 5.1).

Landscape as a verb. Here, the *OED* description is much shorter. In itself, this comparison is significant, given that geographers have sought to imbue landscapes with more active qualities. The *OED* definition is in three parts. First, landscape (or landscap*ing*) is the process of represent*ing* a tract of land. It is the active practice of sketching, painting or photographing. It involves picturing, depiction and selection – turning a tract of land into a landscape, perhaps by painting it. The second use of the term landscap*ing* is more common. It refers to the formal act of creating a landscape: of laying out a garden, of planning a park, and perhaps of the profession of a 'landscape architect'. Third, and tantalisingly, the *OED* refers to landscap*ing* as an act of "concealment". For instance, it cites the advent of road landscaping in 1950s Britain (see Photograph 5.2). According to newspapers of the day, new roads were to be 'landscaped' into the countryside and not 'stuck on it' (see Merriman 2006). So landscaping is also about hiding that which is not acceptable – whether because it is visually displeasing to some, or socially demonised by others.

Thus 'landscape' can be a noun – something material, like rocks or houses, or something representational, like a painting or photograph. And 'landscape' can be a verb – the act of creating a landscape or of representing it in a text or image. These meanings hint at how cultural geographers have understood landscapes. We therefore expand upon – and query – some of these definitions in the remaining sections of this chapter. We consider landscape as three things: as material; as text (or representation); and

Box 5.1
Landscape photography: ways of seeing landscapes

Landscape photography is something many of us do, especially on our holidays. Certain conventions for taking photographs of landscapes have endured since the invention of the camera in the mid-nineteenth century. In turn, some photographic conventions were influenced by enduring features of landscape painting, such as perspective. Others were not. For, after its invention, the camera was soon considered to be different from other media. The photograph was understood as a technological achievement, a faithful, mechanistic representation of reality (Snyder 1994) – a belief that to some extent still holds in the phrase 'the camera never lies'. However, understanding cameras in this way meant that photographs could also be used to serve certain ideological interests – in the nineteenth-century USA, for instance. Here, high-quality images taken by US photographers highlighted both the progress of American technological innovation and the special (explicitly *not* European) nature of America's wild landscapes (Snyder 1994).

Think about some of the long-standing conventions of landscape photography that many of *us* adhere to. If we are taking a picture of some hills, we usually take the picture so that the camera is horizontal – in what is commonly referred to as 'landscape' format, in fact. We usually try to get a bit of foreground – a tree, a person, a rock, a cute furry squirrel – to get a sense of scale. If we're keen, we try to make sure the light is interesting, maybe by fiddling with a few buttons. Many of these practices have been learnt and shared in tandem both with (largely Western) cultural assumptions about landscapes, and with camera technologies themselves. This is why so many of our holiday photographs (like Photograph 5.1) look so similar!

The way of seeing landscapes through a camera lens also evolves. For example, slight changes occur as a result of developments in camera technologies. Take the change from 35mm to digital cameras. In the past, people placed their eye against the viewfinder so that the view was seen only through a small piece of glass. Now, people lean back and look at the digital image displayed by the camera itself – and sometimes at the view itself too – so that they are framing the image in a very slightly different way. They also get an instant result: if they don't like one particular landscape image, they can delete it, and take another one, until it's just right. These are important – if routine – ways in which we may see, select and frame landscapes.

Photograph 5.2 An embankment on the M1 motorway, UK.
Source: Press Association Images/PA Archive.

as performance. As will become clear, these divisions are slightly arbitrary (there are crossovers between each section), but they provide a good indication of how cultural geographers have tried to understand landscapes.

5.3 Landscape *as* . . . material

The word 'material' conjures up various images. We deal with the complexity of this term elsewhere (see Chapter 10). In this section, we see how geographers have *applied* different 'material' approaches to the study of landscape. And, as we will see, geographers have used very different understandings of the word in their research.

When discussing landscape, it is customary to begin with a brief introduction to the **Berkeley School** and Carl Sauer. As Wylie (2007) notes, it is then customary – in the UK at least – to offer a line of sometimes dismissive critiques of Sauer's brand of cultural geography, before moving on to outline **new cultural geographies** (see also Chapter 1). You will no doubt come across this conceptual manoeuvre if you read some of our recommended texts about landscape. We want to take a slightly different approach. Like Mitchell (2000), we want to emphasise the *continuing* importance and interest of some elements of the Berkeley School's approach. This is partly down to Sauer's influence on later geographical research, especially from the 1920s to the 1970s. But (in our view more importantly) this is also because his work was instrumental in requiring geographers to simply look at landscapes – and especially the materials that constituted them.

For Sauer (1925), a landscape was the key unit of geographical analysis. A landscape could be more or less defined and bounded – a particular 'tract' of land in our definition above. A landscape was home to a particular cultural group. A region containing relatively similar landscapes and cultural groups was a 'culture region'. The task of the geographer was to demonstrate the effects of a cultural group on its local landscapes. The word 'effects' is critical here. Unlike later cultural geographers, Sauer was not so interested in how or even why a culture region had produced the landscapes it harboured. Rather, he was concerned with the physical, material effects of a culture upon the land. This really is an understanding of landscape as a *noun*.

Two key words describe Sauer's approach to landscape. First, it was *idiographic*. This meant that it was not concerned with making general laws or models about how landscapes came into being. Rather, Sauer showed what made places unique. In other words, he and his many students looked for the material elements of one landscape that signalled difference from others – things like house types, religious buildings, barns, artefacts, agricultural technologies (Kniffen 1956; Zelinsky 1973: see Chapter 4).

Second, a landscape was viewed as a *palimpsest*. In other words, a landscape was built up over time. Different cultures brought different 'layers' with them when they settled – different house types, or perhaps subtle changes to road layout. The idea was that a landscape could be 'read' by looking for material evidence of these layers. Observation and description were key. Sauer and colleagues built an inventory of stuff in different landscapes in their detailed field notebooks – a type of document you will no doubt have to produce in the course of your studies.

Since the 1950s, various cultural geographers have revised the Berkeley School's approach. With conceptual roots in Marxism, they sought to restate the importance of material things in understanding landscape. This move was important. A number of critics expressed dismay with the ways in which 'culture' has been used by cultural geographers of different persuasions (see Barnett 1998). In particular, Mitchell (1995, 2000) argues that both Sauer and new cultural geographers have a rather thin conception of culture. Whereas Sauer simply assumed that culture referred to a cultural group or way of life (a thing, essentially), later geographers used culture as a 'level' or layer of explanation. The problem with both is that they assumed culture was actually an empty, already existing abstraction. Because it was abstract, this meant that they ignored the ways that culture was produced – the social processes, contradictions and, especially, the sheer *work* involved in producing culture went ignored.

Mitchell (2000), then, calls for a different understanding of culture. His work also provides a different way of thinking about the materiality of landscape, in two ways. First, because it sits in a tradition of Marxist thinking. This emphasises the socio-economic relations between groups of people, focusing in particular on the value and distribution of resources (and hence materials). Second, it is materialist because it retains a focus on the things that make up landscapes. But in the case of Mitchell's work, it is more processual and active. Landscape is a verb *and* a noun.

So landscapes are not a done deal. They are never finished, ready to be simply read, or visualised, or represented. Landscapes are *produced*. Landscapes are the outcome *of* class relations and capitalist accumulation. They are also represented *by* the powerful actors of capitalism (managers, politicians, business leaders). As Mitchell (2000: 94) says:

> the landscape itself is an active agent in constituting . . . history, serving both as a symbol for the needs and desires of the people who live in it (or who otherwise have a stake in maintaining it) and as a solid, dead weight channelling change in this way and not that.

As an example, Mitchell (2000) traces the ways in which Johnstown, California, has been regenerated after deindustrialisation. Regeneration planning has drawn selectively on the town's industrial past, and a rather spectacular flood that devastated the town in 1889. Whereas numerous labour disputes have been written out of the town's revised culture, the flood and a related but rather inflated heroic sense of the town's industrial past have been written in. Critically, planners wanted to stress the town's industrial heritage as part of a (touristic) regeneration of the town. They sought to *use* the town's declining landscape – the empty mills, the ruined factories – as a source of economic development, and as a seemingly accurate representation of the town's history. In other words, the landscape was made to *work* again. The material elements of the landscape – stones, bricks, wood of factories and mines – were no longer in the service of industrial development, but of post-industrial heritage regeneration instead. Landscape is the stones, bricks and wood: these both represent and embody the contested cultures of a place.

In ending this section we want to highlight how geographers influenced by poststructuralism and postcolonialism have suggested other ways in which landscapes are material. We look first at poststructural approaches to landscape, and then at postcolonial approaches.

A number of poststructuralist geographers have been influenced by **actor-network theory** (ANT). As we detail in Chapter 10, ANT proposes a different way of thinking about the boundaries between objects and people (Bingham 1996; Latour 1999). ANT argues that we need to consider the active role of *non-human* agents in creating places – agents such as animals, plants, technologies, even buildings (Jenkins 2002; Whatmore 2006). Doing this breaks down the boundaries that many of us think exist between 'humans' and 'everything else' – including material landscapes that are 'out there'. ANT requires a way of thinking about the world that focuses on the relations between different agents – looking at connections rather than the things themselves. This means that landscapes are not only created by the ways we look at them. Rather, if we accept that many different agents work on the landscape, then we have to consider landscapes as things that are always being created, and which are often out of 'human' control (literally, in the case of the Johnstown flood and others since). ANT actually challenges our assumptions about landscape (for more detail about this, look at Box 5.2).

Postcolonial geographers have taken a different approach to the materiality of landscape. They emphasise how certain bits of material culture have particular meanings, and can be seen as key parts of the ways in which we imagine landscapes. Material objects – in the example below, the Muslim veil – can be considered one element (a very important element) in a collection of objects, images and imagined places that together make up a group of ideas that we can loosely term an imaginative landscape. Such seemingly everyday objects may also link up with broader controversies and differences about place. They may in fact symbolise profound differences in attitude about a place that are more powerful than any identifiable land-'scape' in the sense that we described in our section on definitions.

Young (2003) shows how the Muslim veil has often been represented as a symbol of difference between the Western and Muslim worlds:

> For Europeans, the veil used to symbolize the exotic mysteries of the east. For Muslims, it signified social status. Today, the meaning of the veil has changed dramatically. For many westerners, the veil is a symbol of patriarchal Islamic societies in which women are assumed to be oppressed . . . On the other hand, in Islamic societies . . .

Box 5.2
The challenge of ANT to 'landscape'

John Wylie (2007: 204–205) argues that actor-network theory (ANT) poses a number of challenges for previous ways of thinking about landscape. First, ANT prefers ways of thinking about the world that stress change, fluidity, and mobility. This contrasts with most notions of landscape – like Sauer's – which are generally static, looking as they do at the material *effects* of cultural processes on a landscape, at a certain point in time.

Second, ANT starts with relations, rather than specific or named places, scenes or representations. Take the example of Johnstown again. ANT theorists would not see 'landscape' as the result of work, or even as work (like Mitchell). That's because starting with 'landscape' brings with it a whole set of aesthetic and political assumptions. Instead, ANT would start with social and technical achievements (such as the factories) themselves. ANT would query the involvement of many different technologies, techniques, materials and professional practices in the construction of a factory – the work, if you like, of a much greater range of agents – not just human agents, but non-human too. It would also ask what relationships broke down – what went wrong – when the factories closed down. But this work might not simply just give us a landscape. The outcome might be a quite different metaphor or idea. So the idea or 'fact' of a landscape is not assumed if we look at particular sets of relations in a place.

In these two ways, thinking about the materiality of landscapes (in this case via ANT) can also pose challenges to there being any value to the idea of landscape at all. ANT asks us to fundamentally rethink what we mean by 'landscape' – is every bit of land really a landscape?

the veil (*hijab*) has come to symbolize a cultural and religious identity.

(Young 2003: 80)

Hence the veil's changing status says a lot about the changing relationship between the West and postcolonial cultures. It also refers to the way in which we imagine the peoples and landscapes associated with the veil. Certainly, Western ways of thinking about the veil were part of a whole range of ways in which non-Western countries were exoticised in quite generalised ways (see also Gregory 1994; McEwan and Blunt 2002). Hence in all manner of orientalist representations (see Box 2.3) images of the veil, of Muslim women, were for instance synonymous with the bright colours, the pungent, spicy smells and the bustling sounds of a (probably unspecified) Asian city. Other postcolonial work has shown how material cultures are a key tool for migrants to come to terms with unfamiliar spaces while retaining a connection to their home (e.g. Hall 1990; Slymovics 1998; Tolia-Kelly 2004a). That work is important, because it shows how people's connections to landscapes are more multiple and complex than the generalising, exoticising images of 'oriental' landscapes represented in many popular discourses.

In this section, we have shown that landscapes involve materials and material cultures in all kinds of ways, and are *not* simply about views, surfaces or images. Quite ordinary landscapes have material things in them – houses, barns, roads. But cultural geographers have continued to debate how we should understand those everyday, material things. Are they, as Carl Sauer thought, 'effects' that are left on the land, and which make up a landscape? Are material things, as Don Mitchell argues, a part of the work of landscape – more than simply another layer of representation? Are they – as ANT theorists suggest – part and parcel of relationships that question the definition of 'culture' and 'nature', and which therefore question what we mean by landscape? Or are material things – as postcolonial geographers argue – a symptom and sign of contested, political images about the very different places that make up our earth? The short answer is: material landscapes can be and do all of these things, and all of the above approaches continue to be used by cultural geographers. But as we see in the next section, landscapes can also be more than material.

5.4 Landscape *as* . . . text

In Chapter 4 we introduced the idea that **architecture** – even ordinary domestic and public buildings – could be read like a text. This section of the chapter works from a similar premise. That is, that landscapes can be treated as texts, albeit in slightly different ways from buildings. It is worth stressing the links between landscape as material (the topic of the last section) and landscape as text. When you read about landscapes, you will probably come across both terms in the same book or journal article.

The links between landscape as material and landscape as text go back further than recent geographical work. From the 1960s onwards, American geographer J.B. Jackson developed a large body of work about the American landscape. His work had echoes of the Berkeley School. It was concerned with everyday – what he termed 'vernacular' – landscapes. It was concerned with an approach rooted in culture as a way of life, rather than culture as a work of art (with 'low' rather than 'high' culture, if we can put it that way). Jackson used the term 'landscape' to refer quite specifically to the material worlds of everyday towns and cities – houses, cars, shops, workplaces.

In stark contrast to the Berkeley School, Jackson also did something more in tune with later, 'new' cultural geographers. He also explored *symbolism*. He sensed that landscapes were a source of myth, imagination, and symbolic value to the people who lived in them. As humanist geographers soon also argued, landscapes come to hold deep-rooted meanings for both individuals and social groups. Indeed, Jackson worked with humanist geographers to explore the everyday, material and symbolic meanings of 'ordinary' landscapes (Meinig 1979; Olwig 1996). For instance, Jackson focused on differences in agricultural and settlement patterns between northern Mexico and the southern USA. He argued – unlike Sauer – that those differences were not just down to the 'effects' of different culture groups living in those regions. Rather, they were also the effect of distinct sets of cultural, economic and political values. Quite simply, those different patterns were down to the different ways that those groups felt about social organisation, land ownership, and land value. The landscape reflected those beliefs.

While Jackson's studies of vernacular landscape acknowledged symbolism, cultural geography in the 1980s went far further. It developed sophisticated tools for understanding landscapes through such concepts as 'text', 'discourse', 'representation' and 'metaphor' (see Barnes and Duncan 1992). Originating in the UK, new cultural geographers developed these tools in direct response to what they saw as the many problems of US cultural geography, and especially the Berkeley School. Their critiques were many (see Box 5.3).

The work of new cultural geographers was (and is) diverse. It is a bit unfair to generalise about what they did, but we can see a common point of departure. That is, new cultural geographers picked apart the assumption that landscapes are 'just there' – that they are taken for granted, in the background. In other words, the Berkeley School had opened our eyes to the fact that landscapes were the effect of cultural processes. But then new cultural geographers sought to look again, to look in more detail, and to look behind the façade, behind those effects. They sought to explore what were the debates, tensions, ideals and morals that were operating within the material effects seen by Sauer and his colleagues. They sought to see what landscape was hiding (see our last definition of landscape as verb).

So the point is that landscapes reflect particular meanings and values that people hold about a place. More particularly, they tend to reflect the values of the most powerful people in society. In feudal and early industrial societies, this meant the aristocracy. In this way, the owners of large, English houses had the money and education to rebuild their parkland estates according to the latest fashion. During one period, they would incorporate elements of Classical mythology; later, they would rebuild their gardens to look 'natural', following tastes for the so-called 'Picturesque' and then 'Romantic' periods in landscape taste (Daniels 1993; Photograph 5.3).

In capitalist societies, the powerful tend to be the middle and upper classes, as well as professional planners. For instance, Judith Kenny (1992) critically reads a plan for the future development of Portland, Oregon, written in 1976. The document indicates the importance of the landscape for reflecting the way of life and character of the city and its inhabitants – as surrounded by nature, as 'dynamic', 'liveable' and 'diversified' (Kenny 1992: 176). Yet the document does not necessarily encapsulate the values and experiences of the diversity it mentions. It does not reflect the interests of the many different groups living in Portland – people from different socio-economic

Box 5.3
Critiques of the Berkeley School approach to landscape

The Berkeley School was criticised for:

- over-emphasising rural, 'vernacular', folk landscapes at the expense of urban ones;
- treating culture almost as a living thing, something independent of human intervention or action (this approach has been termed 'super-organic');
- downplaying the role of human agency in creating, resisting or working on a particular culture;
- being too descriptive, and not explaining the patterns they documented;
- essentially ignoring the effects of different kinds of tension and exclusion (such as the uneven effects of commercial development, or the different ways that men and women 'accessed' and worked on landscapes);
- looking at landscape in a static way – compare the Berkeley School approach with the ANT approach, and look at the contrasting words (e.g. 'effects' versus 'fluidity');
- downplaying the many representations and images of landscapes (such as poems, novels, paintings, photographs) that are an important part of what landscapes mean.

Photograph 5.3 Stowe Landscape Gardens, Buckinghamshire, UK. An excellent example of a landscape garden surrounding a large country house, which incorporates Classical design (the bridge) and the 'Picturesque' (the rolling, 'English' hills).

Source: Peter Kraftl.

classes, ethnicities, religions – indeed any that do not conform (whether by chance or by design) to the ideals expressed in Portland's planning document. Rather, Kenny (1992: 176) argues that "[a] planning document, possibly more than any other written text, articulates the ideology of dominant groups in the production of the built environment." So landscapes are not neutral; they very often reflect the interests of a dominant group, and have done throughout history.

But Kenny's argument also leads us into a discussion about texts (also Short 1991). In fact, it makes us think about landscape as text in two ways. First, we can think of the texts that are written about landscapes. In the case of Portland, this would be the official planning document. In other cases, this would be a poem, a piece of music, or a painting, which we could read in a similar way to a **text** (for detailed examples, see Chapter 6). Some geographers have suggested that we can 'read' elements of paintings by famous artists such as Turner or Constable (Daniels 1993; Cosgrove 1997). We can look at the component parts of a landscape painting (like that in Photograph 5.4) and try to 'read' it as a **cultural product** (see Chapter 2) by asking certain questions:

- What and who are shown in the picture, and why?
- What and who are *not* shown, and why?
- Why did the artist paint *this* scene, and who commissioned them?
- Does the painting refer to a particular style, or to other paintings in that style? (This is a simple question about the idea of 'intertextuality', which we look at below.)
- And, perhaps most importantly, after answering all of the above: does the painting show certain beliefs about a landscape or a society – perhaps by depicting people, things or places in a way that we would associate with a particular religious or political group?

The image in Photograph 5.4 is one you may well have seen, even if it is in other geography textbooks about landscape! It shows what for many would be a typical English, rural scene: a picturesque cottage, a mill, and, in the background, a working agricultural landscape. Yet the emphasis on rurality and work

Photograph 5.4 John Constable's *The Haywain*, painted in 1821, and an iconic image of quintessentially 'English', rural landscape.
Source: Alamy Images/Ivy Close Images.

also holds *hidden* meanings. At the time, landowners had a particular understanding of nature – one that focused on the working land and the labour of the rural poor. With this model, landowners could stress the aesthetic qualities of a landscape such as that in Photograph 5.4 – because it was 'natural'. But they could also justify the low wages they paid, which kept farm labourers in a cycle of poverty (and under their power). As a result of this model of the 'working landscape' – reinforced in paintings such as Constable's – landowners maintained their privileged lifestyles and their grip on power. Indeed, this very organisation of society was seen as natural; Constable's famous painting is an example of how this economic and aesthetic way of thinking about landscape was reflected back to the ruling classes in the nineteenth century. With this knowledge, the different elements of the painting – and of similar paintings – can be read like symbols or words that indicate the above values. With its emphasis on class, it also points out the clear Marxist influences on new cultural readings of the landscape (see Chapter 1 for further explanation).

This brings us to the second way of thinking about landscapes. That is, that material landscapes themselves (rather than paintings or poems) can be read as texts. This requires perhaps an even greater leap of faith than thinking of paintings as texts. Generally, most of us do not see landscapes in this light. But, this concept should make us think about the processes and tensions that go into making up a landscape (Mitchell 2000).

As J.B. Jackson demonstrated early on – and Don Mitchell (2000) much later – even very ordinary landscapes can be subjected to this kind of reading. But, in this way, geographers have explored the symbolic values that are 'inscribed', as we say, into landscapes. On the one hand, they have explored spectacular kinds of symbolism: expressions of power in Renaissance Venice or Nazi Germany (see, Cosgrove and Daniels 1988; Hagen and Ostergren 2006). In the example of Nazi Germany, powerful (and very different) sentiments about nationalism, ideology and the display of political-economic strength were symbolised in knowingly extraordinary landscapes. These landscapes included enormous arenas for Nazi party rallies and imposing political buildings, which combined to position Berlin as a symbolic centre of power for the party. The landscape of the city could be read as a series of monumental reminders of the party's political power – reinforced by the fact that the architectural style of the buildings referred back to 'Classical' civilisations (such as the Ancient Greeks) whose cultural influence Adolf Hitler and his architects held in high esteem. Indeed, it could perhaps be concluded that Hitler saw the resurrection of Classical design as a key part of his dream for a new German empire that would at least live up to that of past civilisations.

On the other hand, geographers have shown how more everyday landscapes of gentrification and gated communities express particular ideals about their intended inhabitants (Mills 1993; Till 1993). For instance, the design of such communities constantly refers to middle-class ideals: the detached house for a nuclear family (two parents and two-point-four children living as a separate family unit); having a plot or garden for the children to play in; mimicking previous, 'traditional' town layouts to instil a seemingly long-lost sense of community. We can read very particular features in these landscapes, just as we might pick out 'key words' when reading a textbook. Looking around neo-traditional communities in the USA or the UK, for instance, we might pick out the white picket fence, the neat patch of lawn and flowerbeds in the front yard/garden, the recurrence of half-timbered, half-plastered walls, or a town square or 'village green' where organised community events take place (Photograph 5.5).

We see, then, that landscapes can be seen as texts in two ways. Landscapes can be represented *by* texts, and landscapes can be read *like* texts. As might be becoming obvious, these two ways of treating landscape overlap. There is a constant batting back and forth between the two. Just as landscapes are created like texts (in the examples of 'gated' communities and Disneyland, for instance), they are represented in other texts (in planning documents and adverts, for instance) and then further represented by still more texts (critical books about gentrification, or holiday photos, for instance), all of which may then affect how the material landscapes are rewritten, redeveloped and re-read. This is the essence of the constantly 'intertextual' nature of landscape (Duncan and Ley 1993).

As we mentioned in Section 5.2, landscape is also a way of seeing. Cosgrove (1985) argues that the basic way in which we see landscape is through a linear perspective. That is, we understand there to be a 'vanishing point' in every view – towards that point, things get smaller until they disappear. A road or railway that gradually narrows to a point in the distance would be a common example of a perspectival view. Although it seems natural to think about our viewing of landscapes in this way, the idea of

Photograph 5.5 A new community in Northamptonshire, UK. Finished in 2012, it uses local stone and building styles to mimic other villages in the region, in order to create a traditional 'village community' that will attract families to buy homes.

Source: Dr Sophie Hadfield-Hill.

perspective was developed in Renaissance European painting and science – it is not just 'natural'. Before the Renaissance, painters simply represented objects, people or other items in a landscape according to their importance. So, for instance, a person of greater religious or political importance would appear larger and perhaps closer to the 'front' of an image. Over time – drawing on mathematical and graphical developments from fifth-century Greece onwards – artists used perspective to mimic the effect of distance on the human eye. In other words, they mimicked the idea that objects seem smaller when they are further away, and that lines (like train tracks) seem to converge on the horizon, if we can see that far. The chief reasons for doing this in paintings were twofold, directly mimicking the search for reasoned, scientific explanations of the world during the Renaissance. First, the use of perspective provided the illusion at least that paintings were more 'realistic'. Second, the use of perspective in paintings conformed with new explanations of the world (indeed of space itself) that relied on geometry – on measuring distances, angles and volumes. Hence the use of perspective actually reinforced the development of Renaissance ideas and ideals about the world, and particularly about scientific representations of it.

So when we think about landscape, we almost always think in terms of perspective – a particular, if everyday, way of seeing. But – as Cosgrove argues – there are many problems with the perspectival way of seeing landscapes. First, they are really only an *illusion* of reality: the vanishing point of a picture must always be chosen, and, as anyone who has ever driven along a straight road knows, the 'vanishing point' of that road paradoxically never appears! That is, perspective is simply *one* way of representing how the brain understands both distance and the spaces surrounding it. There are many other ways of seeing landscapes, as we see in the final part of this chapter – where we also explore how non-visual senses also matter to how we experience landscape. Second, that this particular way of seeing is now so commonplace is a function of how a perspectival view brings with it certain political and moral values, such as authority and control. Quite intentionally, the perspectival view emphasises the power of the person at the centre (usually not *in* but looking *at* a landscape). The viewer (whether artist or art viewer) holds the power to gaze out over a landscape that has been ordered by the rational power of science, geometry and class-based privilege. For we must think about who is even *able* to do the representing in a painting (the trained artist), and who has enough cultural and economic power to view that painting. On the latter, looking again at Constable's *Haywain* (Photograph 5.4), it is quite clear that it was the landowning class who would be doing the gazing at the rural labourers in the landscape, not vice versa. So the idea of perspective is an invention – an illusion, even. It is a particular way of seeing that developed over time and, if we look a little closer, has associations that can be understood to be problematic and even oppressive.

Somewhat differently, postcolonial geographers have demonstrated how the 'gaze' of the coloniser (for instance, the English plantation owner) placed the colonised (the plantation and its workers) into a subordinate position. This postcolonial analysis has been extended to other domains, to explore how the 'West' views and exoticises landscapes that are 'Other', thus subordinating those peoples and places (Gregory 1994). These ways of seeing are of critical importance to the texts we produce about those places – whether popular myths about Eastern Asia, or photographs of Muslim women wearing the veil (Young 2003). Box 5.4 shows other ways in which we have 'learnt' to see landscapes over time, with reference to the Grand Canyon, which appeared at the very beginning of this chapter.

It is important to remember, then, that cultural geographers have thought about symbolism for some years. They have tried to get beyond thinking about the effects of a culture on the land. A key way for doing this has been to think of landscapes as texts. This opens up a number of debates, all of which mean we cannot just take landscapes for granted. In particular, we need to understand how landscapes very often reflect the interests of the dominant classes, such as wealthy landowners. These are the people who have the power to represent landscapes as texts in two overlapping ways. First, through different kinds of text *about* landscapes – paintings, novels, music, town plans. Second, through writing or 'inscribing' the landscape *itself* with particular symbols – be they the marks of consumer capitalism at Disney, or the preferred lifestyles of the middle classes in gated communities. Finally, we must remember that

Box 5.4

Seeing the Grand Canyon

Look again at the photograph of the Grand Canyon that started this chapter. It's a famous place. Most people know what it looks like. Most people who go there are awestruck by its impossible size (it's an average of a mile deep and ten across). Most people visit, stand, stare, take a photo or two; fewer stay, walk or raft. Most just look. But what are they looking at? And why are they looking? And why are they looking in the *way* they are looking? To answer this, we need to look at a timeline summarising how the Canyon has been seen.

- Pre-sixteenth century: Native Americans inhabited parts of the Grand Canyon. They used water as a resource, and – as for other landforms – attached rich spiritual meanings to the landscape.
- Sixteenth to seventeenth centuries: Spanish explorers saw the Canyon, but did not appreciate its size. They thought it was just a gash in the ground, and that the River Colorado running through it was only a few metres wide (it's actually over 100 metres wide).
- Eighteenth century: fur trappers moving through southern Utah and northern Arizona ignored the Canyon – later explorers thought it was a useless wasteland. Seems odd, looking back now?
- Nineteenth century: the first geological expeditions by John Wesley Powell and his team. This was the first time anyone had travelled the full length of the Canyon, viewing it from below.
- Later nineteenth century: Americans began to forge their sense of national identity using the country's spectacular landscapes. The Grand Canyon was a key landscape in this process. It became 'useful' and worth looking at – perhaps for the first time in non-Native history.
- Still later in the nineteenth century: the Canyon became a destination for a railroad, and began to attract tourists. Early tourists entered the Canyon, and so the early 'classic' view was from below, or halfway down. As in Europe, spectacular landscapes were becoming acceptable, even beautiful, whereas they had rarely been considered as such before.
- Beginning of the twentieth century: the Kolb brothers opened a photography studio on the Canyon's rim. Since most tourists were not able to walk into the Canyon, they produced more and more images of the Canyon from the rim. This gradually became the 'classic' view. That way, they could sell more photos, and the railroad companies could attract more visitors!
- Later in the twentieth century: the Grand Canyon became a National Park. This happened quite late compared with other US National Parks. This sealed the Canyon's fate as both a tourist destination and a landscape 'worthy' of protection. Most development occurs on the Canyon's rim – hence most of the visits, views and photographs take place here.

All of this history gets subsumed into the particular visits people now make. We have, essentially, learnt to see the Grand Canyon as spectacular and awe-inspiring. We have also learnt – partly out of practicality, partly out of commercial interest – to view the Canyon in particular ways. Without this history, we might not see the Canyon as we do today.

texts gain their significance through certain ways of seeing – through gazes that assert power, control and authority.

5.5 Landscape *as* . . . performance/feeling

In this final section, we want to emphasise the work, the practice and the sheer emotional energy that go into landscape. This idea is actually hinted at in the previous sections. For instance, Don Mitchell suggests that we need to explore how landscapes are produced. We argued earlier that his approach is in many ways material; but, because he looks at the work that landscape does, we might say that he also looks at how landscape is a performance.

W.J.T. Mitchell advocates an equally *active* way of thinking about landscapes. His edited (1994) collection on *Landscape and Power* is not a 'cultural geography' text by name. Having said that, it deals with many of the issues considered by cultural geographers. In particular, the essays in the book question whether 'landscape' should be a noun or a verb (as per Section 5.2). Mitchell (1994: 1) is clear from the very first line: "[t]he aim of this book is to change 'landscape' from a noun to a verb. It asks that we think of landscape, not as an object to be seen or a text to be read, but as a social process by which social and subjective identities are formed."

W.J.T. Mitchell and the other contributors to the volume ask how landscape works as a kind of cultural practice. That is, how landscape has a power of its own – sometimes even independent of human agency. This statement could be read conservatively – like the Berkeley School idea that landscape is an external reality waiting to be studied. But Mitchell's is a more radical understanding of landscape, which we actually prefer to align more closely with ANT (although there are *no* direct links between his work and ANT theorists), because here is a landscape theorist arguing that landscapes actually have agency – they can act.

The point is that, for Mitchell and others, landscapes *do* two very specific things. First, they hide or mask the fact that they are socially constructed. As we mentioned in our section on definitions, landscaping involves hiding things we do not want to see (such as roads), as much as drawing attention to stuff we do (see also Merriman 2006). Landscapes can even make things seem 'natural' or 'just so' when in fact they are not.

Second, landscapes make those representations *work* somehow. Mitchell (1994: 2) calls this a kind of 'greeting'. Landscapes make us feel, act, or speak in a particular way – they do particular things to us when we look at them. For instance, in the same collection, Adams (1994) describes how seventeenth-century landscape paintings were not just 'inter-textual' texts to be read alongside others (like the Constable painting in Photograph 5.4). Instead, they were a part of many contemporary economic and political practices – they were a part of the making of history, because they helped certain people to make particular identities. Actually, we can read Constable's paintings in the same way – they actively helped to *create* (or at least reinforce) the distinct class identities of landowners and working labourers, as they represented the 'natural' landscapes of eastern England.

There are, however, a number of other ways in which landscape can be considered as a performance – or in which landscapes can be performed. All of these approaches continue to chip away at the tropes (common terms) of empiricism and representation when thinking about landscape. When we talk about **performance** (see also Chapter 7), we mean it in a number of ways, but are principally thinking of what people do (or don't do) in and within landscapes.

Some cultural geographers like to think about **bodies** (Chapter 12). David Matless (1998), for instance, discusses national identity. At first glance, he does so by reflecting upon the kinds of text and symbol in the English countryside that make it 'English'. But he also looks at the way in which the English landscape mattered when it came to thinking about English *bodies*. Focusing on the period between the two World Wars, he looks at health initiatives that were supposed to improve the nation's fitness. Wherever you are reading this, you can probably think of recent examples of where a government has tried to improve a nation's fitness – particularly children's (for a US example, see Gagen 2004). But the key thing was that healthy English bodies were meant to be out in the English countryside. But not doing any old thing – because the audiences for this campaign were working class. They had to learn to behave, to act, to enjoy countryside landscapes 'properly'. This meant learning to read a map and a compass; not making too much noise; not dropping litter; and doing nice, inoffensive activities, like hiking.

Matless's is a story that might *seem* quaint and amusing. But the point is that these issues mattered, profoundly, to the ways in which the English viewed their health, their identity and their senses of citizenship. More to the point, these bodily practices were at the very heart of serious debates about the English countryside – about who and what it was for – in the context of the gradual opening up of that countryside for public access. Indeed, tensions about truly 'open' access to rural landscapes in the UK are still unresolved as ramblers' groups demand greater access beyond marked footpaths, and as farmers and other landowners try to limit the effects of open access upon their livelihoods.

Another way to think about performance and landscape is to think about *being* in a landscape. It is to ask: What is it like to live in a place? What is it like to walk through a forest? How does it feel to climb a mountain? What energies are exerted, what senses are triggered, what emotions are touched, when we walk on a windswept beach? What memories are evoked when we look at a photograph of a far-away, much-loved place (Tolia-Kelly 2004a, 2004b)?

Some humanist geographers focused on what it is that makes us 'love' a place – what makes us find some kind of attachment with it. Yi-Fu Tuan, for instance, made relatively early (1974) attempts to address this question. He questioned what it was about the relationship between material spatial structures (slopes, frames, internal and external spaces) and the human consciousness (senses and perception) that made us feel stronger bonds for some places than others. Humanists drew on poetry, painting, architectural design, memory and the physical processes involved with simply *being* in a landscape. Individual experiences of a place mattered – as did the ways in which bodies moved through landscapes in all their detail. Think for a moment about what it is that makes some places special to you, that makes you think back happily (or perhaps not) to being there . . .

There are many connections between humanist geographies of landscape and more recent, non-representational geographies of landscape. Some of these draw on ANT, as we mentioned above. Others share their philosophical roots with humanism in phenomenology. We are not so concerned with the formal definition of phenomenology here. Rather, we focus on how contemporary geographers are thinking again about what it is like to *be in* or *move through* a landscape. Once again, here, we are concerned with everyday experience, but not quite of the kinds of vernacular geography so favoured by Sauer or even J.B. Jackson. Instead, the emphasis is upon de-centring *vision* as the most common way of thinking about landscape.

It goes without saying that we experience landscapes through our eyes; and that this is the predominant sense through which most of us record the landscapes we travel through. Yet, seeing is associated with a range of problematic values – such as being rational, or objective, or having authority. Seeing is also, clearly, not the only way we 'sense' landscapes. Seeing actually distances us from the world (or at least it does if we think about perspectival and colonial kinds of gaze). But bodies are not detached from the world. The practices we do in landscapes are in effect a part of those landscapes. So to recognise the importance of other senses – arguably closer to us – is to recognise that landscapes are *closer* than we usually think. There is nothing wrong, then, with thinking about how landscapes matter in terms of touch, taste, smell and sound (Paterson 2011). This means that we pay more attention to our bodies as part of the landscape – how we *dwell* in it rather than construct images about it (see Ingold 2000 for much more on this idea).

This kind of acceptance of the body challenges the idea that landscape is 'out there', beyond the human body. This has allowed cultural geographers to consider landscape in a whole new set of ways. For instance, Jones (2003) discusses whether and how adults' memories can help us to access the landscapes and places that mattered to us as children (see also Philo 2003; Horton and Kraftl 2006b). For Jones (2003), the landscapes of childhood – self-built dens, small woodland areas, forgotten or neglected corners of cities – are accessible only through being a child, and not through the more 'rational' lens of adult memory. To children, those sites have a meaning and importance that is simply not accessible to adults. For Jones (2003), this occurs in part because adults have developed psychologically and emotionally, such that they have learnt to block out the imaginative, fantastical, playful possibilities of those places. Adults learn not to see the creative potential of such places and, rather, choose to make use of places that are more properly 'functional' (for work, learning, or adult versions of 'leisure', for instance). So childhood is an almost dark – what Jones calls 'other' – place. So the size and ability of a body (in this case a child's body) to sense and perceive the landscapes around it matter profoundly to the meaning of those landscapes – meanings that might be lost to an adult.

Elsewhere, in a set of innovative papers, Wylie (e.g. 2002) writes non-representational geographies of walking through landscapes like Glastonbury Tor in Somerset, England. His accounts remind us of the ineffability of being in landscapes – that is, how hard it is to write or speak about everything that goes on during a walk in the countryside. Indeed, the papers do not discount the possibility of writing about or seeing the world. Rather, seeing – gazing – is understood as something the body and landscape *do together*. On a walk, we *do* seeing the landscape as much as the landscape touches us, affects us. For instance, Wylie, writing about the climb up Glastonbury Tor, explains how embodied movements combine with vision to produce feelings and sensations that embed humans in the landscape as they move through it:

> What is surprising now is the height the path has already reached. Already the width and spread of the [Somerset] Levels are distinctively below, and we are above. From the shadowed glade, the view is a vivid sudden entrance into depth, into 'the blue sky, the pearly clouds and the green plain, the darker mountains in the background, like looking through a television screen. It stops the rambling of your view, gives you a focal point'. The view is a moment of focus, and comparative stillness: you compose yourself by taking in what's around you. And this *composure* is an essential aspect of the experience of ascent. . . . . With the view, you are composed.
>
> (Wylie 2002: 449, original emphasis)

The feeling we get is not just about the emotion we have for a place. Instead, it is about a kind of atmosphere, or what Crouch (2003) aptly calls 'the feeling of doing'. It is a constant mixture of body and landscape that creates these feelings: what geographers have come to term '**affect**' (Chapter 11). Our being in the landscape is so much more involved – although at times, we may just stop, and stare.

Spinney's (2006) account of cyclists racing in France is similarly involved. He argues that meanings about a place are made as we move through it. Meanings are made in the very act of engaging with a place; in this case, the act of cycling produces particular rhythms, speeds and senses (including sight) that make the surrounding landscape appear in particular ways. Perhaps blurry. Perhaps bumpy. Perhaps smooth. Perhaps exhausting, or dehydrating. It is the sensation of being on a bike, moving through a landscape, that provides a different feel from, say, walking through it.

This part of the chapter, therefore, shows more than any other part how landscape can be seen as much as a *verb* as a noun. A number of geographers have tried, in different ways, to de-centre the overwhelming focus on seeing and reading/writing landscapes. Some have suggested that we watch what landscape *does* – whether part of identity creation or in the service of a dominant class. Others have tried to link an interest in symbolism and national identity with the kinds of thing that people's bodies are *meant* to do in particular landscapes, in order for them to be 'English' bodies (or otherwise). Humanist geographers stressed the place attachments that most of us forge with particular places, trying to find the links between material places and human perception and emotion. Finally, recent geographers have tried – like ANT – to break down the boundaries between 'human being' and 'landscape'. This has been done by trying to incorporate all of the human senses. Then, we can observe how different ways of *being* in landscapes produce very different feelings about them.

Summary

- Landscape can be defined as both a noun and a verb. It 'is' particular things and it 'does' particular things.
- Landscape has many meanings. It involves material tracts of land, representations and ways of seeing that land, and ways of being or acting with/in that land.
- Many geographers argue that we need to explore 'material' landscapes. This can help us to understand the impact of a culture upon a landscape. It can also show the ways in which cultural processes interact with material things (such as houses) within the landscape.
- New cultural geographers argued that landscapes were not just 'material', or just 'there'. Landscapes are not neutral, but are created and represented to support certain interests and exclude others. Reading the 'symbolism' or 'textuality' of landscapes (whether in paintings or the landscape itself) can help us understand those interests.

- The idea of 'performance' has been a significant part of geographical research about landscapes. Increasingly, it has been shown that what we do in landscapes breaks down the boundaries between 'human' and 'non-human'. In particular, we do not just see landscapes, but we feel them – and feel strongly about them – through all of our senses.
- The differences between the various approaches here (such as the Berkeley School, new cultural geography, humanist geography and non-representational theory) are not as large as it sometimes seems. Almost all of these theories have something to say about landscapes as material, as text, and as performance. But they do say different and often conflicting things about these issues.

Some key readings

Duncan, J. and Ley, D. (1993) *Place/Culture/Representation,* Routledge, London.

An edited collection that could be placed quite easily into the new cultural geography approach to landscape. There are examples about home, gentrification, leisure and multiculturalism. This particular collection also says much about landscape as work and practice – it shows that the differences between this and later work are not as sharp as they are sometimes touted to be!

Mitchell, D. (2000) *Cultural Geography: A critical introduction*. Blackwell, Oxford.

More difficult than the other texts on this list, but provides a stinging critique of new cultural geography in Chapter 2. Chapters 6–10 are particularly good for showing how landscape is entwined with a number of 'big issues' such as sexuality, gender and ethnicity.

Short, J. (1991) *Imagined Country,* Routledge, London.

Still one of the most readable texts on landscape, which explores the social meaning of landscapes through myths, ideologies and texts. There are examples about Western films, Australian painting and the English countryside.

Wylie, J. (2007) *Landscape,* Routledge, London.

An excellent book that delves into the details of what we have only summarised here. Especially good at linking theories of landscape with examples. Try Chapters 5 and 6 if you want to read about contemporary approaches to landscape from an author who has been at the forefront of developing them. Chapter 2 is an even-handed treatment of earlier traditions in cultural geography, such as the Berkeley School.

Textual geographies

Before you read this chapter . . .

There are many kinds of 'text': poems, newspaper articles, policy documents, diaries, works of fiction . . . to name but a few. Before we discuss some of these texts, we would like you to consider the relevance of texts to your own life.

- How important do you think texts are to your everyday life?
- What kinds of text feature in your everyday life, and in what ways?

At different points in this chapter, we shall reflect upon these questions, not least because they bring to mind the ways in which texts gain meaning in different geographical contexts.

Chapter Map

- Introduction
- Spaces/texts: changing approaches to textual geographies and the poststructural challenge
- Geographies of fiction
- Policy texts and discourse analysis
- Writing worlds: maps, feminism and the stories that geographers tell
- Concluding reflections

6.1 Introduction

The *Oxford English Dictionary* indicates that the word 'geography' has its origins in the Greek *geōgraphia*: "from *gēo* 'earth' + *-graphia* 'writing'". This is often translated as 'writing the earth'. Pick up any introductory geography textbook and you will probably find a variation on this definition.

We start with this definition not for some pedantic excursion into wordplay, but to highlight something that may surprise you: the discipline of geography has an intimate, complex but enduring connection with *texts* of all kinds. The later chapters in this book look at all kinds of theoretical approach that seek in many ways to go 'beyond texts': at **bodies**, **practices**, **materials** and so on. One would therefore be forgiven for thinking that cultural geographers have become increasingly distanced from (or, at least, sceptical about) textual sources. However, this chapter paints a rather different picture. We stress the importance of texts both as objects of enquiry for cultural-geographical research, and for their part in the ways in which the discipline of 'geography' has been understood. We shall also try to provide a flavour of the ways in which cultural geographers have studied texts.

In this chapter, we focus on quite literal and – perhaps – conventional understandings of what a 'text' is. We look at texts that are dominated by *words* – usually written, sometimes spoken. In so doing, we introduce you to a necessarily partial selection of texts, including fiction, oral history, maps and public policymaking. Eagle-eyed readers will notice that this runs somewhat against our use of the term 'text' elsewhere in this book. We have used the word 'text' in several other chapters, where the implication is that a 'text' can take many forms besides the written or spoken word. For instance, we look at **landscapes** *as* texts in Chapter 5; and, in Chapter 2, we look at the **cultural production** of news media (especially in Box 2.1).

We adopt a narrower definition of texts in this chapter, for three key reasons. First, to highlight disparate but significant work by cultural geographers that focuses on written and spoken texts. Second, to foreground a series of debates about representation in cultural geography (see Sections 1.4 and 1.5) and, in particular, the important challenges made to traditional forms of academic representation by feminist and poststructuralist geographers. And third, we draw attention to the ways in which texts have been re-imagined in the light of increased interest in what we might term 'more-than-textual' or **non-representational** geographies of **performance, affect** and **embodiment** (Chapters 7, 11 and 12). We begin, though, by encouraging you to think about the status of texts in cultural geography.

6.2 Spaces/texts: changing approaches to textual geographies and the poststructural challenge

In the box that began this chapter, we asked you to think about the texts that feature in your everyday life. Look again at these questions: what sprang to mind? Perhaps your answers fell into one or more of these three ways of thinking about a text:

- a way of communicating – in emails, text messages, oral presentations or blogs;
- a source of information or entertainment – in websites, newspapers, magazines or books;
- a mode of expression – using slang or dialect to express identity, or joining a social networking group to demonstrate solidarity with a political cause.

In each case, texts do different things. Moreover, in each case, texts are surrounded by particular contexts – from the technological and social forces that have produced 'textspeak' (text message shorthand) to the geographical contexts in which regional dialects evolve. And, finally, in each case, texts have a different relationship with what we might – for the sake of simplicity – call 'reality'.

These observations encourage us to spend a little time thinking about the relationship between human geography's core ideas – space, place, scale, etc. – and texts (Hones 2008). They raise a series of questions.

1. How much can texts tell us about 'real' places and spatial processes?
2. To what extent do texts create, rather than represent (or mirror), reality?
3. What kinds of thing do texts miss out – and what else do we need to know about texts to gain a 'fuller' picture?
4. Are particular groups of people placed at an advantage or disadvantage by different textual practices?

Each of these questions has prompted an enormous amount of deliberation in cultural geography and beyond. Yet, in order to prompt critical reflection upon the selected 'textual geographies' that follow this section, it is worth looking briefly at how cultural geographers have sought to deal with these questions.

Question 1: How much can texts tell us about 'real' places and spatial processes?

The immediate answer to this question is (and has been) that texts can tell cultural geographers *something* about the kinds of spatial formation and process that we discuss elsewhere in this book. As Kitchin and Kneale (2001: 19) point out, cultural geographers have held an enduring interest in literary texts as sources of 'geographical data' and for what they can tell us about "the subjective experience of place". Most famously, humanist-inspired geographers like Tuan (1977) drew upon literary texts for their powerful, often poetic evocations of particular **landscapes** (Chapter 5; also Porteous 1985). A more recent example of this approach – set in Japan – is discussed in Section 6.3.

Yet – as Tuan and others were aware – there are limitations to what texts can tell us about 'real' places. Drawing upon ideas from literary theory and philosophy, cultural geographers now routinely accept the idea of a 'crisis of representation' (Chapter 9). If rather melodramatically put, the 'crisis' is that "a gap is opened between words and things, which will never be closed again. In the process, representation becomes . . . a questionable and active fabrication instead of a source of certainty" (Söderström 2005: 13). The implication is that it is impossible to translate the world around us into words, because words have trouble 'signifying' (conveying the exact meaning of) reality. To understand why this might be the case, we need to explore the remaining three questions posed above.

Question 2: To what extent do texts create, rather than mirror, reality?

A key reason why texts cannot simply represent 'reality' is that it is often hard to distinguish between a text and the reality that is purportedly 'outside' it. An example taken from work by the sociologist Stanley Cohen (1967) illustrates a little of what we mean. He looked at the now infamous 'battles' between two subcultural groups – the Mods and the Rockers – on the English South Coast during the 1960s (Photograph 6.1). Part of his analysis was based on local and national newspaper reporting; part was based upon in-depth research that involved talking to young people in both subcultural groups. Newspaper reporting took the form of a 'disaster sequence', predicting war-like scenarios such as the 'Battle of Brighton'. Reporters predicted that violence *would* take place, suggesting that 'reinforcements' were coming from each group, and calling upon extra policing from neighbouring areas. The majority of young people did not go looking for trouble, but the newspapers' predictions came true: fights broke out, reinforcements came, and the presence of extra policing simply served to inflame the situation. Cohen's point was that the print media were in part responsible for the opposition between the two groups, and for creating – not merely

DAILY EXPRESS

No. 19,894 MONDAY MAY 18 1964 Weather: Sunny Price 3d.

'Wild Ones' at it again

NOW MARGATE SEES A MODS v. ROCKERS BATTLE ON THE SANDS

TRUNCHEONS OUT!

EXPRESS STAFF REPORTER

POLICE were ordered to use their truncheons as Mods and Rockers rioted through Margate yesterday. Family holidays were ruined when part of the beach became the centre of a vast pitched battle.

Parents hurriedly collected their children from the sands and made for the safety of their cars. Many did not bother to stop as they arrived to find streets swarming with youths and police.

[illegible]

MISSILES

[illegible]

Mods hurling bottles and smashed deck chairs charge a Margate pavilion held by Rockers yesterday

Dingle Foot in prison row over 'secret tapes'

Brabazon dies

Pet bear turns

66

Black Whitsun's death toll so far

Express Staff Reporter

LONG queues of cars crawled back to London and the big cities last night from the coast.

[illegible]

Doubled

[illegible]

LATEST

Photograph 6.1 Newspaper front cover showing reporting of the skirmishes between Mods and Rockers in 1964 that were in part exacerbated by press reporting such as this.

Source: John Frost Historical Newspapers.

representing – the series of 'battles' that took place. He insisted that, without said reporting, the rivalry between the two groups would have been far less evident, and the ensuing violence might never have happened.

Cohen's is just one example of many approaches that *question* textual representations and their purported legitimacy to simply represent reality. In many ways, textual geographies are all about questioning who and what might (or might not) be represented. Thus, as we highlighted in the introduction to this chapter, an important source of conceptual inspiration for textual geographers has been poststructural philosophy and, in particular, the idea of 'deconstruction' that was central to the work of the French philosopher Jacques Derrida. Doel (2005) argues that geographers have used deconstruction in three ways:

- *to deconstruct*: deconstruction is about the disassembly or dismantling of terms (such as 'identity') that hold more-or-less assumed meanings;
- *as a method*: in order to dismantle a more-or-less assumed term, one must "enquire how it has *taken* position and how it has sought to *secure* its position" (Doel 2005: 247, original emphases);
- *as a recognition of instability*: deconstruction is a way of demonstrating that some terms hold particular meanings when they could have held others; it is also a way of showing that meanings are always provisional, and that those ideas that persist (like ideal gender roles) probably do so not because they are more accurate, but because a dominant force (like men) have the power to ensure that that meaning is perpetuated.

Deconstruction is, if you like, an invitation to question some of the most taken-for-granted terms that make a very particular kind of order from the chaos of reality. Drawing on these insights, we explore some specific examples of how texts create reality in Sections 6.4 and 6.5.

Question 3: What kinds of thing do texts miss out – and what else do we need to know about texts to gain a 'fuller' picture?

As we discuss elsewhere in this book, poststructural theories – such as deconstruction – have been part of a broader groundswell of work in cultural geography that has questioned the legitimacy of textual representations. Since 2000, much of this work has been grouped under the term **non-representational geographies** (Chapter 11). The term '*non*-representational' might seem to imply a move beyond using textual sources in cultural geography. Indeed, textual sources have been downplayed in some work on **embodiment**, **performance** and **affect**. However, it is far too simplistic to assert that non-representational geographies draw a simple line between 'textual' and 'non-textual' sources. Non-representational approaches *challenge* the ways in which texts can be understood, to be sure; but cultural geographers have shown that there are several reasons why such approaches can extend our ways of doing textual geographies.

- Textual practices – like written or spoken words – can form part of the symbolic meaning that is central to some embodied performances. For instance, Nash (2000) shows how it is impossible to appreciate the full richness of different styles of dance without understanding the historical narratives and collective meanings that surround them.
- As Saunders (2010) argues, a balanced account of a piece of literature needs to be aware of its context. This might mean accounting for the social or economic conditions that prevailed when a text was written. However, it might also mean accounting for the kinds of bodily practice – the very stuff of non-representational geographies – that give texts their meaning in any given place. How, for instance, are you reading this text? Perhaps in silence, in a library; perhaps in a cafe, with friends, underlining bits you think are important; perhaps in the park, lying on the grass, lolling in the sunshine in a semi-awake state of dreamlike contentment . . .
- It is an inescapable fact that the vast majority of non-representational geographies rely on texts at some point: they are published in academic journals or spoken at conferences; they draw upon the weighty words of poststructuralist theories; they sometimes draw upon archival sources or policy documents for their 'data'. If we accept that words are *part of* the flow of everyday life, we realise that even seemingly stuffy old archival documents are subject to particular kinds of practice – ordering, reading, storing, preserving, etc. (DeSilvey 2007).

Question 4: Are particular groups of people placed at an advantage or disadvantage by different textual practices?

Doel (2005: 248) argues that an "affirmation of difference, otherness and alterity . . . has become a mainstay of Human Geography" (Chapter 8 details how this has been the case). While Doel is troubled by the ways in which deconstruction has been assumed in this work, the broader point here is that an acknowledgement of difference has been a further reason for *questioning* the authority and accuracy of textual sources. For this reason, geographers inspired by postcolonial and feminist approaches have demonstrated how texts can support the interests of dominant social groups, and/or how they may exclude, patronise, disempower or offend other, less powerful groups. For instance, feminist geographers (like Rose 1993) observe that academic geography itself is a male-dominated discipline, that women's voices have been under-represented in academic research, and that the discipline has excluded ways of writing about the world that are assumed to be female (like the inclusion of **emotions** in research). We look at these arguments in more detail later in the chapter.

We began this section by asking you to reflect upon what texts 'do' in your own everyday life. This led us to think about the relationship between texts and the world: textual geographies are as much about *questioning* the legitimacy of any text as they are about simply excavating its 'geographical' content. Therefore textual geographers are also interested in the contexts – historical, economic, performed, material – that make and are made by a text. We want to bear these questions in mind as we turn to a series of examples of work by cultural geographers.

6.3 Geographies of fiction

In this section, we look at works of *fiction*, which, as Saunders (2010) argues, have predominated as sources for work on textual geographies (also Sharp 2000; Ogborn 2005). As noted above, many cultural geographers have turned to fictional texts as sources that can tell us about subjective experiences of particular places and landscapes, as well as a series of broader political, social and cultural processes. We examine two themes: first, focusing on how fictional texts can bolster our understandings of landscape; and second, analysing how fictional texts help us understand spatial processes such as urban change.

Fictional geographies of landscape

While we deal with **landscape** in another chapter of this book (Chapter 5), we want to explore here what kinds of insight works of fiction can provide about landscapes. We do this through a novel that, although perhaps not well known among Western audiences, is famous in Japan (where it is set, and where its author originated). The novel is Yasunari Kawabata's (2011 [1956]) *Yukiguni (Snow Country)*, which was analysed by the cultural geographer Iraphne Childs (1991). Our focus is on the relationship between the characters in the novel and its landscape, and between the treatment of landscape and particular (enduring) ideals in Japanese history and written style.

Snow Country is about the relationship between a wealthy businessman from Tokyo and a young *geisha* who lives in a small mountain town in the western region of the Japanese Alps. The region – the Yukiguni – contains steep, forested mountains and picturesque towns, and is often used by Japanese visitors for rest, retreat and honeymooning. According to Childs (1991), Kawabata affords the natural landscape a far more prominent role in his novel than do many Western novelists. For instance, the changing seasons allude to changes in the emotional relationship between the businessman and the *geisha*. Kawabata also uses the landscape to define the characters in the novel. The *geisha* is often portrayed through "the adjective *seiketsu* (clean and pure) many times . . . as if she embodied the mountain snow" (Childs 1991: 11). A longer example of this use of landscape can be found in Box 6.1. Childs (1991: 3) argues that "[i]n this respect, the novel exemplifies the view of traditional Eastern landscape art which invariably depicts human beings blending in as part of the scene, as opposed to Western landscape art which usually takes the perspective of the human being looking on from the outside." Childs (1991) argues that this is a tradition of viewing landscape that is more than 1,000 years

Box 6.1

Landscape and poesis in *Snow Country*, by Yasunari Kawabata

Kawabata's novel contains several stylistic elements that differ quite dramatically from many Western novels. Two central components that have made his work so famous are his treatment of landscape, and the poetic form of his novels. We have included two slightly extended extracts from *Snow Country* to provide a sense of these two components (see Childs 1991).

As we indicate in our discussion, Kawabata uses landscape to provide an evocative sense of the personality of the key characters in the novel. Here, he describes Yoko, another girl from the Yukinugi, who "embodies the traditional aesthetic principle of *yugen* (subtle mystery and depth)" (Childs 1991: 12):

> The girl's face seemed to be out of the flow of the evening mountains. It was then that a light shone in the face . . . The light moved across the face, though not to light it up. It was a distant, cold light. As it sent its small ray through the pupil of the girl's eye, the eye became a weirdly beautiful bit of phosphorescence on the sea of evening mountains.
>
> (Kawabata 1956: 16)

The second component we noted was Kawabata's use of haiku to convey motion and sound. The following extract exemplifies this combination:

> It was a stern night landscape. The sound of the freezing snow over the land seemed to roar deep into the earth . . . the high mountains near and far become white what the people of the country call 'the round of the peaks'. Along the coast the sea roars, and inland the mountains roar – 'roaring at the center', like a distant clap of thunder. The round of the peaks and the roaring at the center announce that the snows are not far away.
>
> (Kawabata 1956: 129)

old in Japan. The novel therefore uses symbolic images rather than a simple account of the action: it tells its story *through* the natural landscapes.

Snow Country is also exemplary for its attention to poetic form: to the ways in which written "form speaks and creates; for listening not only to what this voice is saying but to how it is saying it opens up the place" (Saunders 2010: 441). Thus it can be argued that a text's meaning becomes more intriguing when attention is paid to the style of its delivery. In a simplistic sense, most of us know that the same words delivered in a different tone of voice can have a different impact on the listener. The same is true of *Snow Country*, which uses traditional Japanese *haiku* forms to convey the relationship between its characters and landscape. In particular, Kawabata uses what Childs (1991: 12) terms "a peculiar turn of phrase . . . to convey the fusion of motion and sound" – two central elements of *haiku*. Box 6.1 includes an extended example of this turn of phrase.

Kawabata's novel (and Childs' analysis of it) is significant, because it affords a glimpse into understandings of landscape that have tended to be marginalised in many European and North American cultural geographies of landscapes. Works of fiction are clearly not the only ways to access these understandings; however, texts work through their very textural qualities to connect landscapes to broader, cultural principles (such as that of *yugen,* mentioned in Box 6.1). As Saunders (2010) argues, this may mean that in focusing on the very details of a text – single words, sentences or paragraphs (as in Box 6.1) – textual geographies appear a little self-indulgent. But, even if we accept these criticisms, the point is that, just like music and dance, works of fiction may work to create imaginative spaces that can be profoundly evocative. (We suggest you compare this point with the example of French regional folk music in Chapter 7 on **performed geographies**.)

Fiction as a 'critical mirror' to spatial processes

While fictional texts can tell cultural geographers a significant amount about particular landscapes and how humans interact with them, they can also help us to critically analyse spatial processes. In other words, works of fiction do not just tell us about landscapes. Rather, they can offer up what we are calling a 'critical mirror' to some of the broader issues that interest many cultural geographers (indeed, human geographers in general) – from the dynamism of urban forms to the functioning of political-economic systems. By this, we are returning to the idea introduced in Section 6.2 that textual geographies are all about *questioning*. We discuss two ideas raised there. First, the relationship between a 'text' and 'reality' and, in particular, the idea that a text can have an effect upon a reader's perception and understanding of what Kitchin and Kneale (2001: 21) term "the present condition" of the world. Second, we explore what works of fiction can tell us about the ways in which spaces – especially urban spaces – reflect social difference.

In order to discuss these two ideas, we focus upon what Kitchin and Kneale (2001: 19) call the "imaginative geographies" of two overlapping genres of fiction: science fiction novels and utopian novels. Science fiction (SF) novels often take existing realities (especially in terms of science and technology) and extrapolate them onto the past or future of a society. They offer a detailed picture of another world – or, more accurately, they often present an image of *our* world as it could have been or could be otherwise. A key feature of SF novels is their plausibility: SF authors draw upon more-or-less accurate, existing scientific knowledge – about genetic cloning, communicative technologies, or energy production, for instance – and imagine worlds where those plausible forms of knowledge have given rise to alternative versions of our world (Kitchin and Kneale 2001).

Kitchin and Kneale's (2001) analysis of 38 SF texts is particularly helpful, because it highlights both of the ideas we want to discuss here. First, Kitchin and Kneale discuss the *effect* on the audience of an SF novel. For them, SF is inherently spatial – and it is its treatment of spatial processes that is key to the effect upon its readers. SF causes what they call 'estrangement', because it combines the plausible with the implausible. It makes readers stop and think that the world could have been or could be otherwise. It asks them to consider what might happen if this or that piece of technology was adopted in a particular (perhaps extreme) way. It encourages readers to take a critical stance on their world – thereby holding a 'critical mirror' to existing spatial processes. As Kitchin and Kneale (2001: 21) put it:

> Here it is recognized that estrangement is bound within spatial metaphors such as being 'out-of-place'; the invasion of the bounded space of the self by strangers; the construction of new, unfamiliar spaces; and the disruption of territorial identities.

As we highlighted above, Kitchin and Kneale are talking about the very stuff of cultural geography (indeed, of many strands of human geography). In particular, they suggest that so-called 'cyberpunk' SF – like William Gibson's *Neuromancer* and Bruce Sterling's *Islands in the Net* – construct prescient visions of future urban places. In *Neuromancer*, Gibson talks of a future 'Boston–Atlanta Metropolitan Axis', the result of decentralised urban sprawl that has swallowed up the large area of land between those two cities. In *Islands in the Net*, Sterling "writes of a futuristic Singapore where [buildings] are virtualized through the incorporation of computer networks which render them 'smart'." And, in many novels, SF writers chart what geographers like Michael Dear (2000) term the 'postmodern urban condition':

> a world reordered by libertarian capitalism . . . reshaped at all spatial scales through the sociospatial processes of globalization . . .; a world dominated by a few large multinationals . . .; a world where the middle-class has been eliminated with the population neatly tied into have and have-nots; a world of fractured and fragmented cities . . . where the wealthy live in private and defensible spaces (where public space is eliminated) and the poor are left in ungoverned, anarchic, lawless spaces.
>
> (Kitchin and Kneale 2001: 25)

SF texts, then, expose and extrapolate some of the potential fault-lines of late-capitalist societies, drawing attention to stark dividing lines in society that for many commentators are becoming ever-more real (for instance, Davis 1990). Thus these texts not only draw attention to the potential future of contemporary urban forms, but also ask their readers to question the effects of these forms upon different social groups (and, especially in Davis' case, deprived immigrant communities living in cities).

We turn next to utopian fiction – which we include here because of increasing interest in utopias among cultural geographers (Harvey 2000; Anderson 2006; Kraftl 2010). In many ways, SF is a kind of utopian writing (Moylan 1986). Yet utopias take a more diverse form than SF, concerned as they are with writing about the 'good place' that is 'no place' (Kumar 1991; Pinder

2005b). In this way, Claeys and Sargent (1999: 1) define a utopia as:

> a nonexistent society described in detail and normally located in time and space [which] the author intended a contemporaneous reader to view as considerably better than the society in which the reader lived.

Many of us – if not all of us – take part in forms of dreaming that transport us to a desert island, an ideal home, or a perfect garden of plenty. But utopian writing has at its heart the idea that virtually every aspect of human life can be brought under human control and rendered as good (even perfect) as possible. Strikingly – like the SF novel – utopias take the form of an imaginary or ideal city, of which Thomas More's *Utopia*, published in 1516, is the most famous early example.

Withers (2006) argues that a key way to understand literary geographies is to understand the function of a text – how it is to be received (indeed, how it is received) by its audience. Thus Kraftl (2007) demonstrates that we need to more fully understand the emotional impact that a utopia sets out to have on its reader. Many utopias – like Aldous Huxley's ideal *Island* – are what we call 'compensatory': they offer escapist fantasies that transport their readers to another time and place. They are, as Kneale and Kitchin put it above, 'imaginative geographies' that offer, if you like, a conceptual space for readers to momentarily get away from the troubles of everyday life. However, Kraftl (2007) also argues that many utopias are 'unsettling'. For instance, while William Morris's *News From Nowhere* was an image of a familiar city (London), it was presented in a new light – taken over by unfamiliar forces (wild nature and farming) that were at odds with how most of us think of London. Just like SF, utopian novels like *News From Nowhere* combine the real and the imagined to provoke a reaction that is disquieting for readers, and provokes them to consider whether contemporary London could (and indeed should) be organised otherwise.

We argued above that some texts have the power to create or transform reality in more-or-less powerful ways. A final function of some utopian fictions has been just that: some novels have acted as direct inspirations for real-world experiments in alternative forms of living. They have spawned all kinds of real-world spaces: from the architectural innovations of modernism and postmodernism (Pinder 2005b) to alternative or 'autonomous' communities created in pockets of land all over the world (Pickerill and Chatterton 2006). To read more about how one utopian text (which also happens to be classed as SF) inspired real-world spatial experiments, turn to Box 6.2.

In this section, we have highlighted that, despite weighty critiques from non-representational geographers, an interest in fictional texts has been and remains an important part of the work that cultural

Box 6.2
B.F. Skinner's (1948) *Walden Two*: a science fiction utopia made real?

B.F. Skinner's utopian fiction *Walden Two* typifies the science fiction utopia. Skinner was a psychologist, working in Harvard during the middle of the twentieth century, who was interested in discovering the determinants of human behaviour (including environmental determinants). He was a pioneer of so-called 'behaviourist psychology'. But Skinner also wrote a utopian novel – *Walden Two* – where, in the SF tradition, he took elements of his research and extrapolated them to describe an ideal, intentional community. Skinner's work is so interesting because his novel actually inspired some 'real' experiments in alternative living – including Twin Oaks in Virginia (USA) and Los Horcones (Mexico).

A key component of *Walden Two* was that its design – the 'Plan' as it is termed in the novel – helped to predetermine the behaviour of the members of the community (thus inspired by Skinner's own research). This Plan controlled both the spatial and social organisation of the community, evident here as Frazier, the orchestrator of *Walden Two*, describes his work:

> I did plan *Walden Two* – not as an architect plans a building, but as a scientist plans a long-term experiment . . . In a sense, *Walden Two* is predetermined, but not as the behaviour of a beehive is determined. Intelligence, no matter how much it may be shaped and extended by our educa-

Box 6.2 continued

tional system, will still function as intelligence. . . . What the plan does is to keep intelligence on the right track, for the good of society rather than of the intelligent individual. (Skinner 1948; reprinted in Claeys and Sargent 1999: 375)

For many modern readers, this extract may be reminiscent of another famous utopian (or, rather, dystopian) novel that described some of the dangers of essentially policing human intelligence: George Orwell's *Nineteen Eighty-Four*. The relative desirability of Skinner's community is not the matter for discussion here. Instead, Skinner's work highlights the interplay between fictional texts, existing social realities and real-world attempts to overcome or escape those realities. The utopian novel – just like any novel – is contextualised by a complex and ongoing interplay with its readership.

geographers do. We have suggested that fictional geographies can add to our understandings of landscapes, and can offer cultural geographers a 'critical mirror' – an additional tool with which to understand complex spatial processes such as postmodern urbanism. In addition, we have tried to convey something of how the form and poetic content of a text can create effects upon a reader, so that what Withers (2006) terms the 'geographies of reception' matter as much as formal, detailed analysis of texts. To this extent, we have highlighted how novels such as SF and utopian fiction may blur the boundaries between 'reality' and textual 'fiction'. For a further resource on literary geographies, reading lists and discussion points, readers may like to consult the excellent 'literary geographies' blog at http://literarygeographies.wordpress.com/, which was started by Sheila Hones and James Kneale.

6.4 Policy texts and discourse analysis

In contrast to the previous section, we look next at 'policy' texts. By this we mean the (usually) official accounts of governments, charities, religious groups and other organisations that may document their work in guides, manifestos or legal papers. Cultural geographers have examined policy texts because they can tell us so much about the priorities that underpin where and who policies are aimed at (or not). Policy texts can also tell us much about how spatial metaphors – like scale and boundary – are used to justify a government's approach to, say, school-building or obesity.

Cultural geographers have used a variety of methods to analyse policy texts. One key method that has steadily gained currency is called *discourse analysis*. As outlined in Box 2.2, **discourses** can be understood as:

1. the particular, interconnected groups of texts, statements and representations which emerge from geographically and historically specific policy contexts or agendas;
2. the entire array of practices through which such contexts and agendas are constituted, sustained and experienced, [including] documents, agendas, statistics, press releases; via meetings, presentations, briefings, media reportage. (Kraftl *et al.* 2012a: 15)

Because of this broad definition of policy discourses, discourse analysis itself can be approached in a huge variety of ways. However, it is possible to distil four central principles of discourse analysis (Wodak 1996; Titscher *et al.* 2000).

- Like deconstruction, discourse analysis looks critically at the taken-for-granted 'truths' or assumptions that underpin any policy text.
- Discourse analysis seeks – following the work of French theorist Michel Foucault – to uncover the relationship between a policy text and power relations that cause exclusion, oppression or domination of particular social groups.
- Discourse analysis focuses on one or a few carefully chosen, 'key' texts that are thought to be particularly illuminating for understanding a particular political issue or social-spatial process.
- Discourse analysis involves close, careful reading of texts; like some approaches to works of fiction, this can mean looking in detail at the syntax of sentences or the choice of particular words over others (compare Saunders 2010).

Before we move on, compare the four principles of discourse analysis listed above with the points raised in Section 6.2 of this chapter. To what extent do they reflect the broader questions we asked about texts? What kinds of policy text do you think would be of interest to *cultural* geographers?

The rest of this section looks at analysis of policies aimed at children and young people, before drawing out some more general points for consideration about policy analysis and cultural geography. The examples illustrate several themes that have been of increasing interest to cultural geographers who undertake policy analysis. Note that not all of the authors cited below will say they undertook 'discourse analysis' (although many do); rather, as we shall see, they display one or more of the four principles listed above in their work.

Critical geographies of childhood and youth

Policy texts about young people are often fascinating sources, because they tend to set out (quite explicitly) the key political and economic priorities of a country or region. They can also tell us much about how children and young people are viewed in different historical or geographical contexts – what some people call the 'social construction of childhood' (James and James 2004). Crucially, these views often involve some assumptions about *scale,* which cultural geographers have sought to unpick.

Many policy texts discuss children at a *national* scale. In fact, they explicitly equate the future of a nation with the contemporary and future prospects for that nation's children. In doing critical discourse analysis of policy texts, cultural geographers have focused on two interrelated themes in respect of national policymaking for children. The first is that – implicitly or explicitly – such policies seek to intervene in particular places and among particular groups of somehow *disadvantaged* young people. Thus national-scale policymaking also represents certain local spaces as 'in need' of particular intervention. The second is that entire generations of young people – not just the disadvantaged – are targeted and invested with *responsibility* for a nation's future.

We can see these two trends in research on children past and present. For instance, Gagen (2004) argues that city planners in 1900s New York City built playgrounds in deprived immigrant communities in order to foster senses of American national identity. She argues that the playground movement focused on how children would glean gendered American identity traits through play, sport and participation in ritualistic events. Essentially, children had to unlearn 'troublesome' habits that planners thought children had acquired from their parents (and which threatened the future of the nation as understood at the time), and to learn habits 'proper' to the future prosperity of the American nation.

Bethan Evans' (2006, 2010) work on anti-obesity measures in Britain in the 2000s shows striking parallels. She notes that, if one looks at national guidance, it appears that the UK, "like other Western countries, is at war, fighting to defuse a 'time bomb' . . . that threatens the future of the nation in myriad ways" (Evans 2010: 21). In her analysis of the 2004 *House of Commons Select Committee Report on Obesity*, Evans shows how 'fault' for obesity is seen to lie in part with individuals, whose "responsibility for their bodies has increased in salience – health risks no longer located outside the body (contamination by germs) but located within the self (lack of self control)" (Evans, 2006: 262). Evans' work is striking not only because it shows how responsibility is increasingly being located at the scale of individual children and their families. It is also important because – as you may have noticed from the quotation in the previous sentence – it shows how policy texts can also illuminate something of the interests of non-representational geographies in the scale of the **body** – a connection between textual geographies and non-representational geographies that we highlighted in Section 6.2.

Research on youth policymaking has also shown that some policies operate at a scale *beyond* the nation. Ansell *et al.*'s (2012) analysis of the youth policies of Malawi and Lesotho shows that such policies do not arise independently in different countries. Instead, they are the complex result of inter-governmental sharing and international guidance (such as the *United Nations Convention on the Rights of the Child*, published in 1989). They analyse two documents: Malawi's (2011) *National Youth Policy* and Lesotho's (2002) *National Youth Policy for Lesotho*. Ansell *et al.* (2012) point to four features of youth policymaking in the two countries.

- Youth policymaking reflects the interests of international agencies (like the United Nations) and development partners (like NGOs and charities) in Majority World countries having an explicit youth policy to deal with issues as diverse as AIDS/HIV, education and economic development.
- Just as in the examples from other contexts listed above, youth policymaking has been used to produce neoliberal future subjects who will serve and take the responsibility for a competitive national economy. Children are framed as malleable adults-in-the-making: if they can be skilled properly, then the entire nation will be benefit at some (usually undetermined) point in the future.
- At the same time, international principles have been applied and interpreted unevenly in Lesotho and Malawi. Malawi's government is more strongly tied into transnational networks and influence, and reflects 'purer' neoliberal approaches to the education of young people for the future. Lesotho, meanwhile, is less tied to these networks, and the language of its youth policy emphasises the continued role of the national government in taking responsibility for the country's future.
- The differences in international influence can also be seen in the content of each policy. Malawi's youth policy stresses economic development as the driving force of its approach to young people, whereas Lesotho's youth policy emphasises that "aspirations for youth are not confined to productivity, economic growth and individualisation. Human development is emphasised, which includes moral, spiritual, intellectual and social aspects. Rather than individualisation, the policy aims to develop a sense of belonging" (Ansell *et al.* 2012: 53).

So far, we have stressed that youth policies operate at a variety of spatial scales in order to justify particular kinds of intervention, sometimes with particular groups of young people. We have also noted that almost all youth policies, to some extent, display a social construction of young people as 'the future'. We want to pick up on this point about the future because, looking more broadly, cultural geographers are increasingly interested in how policy discourses attempt to prepare for, pre-empt and offer precautions for uncertain futures (Anderson 2010).

One significant component of this work is that it highlights how discourses – like texts – are set into and help to create what appear to be 'non-textual' practices. Away from work on young people, for instance, Amoore and de Goede (2008: 173) look at what they call the "banal face of the war on terror" in North American and European countries (see also Box 11.9). Amidst air strikes, counter-insurgency doctrine, mass media reporting and propaganda, they suggest that security policy also involves scrutiny and data-mining about individuals' everyday practices. They show, for instance, that information gained about credit card transactions and travel habits is used to classify "mobile people and objects into degrees of risk" (Amoore and de Goede 2008: 176). Amoore and de Goede (2008: 174) argue that this makes bodies 'legible', by reducing them to text-like data: they become more readable and, it is hoped, more predictable, so that, ultimately, terrorist attacks can be pre-empted through the arrest and detention of 'suspicious' bodies *before* 'they' can act.

As Anderson (2010) argues, such 'future geographies' are becoming an increasingly common aspect of policymaking on issues as diverse as climate change, obesity and security. These future geographies are also embroiled in a series of geopolitical questions about international relations and the power of particular nation-states (like the USA and the UK) to intervene in places such as Iraq, Afghanistan, Libya and Sudan. There are, in fact, striking parallels in the ways in which policies targeted at completely different goals *verbalise* interventions into the flow of people's lives – to govern life itself. For instance, compare how the scrutiny of people's credit card transactions in the '*war* on terror' resonates with the very different (and equally controversial) context of the monitoring of children's body mass index as a 'pre-emptive strike' in the '*war* on obesity' (Evans 2010; Evans and Colls 2009; see Figure 12.1).

At the same time, the governance of the future may be more altogether more spectacular. For instance, since 2000, several countries (including Britain, Australia and Portugal) have embarked upon ambitious school-building programmes designed to transform and revolutionise the environments in which the next generation of children will learn. In each context, the promise of a revitalised national landscape of brand new, shiny buildings anticipates a future of economic promise and greater social inclusion. The Australian programme *Building the Education Revolution* (DEEWR 2010) was symptomatic of this approach, claiming in its launch document that it would

> boost jobs and invest in Australia's long-term future; the Rudd Government will build or upgrade buildings in every one of Australia's 9,540 schools. *Building the Education Revolution* is a $14.7 billion investment to improve the quality of facilities . . . in Australian schools . . . This historic nation-building investment . . . will not only support jobs – it is also a down payment on the long-term strength of the Australian economy . . . [The program] will also help to support local communities.
>
> (DEEWR 2010: n.p.).

It is noteworthy that these schemes connect ambitious, discursive claims about the future of schooling with the symbolic promise of **architecture** (Chapter 4; also Kraftl 2012; Horton and Kraftl 2012b). Once again, we can see the intended power that a text like this launch document is supposed to have in creating and transforming the real world. In this instance, these new schools will 'boost jobs', safeguard Australia's 'long-term future', be an occasion for 'nation-building' and 'support local communities'! In Photograph 6.2 we have included two images of a building from another similar programme – the English *Building Schools for the Future* programme (DfES 2003), notable in part for its name. You may like to judge for yourself whether you think a building like this can live up to the expectations set out in texts like the one quoted above.

In this section we have provided a flavour of the kinds of issues that can be analysed by looking in detail at policy texts. Many cultural geographers use a kind of discourse analysis to examine the underlying assumptions that drive policymaking. By looking at policies targeted at children and young people, we can see how policy discourses often emphasise particular spatial scales, and especially that of the nation, as well as particular timescales – especially the (near) future. This latter point about timescales also shows how analysis of policy texts overlaps with the themes that cultural geographers have explored in relation to fictional texts (Section 6.3).

6.5 Writing worlds: maps, feminism and the stories that geographers tell

In this section, we turn the lens 'back' onto geography as an academic discipline. As we suggested in Section 6.2, geography (including cultural geography) has been dominated by particular ways of knowing

Photograph 6.2 Two images of Suffolk One, a *Building Schools for the Future* school in Ipswich, England. The angular left-hand edge of the building is intended to symbolise 'direction'; the interior atrium is a 'flagship' space that contains a café, seating areas, a space for musical performances and exhibitions of student work. Such flagship spaces were a feature of many *Building Schools for the Future* schools, and were intended to be an obvious, visual contrast with the dark, enclosed, cramped and 'unfit for purpose' corridors found in the previous generations of England's schools.

Source: Peter Kraftl.

the world, which have in turn been characterised by particular kinds of writing about the world. It is not within the scope of this chapter to provide a detailed history of geography as a discipline (for an example, see Livingstone 1999). However, it is useful, as a starting point, to consider a few features of geographical endeavour since the 1800s (after Livingstone 1999; Withers 2006).

- Modern geography originated, at least in part, in eighteenth- and nineteenth-century efforts to *explore* and map unknown lands outside Europe. These efforts saw writing form part of an impulse to *collect*: to garner knowledge about new worlds, especially for trade; and to collect scientific specimens (animals, plants, rocks, tools) that were returned to museums and archives. Thus early geographers were very much part of the development of 'scientific' approaches to understanding and writing about the world.
- Geographers were also involved in efforts by European countries to colonise new lands. Their newly gained scientific knowledge was put to use as countries such as Britain, Portugal and Spain raced to claim territorial control over previously 'unmapped peoples', places and resources.
- As the discipline developed, geography required certain kinds of **embodied** technique (also Powell 2002; see Chapter 12). Geographers were required to take on certain traits and habits that would aid the production of scientific, geographical knowledge. Lorimer (2003), for instance, tells some of the 'small stories' of events, encounters and experiences that occurred during a visit by 40 Glaswegian schoolgirls to the Scottish Cairngorm mountains in 1951, where they undertook a course of outdoor learning (surveying, sketching and *in situ* observation). He argues that the "situated and personal aspects of educational experience" such as this tell us how some of the imperatives of disciplinary geography were translated into the lives of 'non-specialists' – the girls (Lorimer 2003: 200).
- Geographical endeavour has been dominated by a series of sites and spaces, the traces of which remain both within and without university campuses today. As Livingstone (2003) shows, the sites of scientific geographical knowledge production have included Royal courts, botanical gardens, ships, archives, map libraries, museums and laboratories. As dedicated geography departments emerged in universities during the twentieth century, the buildings in which those departments were located reflected the drive for geography to become a spatial science. Some geography departments bear remnants of this past – for instance, it was only in 2010 that a sink, natural gas outlet and illuminated map table were removed from the office of one of the authors of this book – facilities that had seemingly not been used since the 1990s!

This brief overview should give you a sense that geography as an academic discipline has often been exactly that – an ongoing series of attempts to *discipline* the world through ways of seeing, ways of writing, ways of acting, and ways of designating particular spaces as 'proper' to the production of geographical knowledge. For many academics – especially geographers adopting feminist, queer, poststructural and/or postcolonial theoretical perspectives – those traditional ways of writing geography are inherently exclusionary and oppressive. Importantly, traditional ways of writing (about) geography marginalise many kinds of text produced by *both* academics *and* the participants in research – whether interview material gained with disabled children or works of fiction written by novelists writing from non-Western contexts. This section explores just two groups of criticisms of the ways in which geography has been traditionally written and represented, looking first at maps, and then at feminist geographies. It also provides examples of some alternative ways of writing (and speaking) – including auto-ethnography, poetry and oral history.

What's in a map?

Most people would accept that maps are a distortion of reality. Indeed, maps would not work if they did *not* distort reality. Just like other forms of text, maps must be selective, in order to avoid over-complication and confusion. Maps must be schematic, highlighting particular features (say, roads) to make them look larger than life, so that we can use them to navigate. Maps flatten the world: usually, they convert the spherical shape of the earth – with all its lumps and bumps – into a plane. Maps stop the world: at least in their traditional form, they rarely provide a sense of the dynamism and non-representational liveliness of **space** and place (Section 6.2; see also Chapter 13).

Yet, notwithstanding our common knowledge of these distortions, many of us see maps as broadly accurate, apolitical, non-judgemental representations of the world. Just like statistics and scientific experiments, they evoke a rational, objective view of the world that was particularly prevalent in geography in the 1950s and 1960s, as part of the so-called 'quantitative revolution' (Fotheringham *et al.* 2000). While approaches to space and place have changed somewhat in cultural geography since then, a belief (albeit not unquestioning) in the objectivity of scientific method still pervades some disciplines, and in many mainstream societies and governments. Since most maps are based upon some kind of 'objective' methodology – surveying, schematisation, projection, computer representation – and since many of us rely on paper or online maps, or digital satellite navigation systems, to some extent in our everyday lives, most of us tend to invest in the idea that maps are pretty accurate reflections of the real world.

Herein lies a fascinating disjuncture – between the idea that maps are *distortions* of reality (like any other text) and that they are *reflections* of reality. In fact, this is a disjuncture that we explored in general terms during our discussion of geographical approaches to texts, in Section 6.2, where we also argued that texts can *create* reality. Maps are a great example of this argument. Think, for a moment, about how maps can influence how we think or feel about a place, what we think it might look like before we visit, or how we approach it, literally.

- Since common projections of the globe place the southern hemisphere at the 'bottom', the British frequently refer to Australia as being 'down under'. What might Australians think and feel about this term?
- Throughout history, different places have been located at the 'centre' of a map – from the Holy Land to Nazi Germany. The effect has been to make those places seem more important, or powerful, and to deliberately make those places at the margins seem 'peripheral'. Perhaps you can think of examples of how contemporary maps do this.
- Maybe it is just us, but have you ever looked at a map of a place you want to visit and tried to work out what it might look like in reality – from the topographic contours, the road layout, or the vegetation depicted on the map?
- Since maps tend to be oriented in a particular way (usually with north at the top), and tend to show principal routes to avoid confusion, the way we physically arrive at or leave a place may be structured by that map. More recently, some people have argued that satellite navigation systems encourage people to take a very 'blinkered' view of their surroundings – caught up with following a route rather than understanding the wider context of the place they are in, such that they ignore 'commonsense' understandings of that place, and sometimes end up blindly following their satnav the wrong way down a one-way street, or into a river! So we could say that maps and satnavs can sometimes encourage a particularly narrow, route-focused way of seeing and experiencing 'reality'.

These kinds of observations led Monmonier (1996) to question in a more formal way the many ways in which people 'lie with maps'. Monmonier's argument is actually a little more complex than simply being about 'lying' – a kind of deliberate deception – but, rather, signals a range of ways in which reality can be created or misconstrued through maps. Just like statistics, Monmonier (2005) argues, maps are such powerful deceptive tools *because* many people are invested in the belief that they are 'accurate'. His argument is that academics – such as geographers – can manipulate maps to tell particular stories. For instance, he argues that so-called 'choropleth maps' (like Figure 6.1) work by dividing places (in this instance, US states) into categories. But the selection of these categories can be manipulated so that certain areas appear in particular categories – in other words, the cut-off point between categories can be *chosen* to make values (e.g. birth rates in Figure 6.1) appear relatively higher or lower in certain places. He notes that a map like this provides no sense of which places are on the borderline between categories, or of the differences within categories (for instance, in Figure 6.1, Utah and Texas actually have quite different birth rates, but end up in the same category). The effects of a map like this could be wide-ranging: from

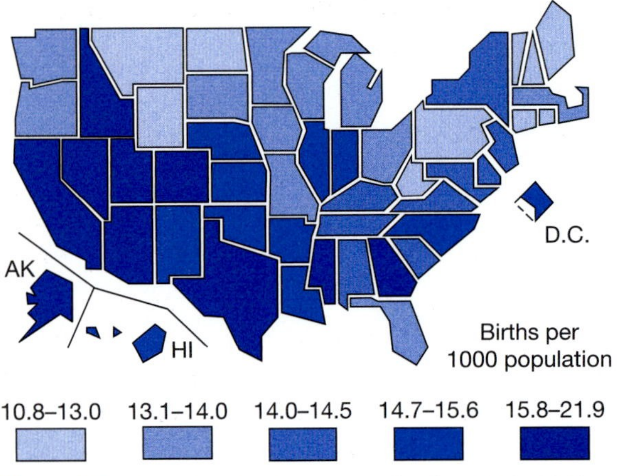

Figure 6.1 Choropleth birth rate map for the USA in 2000, taken from Monmonier (2005).

Source: Monmonier (2005).

manipulation to suit a 'new' argument produced by an academic to the (arguably more serious) effects that such representations could have upon policy-making and the direction of funding to somehow deal with 'high' birth rates. Thus a carefully drawn map can have serious real-world repercussions.

Maps may also be drawn to suit a range of other purposes. Returning to the history of geography, we made the point above that early geographers were a key part of the exploration, mapping and colonisation of large tracts of land outside Europe. In this light, Harley's (2002) work is significant, because, as a cultural geographer, he turns the lens back on those early geographers to critically analyse the maps produced by them. During the 1980s and 1990s, Harley undertook some early critical work on cartography, which drew upon Foucault and Derrida (who, respectively, are foundational authors on discourse analysis and deconstruction, terms that we have already discussed in this chapter). He argued that, during the era of European colonial exploration, maps were deployed to various political ends: to demonstrate the richness of a territory for natural resources and trade (and thereby its strategic importance; Photograph 6.3); to make a tract of land appear devoid of human habitation (even if it was not), and thereby a blank canvas, ready to be exploited; to erase local forms of knowledge, so as effectively to dispossess earlier inhabitants of any power over or claim to a particular territory; to support – using the 'objective' techniques of cartography – particular political and economic agendas, such as the transportation of slaves from Africa to Brazil and onward (Photograph 6.3). Such critical cartographies have been a key case study for new cultural geographies, in which questions about representation have been central (Chapters 1 and 5). Indeed, arguably, maps were a key text through which new colonies were defined, created, imagined and managed. It should go without saying that all of these tracts of land existed prior to colonisation by Europeans, and that they were often subject to complex and intense human habitation. But it was important that such maps propped up the idea that these were 'new' lands, which had been 'discovered'.

To repeat an earlier point, in this way, maps – and the early geographer-explorers – played a key role in *generating* the new 'reality' of these colonies, and not simply representing it. Thus, as Kearns (2010: 187)

Photograph 6.3 Early colonial map of Brazil, showing ships arriving from Africa carrying slaves, and depicting Brazil's interior as a land rich in natural resources, but populated by apparently 'less civilised' peoples.

Source: Getty Images/DEA Picture Library.

argues, early geographers like Halford Mackinder did not merely propagate a "science of empire" through their work, but sought to "promote the cause of empire". Kearns argues that this political deployment of maps is as apparent in contemporary geopolitical projects (like the 'war on terror'; see Box 11.9), as it was in the late-nineteenth-century colonial era.

Turning to the present, one can list a range of uses to which maps can be put – from promotional maps generated by tourist agencies, which channel visitors to particular sites, to maps of National Parks or nature reserves, which delineate the boundaries within which particular kinds of somehow 'special' nature are defined and protected (see Hinchliffe 2008). Admittedly, not all of these maps are produced by geographers any more, but all of these examples demonstrate the key argument for cultural geographers that maps are *social constructions* (Crampton 2011). That is, just like **identities** (Chapter 8), maps are not neutral or natural representations *of* reality, but privilege particular versions of reality, and may constitute those realities in the process. And, just like identities, maps may exclude or oppress the interests of the less powerful as they tell a particular story about a particular place. They are produced by particular members of society, for particular reasons, to have particular effects within particular societies (hence why they are *social* constructions).

For instance, Crampton (2011) assesses how new mapping technologies are being used to calculate risks posed by identifiable populations in specific places. Thus "in Colombia, where security has become a pre-eminent concern, the calculation of spaces of risk is a primary way in which populations and subjects are constituted" (Crampton 2011: 93).

It should go without saying that these 'calculations' are based upon a complex series of judgements that are moral, emotional, geopolitical and (quite often) economic. The point is that, as they have with other texts, cultural geographers have sought to deconstruct maps, to tease out some of these judgements and go beyond the idea that such calculative technologies are rational, neutral tools that simply quantify possible threats. We could say, then, that from the **new cultural geography** onwards, cultural geographers have sought to subject maps (and, indeed, some geographers themselves) to considerably more scrutiny than had hitherto been the case.

Feminism and geography

Feminist geographers have, since the 1990s, also issued a series of wide-ranging and penetrating criticisms of geography as an academic discipline. These criticisms range from debate about the very people *doing* geography – who have predominantly been, and continue to be, men – to the ways in which geographers have reinforced male-dominated ways of organising the world, not least in partial histories of the discipline itself. Some key feminist critiques are as follows.

- Academic geography is dominated by men. In 2008, just 34% of geography academics in British higher education departments were women (Maddrell *et al.* 2008). At the same time, the role of women in the development of the discipline – which has been marked and enduring – has often gone unwritten (Maddrell 2008).
- Geography – as a spatial 'science' – has privileged ways of knowing and writing about the world that admit gender-biased perspectives. For instance, traits like 'objectivity', 'dispassion' and 'rationality' have been viewed as desirable embodied habits for scientists to have. Yet, as feminist writers have shown, there are many other, equally valid ways of knowing the world that have historically been marginalised in a male-dominated discipline.
- Geographers have historically looked at places – at landscapes – through a lens that is typically white, male and heterosexual. As we show elsewhere in this book (Chapter 5 on **landscape**), many feminist geographers have suggested that this 'male gaze' objectifies and oppresses those who fall under it (Rose 1993). In this way, maps can be as equally complicit in perpetuating oppressive gender relations as they can in propping up colonial exploration (as we argued above).
- Perhaps most fundamentally, feminist geographers have pointed to issues such as positionality in geographical research. In other words, scientific approaches would suggest that knowledge

> is authoritative because it seems to come 'from nowhere' (i.e. it is 'objective'). But feminists argue that it is impossible to claim that one's identity, one's personality, one's life history and one's geographical location do not to some extent affect what one finds out about the world (see Haraway 1991).

Feminist geographers have had a diverse and profound effect on the study of geography. Perhaps most important of all, they have not sought simply to *challenge* problematic (masculinist) assumptions about knowledge. Rather, they have promoted a variety of ways of knowing, writing about and identifying with the world that move on from these critiques. Thus early work by feminist geographers like Gill Valentine (1989) sought to foreground women's experiences of place – in Valentine's case, focusing on women's fear of crime in public spaces.

However, this chapter is concerned with textual geographies, and we therefore want to look at three examples of how feminist geographers have sought to *write* geographical knowledge in new ways. Our first example takes us to back to the histories of the discipline – to the stories that we as geographers tell about ourselves. Crucially, as we have suggested, the role of women in the production of geographical knowledge has been underplayed in these stories. Avril Maddrell's (2009a) work represents an attempt to redress this balance. Her study charts the work done by 30 women who contributed to the development of geography between 1850 and 1970. She challenges the notion that geography was an overwhelmingly masculine discipline until the 1970s. She looks at a variety of debates about the role of women in academic geography – from the medals and degrees awarded (or not) by institutions, to women's role in fieldwork and geographical expeditions. Elsewhere, Maddrell (2008) looks at female geographers' contribution to the British effort in both world wars (1914–1918 and 1939–1945) – highlighting the gendered work they did as part of hydrological surveys and the production of naval intelligence handbooks. Rather than draw upon published histories, she uses alternative textual sources – oral histories with female geographers and archival sources – to uncover some of the previously undocumented ways in which women contributed to the discipline and, in turn, to historical events beyond academic geography.

A second way in which feminist geographers have sought to write is through an attention to the body (and, more specifically, their *own* bodies). Indeed, feminist scholars can also be accredited (at least in part) with introducing questions of **emotion** and **embodiment** into cultural geography – theoretical questions that have become commonplace throughout the subdiscipline today. This is also where we begin to get beyond the idea (raised in Section 6.2) that there is a sharp dividing line between cultural geographies of texts and non-representational cultural geographies; there simply is not. Several geographers have been inspired by the work of Luce Irigaray, a French feminist theorist who developed a style of writing known as *l'ecriture feminine* – feminine writing *from* the body. Deborah Thien (2004, 2005) provides an example of this style of writing. However, Thien (2004: 43) seeks "more than simply a theoretical manoeuvre, or word-play" – rather, "to consider what takes place as we *feel* our way through our worldly encounters". Following Irigary, Thien argues that it has been hard for feminist writers to be heard *as* women, rather than as the inverse of men, because women have been encouraged to write *like* men to make their voices heard – using academic standards of referencing and theorising that are measured by male standards. Instead, Thien advocates writing from one's own gendered, emotional experiences as follows:

> Some fifteen years after my first encounter with Irigary, I had picked up her work again and this time I had love on my mind. I was reading and thinking through feminist, philosophical and sociological versions/visions of love, and, not insignificantly, I was also in love. . . [Irigary's work was] a means to puzzle out the gendering and spatialities of emotions more generally, and a feminist ethical map for how to conduct my own love relationships.
>
> (Thien 2004: 44)

Ironically, perhaps, it is hard to communicate in the format of a textbook such as this (which is meant to be 'authoritative') what the full impact of this style of writing might be. Simply cherry-picking a quote like the one above to 'prove' our point rather runs against the grain of writing from the body! Yet this quotation does hint at two features of this style of writing that have been picked up in different measure in cultural geography. First, that geography can be written from the first person – using 'I' – rather than in the seemingly objective, position-less third person. Second, and related, Thien combines a discussion of work *about* emotion with a personal reflection upon her *own* emotions – the two are distinct but intimately entangled in her writing in this short extract. Feminist scholars have inspired many academics to write about their own experiences and even their own personality in their research (Moser 2008). This move has included so-called autoethnographic research, which we look at in Box 6.3 – including an example from our own work, which we hope gives a flavour of what writing from and about emotion (and the body) might look like.

Box 6.3

An autoethnographic encounter with the authors: life going on . . .

You may have come across 'ethnography' – a research method that involves detailed observation of a community or culture. Often, it involves living with or spending a long time taking part in a group's everyday lives. However, autoethnography involves turning the lens back onto the researcher: doing an ethnography of oneself! It is at its heart a form of writing that uses the researcher's reflections on their experiences to illuminate something about broader social trends. Autoethnographic texts may look very different from 'conventional' academic articles – indeed, this is part of the point, as it challenges the idea that knowledge about the world comes from predefined 'scientific' methods of data collection. So, for instance, Melina's (2008) autoethnography of her time with a band called the 'Hot Flashes' is written in the form of a series of diary entries spanning a couple of decades. Melina's paper – published in an academic journal – includes no academic references, but is a provocative exploration of how music and feminist beliefs intersect.

We (the authors) have also published an autoethnographic piece where we reflected upon how our childhood experiences are part of our everyday lives as adults (Horton and Kraftl 2006b). Here are two extracts from our paper, which again was published in an academic journal. Here, we reflect on 'being clumsy' (Horton and Kraftl 2006b: 267–268):

Author one:

> I know I'm clumsy in the way I act and talk and carry myself. There's no point trying to convince me otherwise. I don't know if I've got more or less clumsy as I have 'grown up'. . . I don't think of it like that, I don't think it works like that at all, really. If anything, I'd think of growing up as being clumsy and awkward in a succession of different ways. . . . I've always been a bit clumsy, ever since I can remember. A bit big and ungainly. A bit rubbish at walking in crowds. I can't think of any specific examples: more to the point I can't think of any specific examples of not being all that. It's more of a constant state-of-body and state-of-being thing . . .

Author two:

> I used to be really clumsy. More to the point, I was really hyperactive. Instead of just tripping over things, I would run into them (in the house), crash into them (on my bike), or put my foot in them (by just getting over excited and saying the wrong thing). I would run everywhere. . . I've grown out of being clumsy, and hardly ever trip over any more. But I can be quite hyperactive; I always run around, I like to get things done fast, and can be quite impatient and excitable . . .

The point we try to make in our paper is that childhood is not a distinct phase from which we 'grow up' – rather, that elements of our bodily habits and dispositions are carried forward with us into adulthood and are occasionally evoked in particular situations. We use our own experiences to get beyond rather static views of childhood in geography and the social sciences, and to provide a (perhaps) more lively sense of the relationship between childhood and adulthood.

Several other geographers have used autoethnography as a way of including the voices of non-academics in academic research. For instance, Cook *et al.* (2007) collaborated with some of their students, including their diary entries and own analyses in a paper about commodities. In a very different context, Butz and Besio (2004a) explore the value of autoethnography in postcolonial research, arguing that it can provide colonised groups with the opportunity to represent their lives to colonisers in ways that engage with but differ from the norms of academic representation.

However, many people caution against the use of autoethnography. One criticism – which you may agree

Box 6.3 continued

with, having read our own extracts in this box – is that it is simply an excuse for navel-gazing among academics privileged enough to write and reflect upon their own lives. Another criticism (which is also its strength) is that it lacks a defined 'method', and that it is difficult to distil general trends and therefore compare the experiences of different social groups. A final criticism in doing autoethnography with others (e.g. in Butz and Besio's research in Pakistan) is that, inevitably, some groups in some communities (in this case men) still have greater power to represent themselves. Thus Butz and Besio (2004b) point to how, in the particular context of Pakistan, women struggled to write autoethnographic accounts, even though this method was developed as a feminist research tool.

A third and final way in which feminist geographers have sought to write the world is through creative writing techniques. In this instance, the work of Noxolo *et al.* is particularly illuminating. Noxolo *et al.* (2008) begin with a critique of academic geography that combines feminist and postcolonial perspectives. They argue that geography is an academic discipline that is struggling to become global – in part through research in postcolonial places – but that, at the same, time, it is "marginalising the voices and perspectives that make it global" (Noxolo *et al.* 2008: 146). One problem, they suggest, is that while many geographers have embraced postcolonial perspectives, those geographers are overwhelmingly writing "from northern centres of global power" (Noxolo *et al.* 2008: 147). Ironically, postcolonial geographers may simply end up recolonising postcolonial spaces – perhaps with a slightly more palatable conceptual approach – just as exploratory geographers did centuries before!

In another paper, Madge and Eshun (2012) explore the use of poetry as a postcolonial research tool that has the potential to overcome some of the criticisms that Noxolo *et al.* (2008) note about postcolonial research. In their research, Madge and Eshun undertook in-depth oral history interviews with Akan communities in Ghana about their views on the conservation of monkeys in a local sanctuary. When analysing the interviews, Madge and Eshun noticed that the use of the phrase 'I remember my father telling me . . .' appeared as a way of framing participants' memories. Madge and Eshun (2012) used this to structure a poem in which multiple voices and competing perceptions of the importance of monkeys to the community were condensed:

And I remember my father telling me
About Damoah who beheld a piece of white calico
Guarded by two black and white Colobus
And two Mona monkeys
And that the monkeys are a kismet
. . .
And I remember the early 1970s
The emergence of Satellite communities . . .
And how the congregation of the Saviour Church
Scoffed at the furry taboo and killed monkeys
And how the elders of Boabeng summoned my father . . .
They said he has given ancestral land to monkeys.

As Madge and Eshun (2012) point out, the poem illustrates some of the inevitable contestations around the conservation of monkeys, and the placement of a sanctuary on previously sacred ground for one community. At the end of their paper, Madge and Eshun (2012) question what – in the postcolonial vein – this method might enable geographers to 'give back' to the communities with whom they work, and how it might enable them to speak for themselves (as Butz and Besio 2004a discuss in relation to autoethnography – Box 6.3).

Since undertaking the research, Eshun has incorporated these interpretive poems in the setting up of an NGO in Ghana where local communities perform poetry to tourists, where they use poetry to present their views to local authorities, and which provides a tool through which improvements may be made to their livelihood. However, Madge and Eshun (2012) also caution against poetry being viewed as a 'quick fix' for overcoming the mainstreaming of postcolonial approaches in geography. For it can be easy to assume ourselves as researchers to be in a position of responsibility towards a community, because in so doing it is easy to reinforce the idea that academics have the power to adopt that position. Once again, we return to questions of positionality in the research process that are raised by feminist geographers; the implication is that we need to think just as carefully about how we write poetry or autoethnography (Box 6.3) as we do more conventional, 'scientific' texts.

In this section, we have turned the lens back onto geography as an academic practice that *disciplines* the way we know the world. We have suggested that the stories that geographers tell – about themselves and about the world – constitute a series of textual geographies all of their own. We have highlighted the challenge that feminist geographers (and in this chapter, to a lesser extent, postcolonial geographers) have posed to a discipline that has been dominated by objective, scientific, male ways of knowing and of writing the world. Perhaps the most important implication of feminist critiques has been their questioning of the very foundations of the geographical tradition – highlighting that knowledge is *always* situated, emphasising that writing is positional, embodied and emotional, and calling for care about the textual geographies that are produced from within academia – whatever form they take.

6.6 Concluding reflections

Having read this chapter, take another look at the exercise we asked you to do at the very beginning, and at the questions in Section 6.2. Consider two questions:

- Have we missed out any kinds of text that you thought mattered to your everyday life?
- Do you feel that the approaches (e.g. autoethnography, poetry) and texts (e.g. works of fiction, policy documents) that we have discussed in this chapter have offered you an insight into the kinds of people and place that tend to be marginalised in academic research?

In terms of the first question, we could perhaps reflect immediately that we have not discussed particular textual forms that are a routine element of mass-mediated everyday lives, at least in Western consumer cultures. We have not, for instance, explored text messaging, social networking, websites, email or other aspects of the 'virtual world' (for an introduction, see Adams 2009; see also discussion of digital texts in Chapter 8, on **identities**). So, we wonder, from our particular writing position as (at the time of writing) 30-something, white males living in the UK, what kinds of text have we marginalised in the course of writing this chapter?

In terms of the second question, we hope to have challenged you to think about the ways in which cultural geographers write about the world and, at the same time, how they write about other geographers. Indeed, we have tried to get you to think about how cultural geographers write about themselves – and/or how they write themselves into the research process. Inevitably, every textual geography is partial; it is the very nature of a text that it is selective in its representation of people and places. In addition, writing (and being read) and speaking (and being heard) are practices that are far more accessible to some groups than to others. One key challenge for cultural geographers is therefore to remain open to the diversity of ways in which texts may represent, create and constitute 'real' spaces. A second key challenge is to think creatively but critically about how our forms of writing about the world might offer a conduit for research that at the very least allows the voices of diverse groups of people to be heard.

Summary

- Textual geographies are connected by an effort to *question* what texts can tell us about the world. Texts can represent and/or create reality, and texts have the power to reinforce dominant power relations, marginalising particular people in particular places.

- Fictional texts have been an important source for cultural geographers. Works of fiction can powerfully evoke the meanings of landscape for a national culture. Other works of fiction - such as science fiction and utopian novels - offer a critical mirror to society, and make us think about the spatial processes that surround us.
- Policy texts can be read using discourse analysis, among other methods. Such critical methods can tell us about the assumptions that underlie powerful policy documents, and hence about the assumptions that underpin the organisation of societies and economies. Spatial metaphors - like scale - are often used in policy-making; cultural geographers have sought to question how spatial metaphors are used in policy texts.
- Feminist geographers - among others - have sought to question the ways in which geographers write about the world, and about themselves. They emphasise that any academic research is situated and positional. Feminist geographers have inspired numerous creative and critical ways of writing the world, such as poetry.

Some key readings

Murdoch, J. (2006) *Poststructuralist Geography: A guide to relational space,* Sage, London.

An accessible introduction to poststructuralist thought in geography. Chapter 1 is particularly useful, because it charts the development of work on texts from structuralism to poststructuralism. The rest of the book is useful too; it can form the basis for further reading related to chapters in this book on embodiment, materiality and space/place.

Saunders, A. (2010) Literary geography: reforging the connections. *Progress in Human Geography* **34,** 436–452.

Hones, S. (2008) Text as it happens: literary geography. *Geography Compass* **2,** 1301–1317.

Kitchin, R. and Kneale, J. (2001) Science fiction or future fact? Exploring imaginative geographies of the new millennium. *Progress in Human Geography* **25,** 19–35.

Three journal articles that explore geographical studies of fiction. Saunders' text provides an overview of work in literary geography up to 2010, and explores in depth issues around the form of texts and geographies of reception. Hones provides an accessible and rigorous introduction to literary geographies, which would be especially useful for dissertation and coursework preparation. Kitchin and Kneale's text is based on original research, and provides a great example of the power of such studies to tell us about geographical processes (in this case, urban forms). It also has a handy overview and critique of work on fictional geographies.

Kraftl, P., Horton, J. and Tucker, F. (2012) *Critical Geographies of Childhood and Youth: Contemporary policy and practice,* Policy Press, Bristol.

This edited collection provides examples of policy analysis that extend the examples on childhood mentioned in this chapter. The introduction provides a longer discussion about the concepts of 'discourse' and 'discourse analysis', as well as a guide to further reading.

Women and Geography Study Group (WGSG) (2004) *Gender and Geography Reconsidered,* WGSG of the Royal Geographical Society with Institute of British Geographers.

This collection of essays was published by the WGSG to coincide with the 20th anniversary of an earlier, seminal publication on geography and gender. It was published by the group in order to allow a more flexible approach to written style - which is evident in the collection of essays, which vary in terms of format, length and the positionality of their authors. The essays include the article by Deborah Thien mentioned in this chapter, as well as a series of pieces about sexuality, pregnancy, autobiography and leisure.

7 Performed geographies

Before you read this chapter . . .

In June 2010, Serbian artist Marina Abramovic created what was purportedly the longest piece of performance art ever undertaken. She spent 700 hours in New York's Museum of Modern Art sitting still at a table, staring at visitors, who were invited to sit opposite her and do whatever they liked (Leonard 2010).

Abramovic's vigil rehearses frequent but important questions about what is and what is not 'art'. But in simply sitting still, was she 'performing'? And performing what, exactly? We invite you to return to these questions as you progress through various understandings of performance presented in this chapter.

Chapter Map

- Introduction
- Musical performances
- Sporting performances
- Dance and performance art
- Performing everyday life
- Concluding discussion: performing *what*, exactly?

7.1 Introduction

Practices are arguably at the heart of all cultural geographies. At their basic level, 'practices' imply *doings* of some kind. Practices can enable creativity and communication, and signal the fundamental vitality of life. As Harrison (2009) observes, practices have been understood by geographers as the origin of all meaning – the basic stuff from which social spaces emerge. This is, however, a highly abstracted way to understand the *doing* of cultural geographies. Therefore, in this chapter, we focus on the idea of 'performance' as a way of understanding such practices. In particular, we want to highlight two issues that are raised by studying performance.

First, the notion of 'performance' provides some concrete, specific definition to the rather generic idea of 'practices'. In other words, we can identify a range of nameable performances – be they musical, theatrical, sporting or artistic. At first glance, these performances can be distinguished somewhat from **everyday life** (see also Chapter 9). Perhaps they take place in designated spaces (museums or stadia); perhaps they are carried out by professionals, or by trained amateurs marked out by reputation or skill.

The first three sections of this chapter explore the work of cultural geographers who have considered three kinds of performance: musical, sporting and danced. Since these kinds of performance have become increasingly important to conceptual advances in the discipline – especially, but not only, debates about **non-representational** geographies (Chapter 9) – we highlight some of the broader issues that cultural geographers have explored through their work on music, sport and dance.

Second, however, throughout the chapter, we constantly question what the boundaries are between these specialised performances and everyday life. Thus we include a fourth kind of 'performance': the performance *of* everyday life. This enables us to expand cultural geographies of performance to those aspects of 'normal' social life that proceed via specialised, learnt or otherwise identifiable doings. These include gestures (like a handshake) that conform to societal expectations of how one should act in public spaces in particular contexts. As we will see, many of these acts are difficult to distinguish from those that constitute nameable performances – for instance in the ways that sport permeates many aspects of everyday lives (as the example of football fandom in the pub, in Box 7.2, demonstrates). In the process, we continue to break down the idea that there is a neat division between specialised performances and everyday life. Hence we ask you to continuously return to the question raised in the box at the beginning of this chapter: in performing cultural spaces, *what*, exactly, is being performed?

7.2 Musical performances

Music has formed an increasingly important part of the study of cultural geographies. For some geographers, music is a crucial component of what makes a place a place. Kong (1995a) observes that music in some form is a universal aspect of human cultures and societies; in fact, she argues, music helps us to make sense of our everyday lives. Perhaps we have music on in the 'background' while we study; perhaps we listen to particular artists or genres of music as part of our **identities** (see Chapters 3 and 8); perhaps we play in a band, or practise the piano to relax or keep our minds active; perhaps we travel to concerts or festivals to listen to, dance to or socialise in the context of a live performance. In these ways, as Anderson (2004, 2006) points out, music can accompany some of our most personal memories and hopes (see Box 3.2).

Music, then, is part of the production and consumption of places. Indeed, as a performance, music may be both produced and consumed at the same time. Think, for instance, of how a DJ in a night club plays music that is consumed by dancing clubbers, but also responds to the crowd's reaction as they revise their playlist (see Malbon 1999). However, music is often assumed by geographers to be a 'background' component in our experiences of place, which is then consumed. This may be because music (or at least recorded music) is deemed a part of 'popular culture', and not worthy of serious attention when compared with so-called 'elite' kinds of culture (Kong 1995a). Alternatively, this view may result from geographers' long-standing focus upon *visual* spaces and subjects to the detriment of other senses, including sound. Perhaps it is because, for many of us, music is something to simply be enjoyed; to subject it to formal analysis may be to remove some of the vitality that makes music so poignant and meaningful.

However, music is not just 'there' in the background. Indeed, music is not just 'music': what makes musical styles, performances and sounds *matter* has been interpreted very differently by geographers, whose work on the subject is now quite diverse. The **new cultural geography** (see Chapter 1) witnessed the emergence of a number of geographical studies of links between music, place and space (see Leyshon *et al.* 1998). In this chapter, we shall look at several important ways in which music matters to socio-spatial processes and the experience of particular places, often taking the production and consumption of music as elements that are intertwined.

A long-standing interest in cultural geography has been the relationship between music and culture regions (see Carney 1998, and Chapter 5 on **landscape**). This approach explores how music may be associated with particular places. Specifically, traditional or 'folk' musical styles are often connected to particular cities, regions or nations. For instance, White and Day (1997) used a 'culture region' approach to analyse listening practices associated with 162 country music radio stations in the USA. They found that in the Great Plains, Texas and Oklahoma (the 'birthplace' of country music), country radio stations were very popular. Although country music had 'diffused' to the rest of the USA, it was far less popular in areas with higher population density and higher median household income – especially in cities along the north-eastern seaboard of the country.

While remaining popular, the culture region approach – and its association with the **Berkeley School** – has been criticised by many cultural geographers (see Chapters 1 and 5). Nevertheless, notions of 'tradition' and of 'region' have become the focus for critical attention in geographical research on music, most frequently in debates about globalisation (see Box 7.1). Some researchers have challenged the idea that trends towards the globalisation and commercialisation of popular music have erased the significance of individual places. Rather, music may be a key medium through which local and regional identities are renegotiated (Hudson 2006). One example is the UK city of Liverpool, where rock music has become "a way of life with its own beliefs, norms and rituals" (Hudson 2006: 627). Hudson argues that this 'way of life' is particularly evident in the sheer number of amateur rock musicians, their audiences, and the cultural association of the city and various music venues with Beatlemania and rock music.

A more familiar story about the globalisation of music concerns the movement and mixing of musical styles in creating **identities** (Chapters 3 and 8). This story is familiar both within and beyond academic geography, as it concerns the idea of 'hybrid' musical genres such as British Asian dance music (Jazeel 2005) and 'World Music' (Connell and Gibson 2004). Most commonly, hybrid musical styles are considered to be those that follow the migration of a religious or ethnic group to a new country – especially from the Global South to the Global North. For instance, Connell and Gibson (2004) observe that multiple forms of Caribbean reggae became popular in England during the 1960s, the principal era of transatlantic migration from Jamaica to Britain. But reggae became mixed – 'hybridised', if you like – with other musical forms, scenes and identities in different ways. Some white British groups began to play reggae, and adopt stylistic, musical and production elements (with varying degrees of musical and commercial success) from reggae artists, as the genre became more 'mainstream'; meanwhile, British reggae groups, reflecting in the content of their music the conditions of urban life in 1970s Britain, began to have success in Jamaica. This process – underpinned by international migration – "transformed into new expressions of ethnicity and political motivation" (Connell and Gibson 2004: 347). A more detailed treatment of musical hybridity and identity is provided in Box 7.1.

Box 7.1
Music, tradition and place in France and India

Increasingly, geographers have explored how the consumption of music – especially listening to recorded music – can provide keen insights into contemporary cultural spaces, identities and the bodily practices that constitute them (Anderson 2004). Two examples provide contrasting views on these themes: Saldanha's (2002) study of upper-class youth in Bangalore, India; and George Revill's (2004) critical reflections upon taking part in forms of regional French folk music and dance. Two points stand out.

- *Tradition/modernity*. In Bangalore, young upper-class youth have increasing access to global musical styles and networks such as MTV. Many see themselves as 'global youth', part of a modern India contrasted with the 'backward, hopeless and ugly' images they hold of Indian tradition (Saldanha 2002: 345). They dis-identify with Indian musical forms, favouring Western 'pop' music. In France, Revill contrasts different forms of tradition or 'authenticity' associated with folk music. While attending dances at which folk music was played, he noted that musical styles and practices (dancing) were characterised by *creativity*. Rather than simply conforming to the relatively constricting 'traditional' forms of French folk music (which for some French

Box 7.1 continued

observers would be a recognisable part of French rural life), dance was viewed as "free, fluid, a little risqué, an exciting mix of the urbane and the rustic" (Revill 2004: 206). In both cases, musical 'traditions' are mixed with musical 'modernities' as part of the *creative* (rather than passive) practices of consuming music.

- *Bodily practices/(inter)national symbols.* Both the French and Bangalore examples interrogate the relationship between **representational** and **non-representational** geographies of musical performance (see Section 7.4; also Photograph 7.1). While the 'free', creative capacities of the body are emphasised in Revill's depiction of forms of regional French folk music, he stresses that social and cultural values (especially symbolic or representational notions of identity) are equally important. So, bodily performances and cultural values are taken together to stress the difference between regional genres of French and English folk music/dance, according to the French performers:

> We were too repressed in England, we needed something flirtatious to break through that English reserve. French folk dance gave you freedom of expression; it was less hidebound by tradition than English set dancing and much more sophisticated than 'Morris'.
>
> (Revill 2004: 206)

In Bangalore, upper-class youth also draw upon a variety of bodily performances to make highly symbolic statements about their identities - as modern, global and wealthy. As in the French example, these performances rely equally upon creating obvious markers about their difference from *other* identity groups. Many upper-class youth drive around Bangalore listening to music, driving fast, and treating the city as an 'amusement park' (Saldanha 2002: 342). The sound of pop music and the tingle of air-conditioning arguably turn the heat, dust, mass and (what upper-class youth see as the) misery of the city into something abstract. They also create spaces where upper-class girls reportedly feel safe. Hence youth turn these into their *everyday* expressions of identity. They are knowingly 'normal' but knowingly symbolic, performed in full view (like their 'closed' parties, which can still be seen and heard from the street). Herein, Bangalore's wealthy youth display their identities through musical performances that mark them out as 'modern' and 'Western' compared with those watching them.

Both examples suggest that the consumption of music involves a complex series of performances that are both representational (relying on and revising 'traditional' values) and non-representational (drawing on bodily performances of dancing, driving, and a knowing display of 'normality'). Yet note that, ultimately, two very different kinds of cultural space are created: in France, where 'authenticity' and tradition *do* retain a certain significance; and in Bangalore, where tradition is considered something to move *away from* in the consumption of music.

Photograph 7.1 Traditional regional French folk music being performed in rural France.

Source: Alamy Images/Jeffrey Blacker.

At this point, continuing our focus on identity, we return to a question raised at the beginning of this chapter. That is: exactly *what* is being performed when we write of geographies of *music*? Geographers have answered this question in various ways. First, it has been argued persuasively that cultural geographies of music should attend to their **cultural politics** (see Part 1 and Kong 1995a, 1995b). That is, cultural geographies of music should interrogate the intentions of those who produce and perform music. Here, music may be performed (written, recorded, played) in order to exercise authority; or, music may be a medium through which to display resistance against authority. In her seminal study of Singapore, Kong (1995b) contrasted a book of 'national' songs supported by the Singaporean government to promote national identity with a group of songs in the 'Not the Singapore Song Book'. Whereas the former were supposed to create allegiance among the populace for Singapore's relatively new ruling elite, the latter were a series of tongue-in-cheek songs containing "new lyrics set to popular tunes", performed in live stage shows "as a form of resistance to official cultural representation" (Kong 1995b: 448).

Second, a number of geographers have sought a closer understanding of the form, content and emotional effects of music (see Chapter 6 for a similar argument about **texts**). In this case, the argument is that, too frequently, geographers have seen music as a 'carrier' for lyrics or a symbol for other – perhaps more meaningful – notions of identity or political belief. As Revill argues: "[t]he focus of cultural analysis is always on practices *associated* with music rather than music's sonic qualities" (Revill 2000: 597; also Jazeel 2005). Thus, as part of a broad series of turns to non-representational geographies, some cultural geographers have begun to examine how the intended effects of a particular piece of music – its sound, pitch, rhythm and timbre – are meant to operate on the body of a listener. A particular rhythm may stimulate certain emotions; or it may subdue, calm or pacify a listener (Anderson 2004, 2006). However, Revill (2000: 606) suggests that such bodily effects are neither personal nor simply 'natural'; rather, "[m]usical immediacy is . . . only accessible through socioculturally specific forms." In his example, these forms are represented by particular (paternal, elite and conservative) visions of English national identity during the early twentieth century. This socio-political context formed a kind of 'frame' around the sonic qualities of the music being performed. But the sound of the music as it was *performed* – in this case its "late romantic tonal harmonies and broad folksy melodies"– mattered just as much in making the live performance of this music meaningful to listeners (Revill 2000: 611). Hence Revill (2000: 610, our emphasis) is able to argue that "musical sounds *themselves* played a distinctive part in the cultural politics of Englishness."

Finally, moving to the sphere of formal music production and the music industry, we can consider the dual way in which music is co-performed with cultural-economic processes, right down to the choice of location where music producers base themselves. Recent work has shown, for instance, that the music industry itself is 'performed' (to use the word loosely) in highly particular ways. For instance, the US commercial music industry has become increasingly centralised around three major centres: New York, Los Angeles and Nashville (Florida and Jackson 2010). More broadly, it has been shown that that music is increasingly being enlisted in diverse urban regeneration projects around the world. Hudson (2006), for instance, suggests that as part of a 'turn' to cultural industries (especially the performing arts), cities like Nashville are branding themselves as 'music cities'. Outside the USA, in Sydney, Australia, popular music has been used as part of the promotion of newly gentrified inner city spaces (Gibson and Homan 2004). The city's suburbs had, in the past, been portrayed as a key venue for creativity and musical subcultures. However, from the 1990s, the city experienced a decline in live music venues. Gibson and Homan (2004) explore how the local council in one suburb – Marrickville City – have responded to this decline by funding a series of free live concerts in public spaces managed by the council. In this way, musical performance – like a range of other 'cultural' performances – is incorporated into highly managed, often state-led attempts to render cities more competitive and attractive places to live (Hall and Hubbard 1998).

We have argued that music can be performed in a variety of ways, and that it is suggestive of multiple geographical processes. Musical styles may well be associated with particular culture regions – and these associations may be enduring – but we can equally argue that global flows of people, commodities and practices have led to similarly long-standing tendencies to hybrid musical styles. We may also understand different ways in which music allows the performance of identity – from the symbolic 'resource' that it provides to the rhythms and cadences that may evoke notions, motions and emotions pertinent to national identity.

7.3 Sporting performances

Sporting practices share with musical styles an explicit emphasis upon 'performance', albeit interpreted in overlapping but distinct ways, as we shall see. 'Sports geography' is a broad area of research in geography that encompasses issues as diverse as the mapping of sporting institutions to the constitution of support infrastructures and technologies that provide the location for sporting endeavour (Henry and Pinch 2000; Bale 2003).

In this section, however – and despite the under-representation of sport in cultural geography – we want to consider how sporting *performances* are actually particularly illuminating for thinking about performed geographies. Through sporting performances, we can critically consider a question raised in the introduction to this chapter: what is the relationship between 'marked-out' sporting spaces and performances and those 'non'-sporting practices that constitute everyday life?

We begin with the performance of skilled, athletic practices in 'non'-sporting (or, perhaps, 'everyday') spaces. Examples include skateboarding (Borden *et al.* 2001), ultimate frisbee (Griggs 2009) and parkour or free-running (Mould 2009), which predominantly take place in urban settings rather than at formal sporting venues. When we juxtapose these kinds of activity as 'sport' in 'non-sporting' settings, however, we encounter difficulties in determining exactly *what* is being performed. Take parkour: the practice of running through urban landscapes, turning steps, walls, roofs, railings and other street furniture into 'obstacles' during what is ideally a long, continuous and fluid series of bodily movements (Photograph 7.2). Parkour has been defined variously as an 'art of movement', as a 'sport', and as a way of "learning to be in places differently" that requires physical and emotional training (Saville 2008: 891).

Photograph 7.2 Practising parkour in the city. Interestingly, this image spawned online debate about whether or not the move it shows was actually 'parkour': some commentators suggested that it was not the most efficient, flowing move that could be performed on this piece of urban furniture, and therefore was 'not parkour'. Thus the *performance* (and the *kind* of performance) *matters* to this sport (and to at least some of its exponents) as much as to any other.

Source: Alamy Images/Travelscope.

Geographers' work on parkour and skateboarding is tied together by an acknowledgement that neither is simply a 'sport'; both kinds of performance hold an uneasy relationship with everyday life in the city that professionalised sports rarely do. Marshall (2010: 165) argues that, in Quebec and Paris, parkour is a 'youth movement' that heralds a

> renegotiation of the environment, in which the meaning of walls, parapets, rooves [sic], bollards is transformed in a new urban journey . . . to emphasise the popular energy and meaning-creation which exceed the grids of administration and urban planning.

It is through this 'renegotiation' of urban space that skateboarding and parkour are often framed as transgressive ('out of place') performances, because they often run *deliberately* against the grain of 'acceptable' uses, of movements through, city spaces (Borden *et al.* 2001). As Karsten and Pel (2000) argue in their work in Amsterdam, the association of skateboarding with young people (especially teenagers) tends to heighten controversies surrounding the sport, because teenagers themselves are viewed as 'out of place' by adults.

But the picture is a little more muddled than this. Over time, and in some places, certain *kinds* of skateboarding performance have become acceptable elements of daily public life. In Newcastle, Australia, Nolan (2003) demonstrates that, while skateboarders have an 'underground' image, different kinds of skateboarding performance are acceptable in different kinds of public space. In Newcastle, local councillors differentiate between 'good' and 'bad' skateboarding activities – 'good' being the use of a skateboard for transport; 'bad' being the performance of tricks that damage public or private property. Yet some shopping malls and parks are designated 'no skateboard' zones, so that boarders simply *travelling* through one of those zones are simultaneously 'in' and 'out' of place (Cresswell 1996) – engaging in acceptable behaviour but in the 'wrong' kind of space (Nolan 2003).

So far, we have suggested that activities like parkour blur boundaries – both between sport and everyday life, and between what is and is not acceptable in urban public spaces. We can, however, look at sporting performances in other ways, which equally blur the neat division between 'sporting' and 'non-sporting' performances. We want to suggest that for some people sport *is* everyday life, because it becomes a foundational part of their lifestyles, identities and senses of place.

In this regard, bodily performance matters. In their study of British bodybuilding culture, Andrews *et al.* (2005) argue that geographies of sport need to emphasise the role of bodily performance in producing sporting spaces. They argue that bodybuilding is seen less as a sport and more as an 'extreme fitness activity' – and hence is central to the lifestyle of those who practise it (Andrews *et al.* 2005: 879). They concentrate their work in 'Roy's Gym' in south-west England, to make a number of important points about sporting (or fitness) performance (see also Section Chapter 12.1 on the gym and gym equipment). First, they suggest that the gym is not simply a 'container' for fitness activity; rather, the gym is a "complex social and cultural construction" (Andrews *et al.* 2005: 888). Attendees (who are predominantly male) have to negotiate sets of formal rules, etiquette, advice and help passed on informally from bodybuilder to bodybuilder, and social hierarchies. In the case of the latter, one's 'position' in the hierarchy is governed solely by one's 'physical capital':

> You're judged by the amount of plates you have on the bar. If you squat with two plates (100kgs) nobody gives a s**t . . . You do five and you've got the serious bodybuilders watching you . . . Money is the thing in society ain't it? People who've got more money than you think that it gives them more rights. In the world of bodybuilding its [sic] how big you are that determines how far down the pecking order you are.
>
> (Dorian, bodybuilder, cited in Andrews *et al.* 2005: 883)

Second, the gym – with all its rules and accepted practices – reinforces what it means to be part of bodybuilding culture, and to adopt the identity of 'bodybuilder'. Yet, as we have suggested, bodybuilding is not simply located in the space of the gym. Rather, bodybuilding is a lifestyle that tends to suffuse the performances and spaces of everyday life 'outside' the gym as much as in it. Take Mike, another bodybuilder (cited in Andrews *et al.* 2005: 886):

> I only train three times a week but on the days off I'm preparing for the next day, stuffing my face [with food]. On days off I don't think about work, its [sic] 'OK legs tomorrow night and chest on Wednesday' . . . I just could not have a serious effort towards a serious career and all the thinking that has to go into it, and then expect to walk into the gym

and be totally motivated to building muscles. Bodybuilding has really overflowed into other aspects of my life.

The example of bodybuilding demonstrates how immersed – how taken up, almost obsessed – one may become with certain bodily performances. It also highlights a need for geographers to think critically about what Andrews *et al.* (2005) term 'fitness' geographies as part of a cultural geography of sport that is better attuned to the ways in which bodily performances are central to the creation of meaning in people's everyday lives.

While Andrews *et al.* (2005) make an important point, geographers have also shown that sporting performances may operate in quite different ways as part of everyday *community* or communal lives. A series of studies has shown how sporting activities, sporting clubs and sporting fandom may create 'social capital' (Tonts 2005). Friendships may be forged around sporting endeavour, often across religious, ethnic or class differences. Sport may bond people through a sense of shared purpose, belief or (an often emotional) commitment to a particular place – perhaps through participation or support of a local football or rugby team (Hague and Mercer 1998; Bale 2003).

Thinking about sport in this way enables sporting performances to be placed in their political and economic context. Tonts' work on rural restructuring in Australia is exemplary in this regard (Tonts 2005). He argues that global economic restructuring has dramatically affected agricultural production processes in rural Australia, lowering farm incomes, and causing a steady flow of families to leave farming behind for cities on the country's coast. However, many farming communities remain; Tonts hypothesised that community organisations – such as sporting clubs – may keep communities together in the face of such economic threats. In Tonts' study, one in three residents of rural communities in Western Australia agreed that sport was one of the three most important aspects of social life. In one resident's words, sport kept their community "glued together" (Tonts 2005: 143) – a key process in the creation of social capital (Putnam 2000). Interestingly, the *kind* of sporting performance mattered less than simply being present:

> It doesn't really matter if you don't play. The main thing is to get involved and show an interest. Going to footy [Australian Rules Football] is as much about being with your friends and part of the community as getting fit.
>
> (Male resident of Three Springs; cited in Tonts 2005: 143)

However, Tonts also suggests a 'darker side' to the production of social capital through sport. That is, the binding of community members often involves rules or other criteria that exclude those who do not or cannot 'perform' appropriately. In the case of golf clubs in rural Australia, this may simply be because the sport is too expensive. However, in an area where nearly 10% of the population are Aboriginal Australians, levels of Aboriginal participation in community sport are low, despite appeals to the 'inclusive' nature of those sports. In fact, Tonts suggests that bodily performance is (in part) key to the kinds of sport in which Aboriginals participate – largely netball, basketball and Australian Rules Football. "Not only are these sports affordable, but popular myth suggests that these require considerable speed and athleticism and therefore are ideally suited to Aboriginal people" (Tonts 2005: 146). Thus sporting endeavour may well bind communities together in rural Australia – but implicit kinds of racism stereotyping and exclusion may accompany this process. To read more about sporting identity, team sports and the new cultural geographies of football fandom, turn to Box 7.2.

Box 7.2

The new cultural geographies (and sociologies) of football fandom

During the 1990s, geographers began to turn their attention to the geographies of football. Many studies focused on the players themselves – how they 'migrated' to play in particular countries, and especially major leagues in the UK, Spain and Italy. Other studies explored the social and environmental impacts of football stadia upon local communities. However, Bale (2003) argues that, as the new cultural geography took off during the 1990s, its implications became felt in studies of football that explored its cultural meanings – especially for fans' identities (for more on identities, see Chapters 3 and 8). Since then, there have been many studies in cultural geography and sociology about the meanings and cultural experiences of being a football fan. While the everyday experiences of football fandom are culturally specific (often to specific clubs), we would highlight two broader features that highlight the importance of the cultural *geographies* of football fandom.

First, in a relatively straightforward sense, it has been pointed out that the experiences of being a football fan are often subject to clear spatial rules and boundaries. In most parts of the world, a common experience of attending a football game is that the 'home' fans will sit in one area of the stadium and the 'away' fans will sit in another (Bale 2003). Notably – in part because of the long-standing association between football and violence – this is something that happens less regularly among spectator groups at other sporting events (such as rugby, cricket or athletics).

At an even more micro-scale, in many clubs it is common practice that certain groups of fans from the same club (such as season-ticket holders) will sit together, in the same seats, and see each other almost every week. The straightforward act of sitting together means that a range of shared meanings and practices – from chants and songs to food and clothing – can be performed on a repetitive basis (Photograph 7.3). For instance, Kytö (2011) explores what he calls the 'soundscapes' of Çarşı, a supporters group for the Turkish football team Beşiktaş. Kytö (2011: 77) argues that Çarşı had a range of "marches, chants and sonic rituals" that they performed before, during and after a match. These rituals were culturally specific – in particular because many members of the group were also highly politically motivated (being anti-fascist, anti-racist and generally left-leaning), and motivated as much by current affairs as by football. Although the group's future was uncertain at the end of the first decade of the twenty-first century, they retain in Turkey (and beyond) a reputation for sitting together in a particular area of the club's home stadium, and creating "a sense of shared space" and an "acoustic community" among all the club's fans (Kytö 2011: 77). Thus the cultural geographies of football fandom emerge in part from spatial containment, which is in part enforced (often by the local police): in the act of literally sitting and 'performing' fandom, together.

The second feature we want to highlight is how fans support their teams by *not* being physically present at a football match. Although Mike Weed is a sociologist, his (2007) work attends to this important geographical theme. His research was based on in-depth observations undertaken with football fans in pubs. Unlike the previous examples, he argues that the experience of being a fan was less about 'being there' at a match than about the *shared* experience of supporting one's team with other like-minded supporters. **Emotion** (Chapter 11) is key here. Fans can *feel* close to their team, and to the event itself (the match), because they feel a sense of togetherness: tightly packed into a pub, they wear their team's colours, chant and sing songs, drink and share stories (or whatever is locally appropriate practice for supporting one's team in the pub). Notwithstanding the greater significance that he accords to not necessarily needing to 'be there' at a match, Weed (2007) also argues that, for many fans, there is still a certain cultural kudos associated with *actually* having been present at a match – an experience that, especially among committed fans, tends to live longer in the memory and provide greater cultural capital.

Photograph 7.3 The Çarşı supporters group of the Turkish football team Beşiktaş.
Source: Getty Images/Bongarts.

In this part of the chapter, we have considered the various ways in which sporting performances intersect with everyday life. Sports like parkour – which may not be considered formal 'sports' at all – introduce sometimes radical ways to rethink urban space, but are often considered to be 'out of place'. Other sports like bodybuilding comprise a variety of performances that spill over from the gym to take over many aspects of a person's life-world. Finally, the very act of participating in sport – whether as player or as spectator – may be an important part of how communities in different geographical contexts respond to global economic change. However, those responses may also be exclusionary along class or ethnic lines.

7.4 Dance and performance art

> I want to consider here what is to be gained from the metaphorical and substantive turn from 'text' and representations, to performance and practices.
>
> (Nash 2000: 654)

This part of the chapter provides less a survey of 'cultural geographies' of dance and more a provocation to think about the *kinds* of cultural geography that dance creates. Arguably, dance (like performance art)has been central to debates about the conceptual and methodological direction of cultural geography. In the quotation that begins this section, Catherine Nash (2000) indicates the broad direction of this debate: that is, the respective gains to be made from focusing on performances rather than (or as well as) **texts** and representations. We have already explored some reasons for examining performance – from the relationship between musical style and national identity to the constitution of bodybuilding cultures and spaces of 'fitness'. We, certainly, are in little doubt that performance matters in certain contexts. The question is, then: what does geographers' work on dance add?

It is useful to begin – as does Nash – with Nigel Thrift's work on **non-representational theory** (Thrift 1997, 2000a; see also Chapter 9). Here, Thrift develops two propositions that are designed to challenge the predominant focus of cultural geographers on representations (especially textual and visual kinds):

- Certain elements of life *exceed, elude or occur before cognitive thought* – and hence are difficult to 'represent' in words, images or otherwise.
- The kinds of activity that are hard to represent are precisely the skills and practical knowledges of *'ordinary lives' that tend to be ignored* by those academics and social commentators who define matters of societal and political importance. These 'ordinary' practices could be activities as diverse as walking, playing or gardening, to name but a few.

Nigel Thrift (1997) articulates how dance is exemplary of his non-representational approach. First, dance is playful. Dancers are often portrayed as 'free', somehow, to interpret musical rhythms with bodily gestures – perhaps following recognised 'style rules', or perhaps by what is popularly termed 'giving yourself over' to the music. Second, dance involves complex bodily movements that are hard to represent, and even harder to subject to control. In other words, dancing bodies can simply elude the practices of those in authority, because they do not take place in a format (such as words) that can more easily be categorised (for instance, as 'illegal') (Thrift 1997; see also Box 3.13). Third, dance is an impermanent, constantly changing form of performance, much of whose meaning is experienced in *the moment* of doing or watching, not before (as planned) or after (as remembered, written down or photographed).

Finally, dance can *move us*, emotionally. The example of dance movement therapy (DMT) demonstrates a relationship between bodily *motion* and *emotion* (McCormack 2003). Hereby, movements – such as guiding a blindfolded partner around a room, or play-fighting – are performed as joint, bodily experiments that might create new ways of feeling and being. McCormack uses the notion of **affect** to describe the shared feelings – perhaps brief moments of joy, fear or humour – that emerge from these DMT experiments (see Chapter 11). In a therapeutic setting, such techniques may be used with patients who are uncommunicative: DMT therefore offers alternative – ostensibly non-representational – forms of care that make emotional rather than verbal connections. Importantly, McCormack also develops a range of writing and diagramming strategies through which geographers might record at least some of those affective moments (see Box 7.3).

However, several cultural geographers have expressed qualifications about a non-representational approach to dance (Revill 2004). For example, Nash (2000) cautions that geographers should not overemphasise the pre-cognitive, emotional and non-verbal elements of dance. Rather, dance involves complex, changing combinations of "speech, writing, text and body" (Nash 2000: 656). Thus, as we suggested above in our discussion of music and national identity, the rhythms of any given performance intersect with **discourses** that tend to be expressed in words. As Nash (2000: 657) argues: "[t]his is especially the case when . . . different material bodies are expected to do gender, class, race or ethnicity differently." Moreover, for better or worse, certain forms of dance (such as the Tango, or dance used as part of activist protests) often work only *because* of the symbolic meanings contained in bodily movements, facial expressions and dress – in their evocation of particular times, places, events, gender relations or political beliefs (Kolb 2010; see also Box 7.4).

Peter Merriman's (2010) work on the American landscape architect and environmental planner Lawrence Halprin exemplifies Nash's plea to combine non-representational and representational geographies of dance, as well as some of the political implications of dance. It also indicates a final tension in work on dance performances. During the 1950s and 1960s, Halprin worked on a series of public landscape designs in Illinois, Minneapolis and San Francisco. Using insights from contemporary dance and theatre practice, Halprin and his wife Anna sought to 'choreograph' pedestrian movements through public spaces. Informed by liberal democratic principles, Halprin wanted to create cities that were livelier, freer and more diverse – both socially and in terms of the experiences they afforded (Merriman 2010). He combined architectural design with a series of staged happenings – mixtures of dance and performance art. By mixing formal design with 'informal' happenings, Halprin sought new ways to

Box 7.3
Different 'diagrammings' of performance (from McCormack 2003, 2004)

In his work on dance movement therapy (DMT), Derek McCormack experimented with different ways of witnessing (rather than straightforwardly representing) his experiences and those of his co-dancers (see also Chapter 11 on affective practices). He chose a method that he calls 'diagramming'. As the images below demonstrate (Figures 7.1 and 7.2), diagramming involves, in part, the use of relatively simplistic lines and shapes to evoke a sense of the bodily movements that comprise DMT. Compare these forms of diagramming with other ways of representing or witnessing dance – such as video, your own memory, poetry (Chapter 6) or photography (Photograph 7.4). Which do you find most evocative? Which help to give you a sense of the expressive qualities or culturally specific meanings of dance?

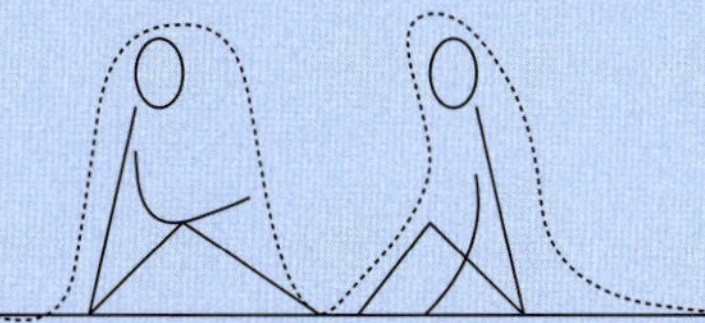

Figure 7.1 An early stage of DMT: participants sit facing each other, each covered by gauze, and try to figure out a 'therapist–client' relationship.

Source: from An event of geographical ethics in spaces of affect, *Transactions of the Institute of British Geographers*, 28, pp. 488–507 (McCormack, D. 2003), p. 492, Reproduced with permission of John Wiley & Sons Ltd.

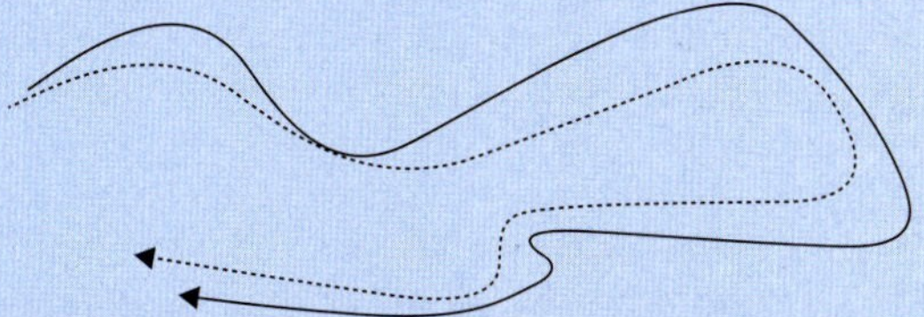

Figure 7.2 Diagramming DMT in action: one participant is led, eyes closed, around a room, following a route roughly mapped in this image.

Source: from An event of geographical ethics in spaces of affect, *Transactions of the Institute of British Geographers*, 28, pp. 488–507 (McCormack, D. 2003), Reproduced with permission of John Wiley & Sons Ltd.

Box 7.3 continued

Photograph 7.4 A blurry photograph of Hula dancers in Hawaii: another way to 'witness' dance.

Source: Getty Images/Alvis Upitis.

Box 7.4
Dance, performance art and politics in India

In terms of the performing arts, India is world-renowned for its lavish 'Bollywood' dance films. However, a less-well-publicised heritage of performance art in India exemplifies how non-representational and representational practice combine in the articulation of political ideas.

Early roots of Indian performance art were evident in the 1970s. In 1971, Bombay artist Bhupen Khakhar staged a performance that mimicked both the rites of an Indian marriage procession and the ceremony to mark the beginning of a new government. While an artistic statement (questioning the role of the 'public' artist in post-independence India), Khakhar's was also a political statement aimed at what some saw as the 'excessive' ritual (including dance) involved in traditional Indian ceremonies.

Since 2000, performance art in India has become increasingly well-publicised - at both national and international scales. A number of artists' associations have emerged since 2000. They provide space for alternative experimentation in India, and forge exchange with international performance artists. Khoj[1] is one of the most high-profile. In 2010 it hosted a project entitled 'Negotiating Routes: Ecologies of the Byways', which, in the context of a potentially destructive national road-building scheme, aimed to combine "research and art creation by artists and local communities, addressing the visible and invisible transformations currently taking place in their immediate environments" (Khoj Workshop 2010). The intention was to engage community participation via dance and performance art, and to link textual and non-textual art forms in the production of a politicised statement. The project aims to produce a community 'route' that reflects hopes for the "regeneration of the local ecology of . . . local cities or villages" (Khoj Workshop 2010). In addition, the project will cut across local and national scales, as each community is encouraged to share its local 'route' as part of an alternative 'road map' that mimics the national road-building plan.[2] Performance art is therefore offered here as a way of practising environmental politics, and as a way to (literally) practise cultural geographies of landscape at different scales.

[1] www.khojworkshop.org

[2] http://www.khojworkshop.org/project/12223

engage with and choreograph the everyday lives of (often ethnically diverse) local communities with their taken-for-granted public spaces.

The example of the Halprins' work provides two qualifications to non-representational geographies of dance. First, as a landscape practitioner and planner, Halprin sought (and was required) to codify his approach in order to address everyday design problems. In other words, of necessity, he had to *think*, to classify, to generalise, in order for the performed movements that constituted his happenings to have any broader design relevance. Second, Merriman highlights the potential *political* implications of dance as it extends into other kinds of performance (such as architectural practice). For instance, Halprin increasingly used staged happenings to promote the participation of local communities in urban planning. In the context of major urban regeneration projects that threatened to displace minority groups, liberal planners like Halprin aligned with civil rights activists to call for more democratic planning techniques. Thus, in forging non-representational geographies of dance, it is important to consider how *both* embodied performances *and* political discourses are frequently drawn together or co-implicated. Some very different cultural and political associations with dance and performance art are explored in Box 7.4, using examples from India. To read more about **architecture** and built geographies, please turn to Chapter 4.

In this section, we have shown how dance – and to a lesser extent, performance art – has been at the very centre of debates about the direction and nature of cultural geography. While, as Lorimer (2005: 557) observes in a telling footnote, "plenty of cultural geography . . . rolls on, unabated" in the wake of non-representational theory, geographers now have an extended repertoire of approaches for understanding where and how performance matters. Indeed, arguably, an interest in dance and performance art has allowed geographers to engage with artist-practitioners in ways not imaginable in the early 1990s (see, for instance, Merriman and Webster 2009 on a collaborative project between a geographer and artist that explores ways of travelling through landscapes). However, at the very least, we have cautioned against a simplistic reading of non-representational theory that values *only* the embodied, emotional elements of dance. Rather, as Nash (2000) suggests, dance combines practices that can be defined as *both* representational and non-representational. The same goes for the performance of everyday life, which we turn to in the next part of this chapter.

7.5 Performing everyday life

In the first three sections of the chapter, we focused on broadly 'nameable' kinds of performance: music, sport and dance. In this section, we explore how the concept of 'performance' has been applied to routines and practices that we consider to be a part of the non-specialist, ostensibly mundane world of 'everyday life'. As the previous section indicated, it is very difficult to maintain a neat distinction between specialist performances and everyday life. In this section, we tackle the issue from the reverse perspective, by asking where and how the seemingly 'non-specialist' performances that we take most for granted gain meaning. We begin by considering everyday routines and rituals that, if you like, 'stitch' together public spaces. Then we consider how the performance of everyday life has become a valuable, visual (or 'aesthetic') project in its own right under conditions of consumer capitalism. We finish the section by reflecting on different political (and artistic) reactions to this trend. To read about **everydayness** in more detail, turn to Chapter 9.

Everyday routines and rituals

Most of us acknowledge that certain performances keep social spaces ticking over in certain ways. Consider the following questions.

- Why (in some contexts) do we wait in line at a bank or a shop (see also Chapter 9 on **everyday spaces and routines**)?
- Why do some of us hold open a door for the person behind us, but some do not?
- How do we negotiate our way to a seat in a busy bus without offending anyone, or taking the seat from someone more deserving?

The sociologist Erving Goffman (1963) provided an insightful explanation of these kinds of process. He suggested that people perform 'roles' in order to present themselves in the best light, and that it is this act of 'self-presentation' that governs how people interact in public. He argued that in different places – the home, the street, the workplace – people act differently in order to create different impressions upon others. While people would draw on social rules and moral codes, essentially, the individual decides how to carry out or *perform* such roles – just like an actor on a stage. Elliott (2001: 32) puts it neatly: "[Goffman's] central preoccupation is with the dramatic techniques by which the self displays agency to others" – the movements, gestures and talk that show off to others who we are (or want to be). A particularly significant development of (and away from) Goffman's approach has been work on the performativity of identity, which is explored elsewhere in this book (see Chapter 12 on **bodies**).

A number of geographers have used Goffman's work directly. For instance, Crang (1994) showed how, when working at a themed restaurant ('Smokey Joe') in south-east England, he and his colleagues had to evoke the ambience of the Deep American South. This was achieved not only through the food and décor but also through the staff playing roles (especially being 'happy'), and encouraging diners to take part. Again in a context of a consumption space, Laurier and Philo (2006a: 196) look at what they call "gesturing amongst the unacquainted in the public space of the café". They show how particular gestures – smiling, turning towards or away from, looking at or interacting "with (and at) faces, tables, chairs, napkins, newspapers and other customers" may produce different kinds of relationship between strangers – from friendliness to rudeness. These may be small gestures and momentary encounters, but they can tell us much about the ways in which people *express* themselves both as and with strangers (Laurier and Philo 2006a: 204; see also Chapter 9 on **everyday** spaces, practices and events).

Important work on disability geographies has developed Goffman's (1963) notions of 'front' and 'back' stages performances. For example, in a study of self-help groups for people with agoraphobia (to generalise: a fear of social spaces) in Scotland, it was found that many sufferers use a variety of techniques to cope with their fears when in public spaces (Davidson 2003). For instance, some try to avoid the kinds of social contact that they perceive as a threat, avoiding large social gatherings or busy public spaces. Others draw on some kind of material assistance to create a 'distance' between themselves and others. Moyra (in Davidson 2003: 117–118) comments on the use of glasses:

> its [sic] partly cause it gives you a wee [slight] feeling of security. Its [sic] partly cause you think, well you know they can see you but they cannae see your eyes. You know how sometimes if you're troubled it shows in your eyes?

These are 'front' stage performances because they involve individuals managing and presenting themselves in particular ways in order to 'fit in' in public view. Once in 'backstage' spaces (like the privacy of the home), one's guard can drop and, usually, such performances are unnecessary. Often, these kinds of act may be about the avoidance of 'stigma', manifest in staring, naming, feelings of shame, or being marked as 'different' (Parr 2008).

Everyday life as aesthetic performance

We have suggested that performance is actually central to the ways in which everyday life proceeds, especially in public spaces. Earlier in the chapter, we suggested that sporting or artistic performances are also blurred with and constitutive of everyday life. In this final section, we come 'full circle'. Our key argument is that some theorists view everyday life itself as a *specialist* kind of performance: an aesthetic (artistic, visual) project, if you like, that has become increasingly prevalent in Western consumer cultures since the beginning of the twentieth century. We want to highlight two contrasting ways in which everyday life has been aestheticised through performance.

First, a variety of thinkers and artists have tried to blur the boundary between everyday life and art. These 'avant-garde' artists included Dadaist and Surrealist performers who took theatre into outdoor public spaces and tried to break down formal boundaries between audience and performer (Bonnett 1992). They also questioned the increasing prevalence of mass-manufactured goods, turning 'ordinary' everyday objects into resources for creative practices. These were deliberate attempts to disrupt the repetitive – for some, alienating – effects of everyday life. While they were knowingly aesthetic projects, the distinction here is that one *begins* with the everyday in order to create meaningful artistic performances. Inspired by the work of French theorist Henri Lefebvre (see Chapter 9 on **everyday geographies**) and the Situationists, cultural geographers have explored the implications of avant-garde performances of everyday life. In particular, they have explored the potential for *doing* cultural geography through walking as a particular form of performance art (Box 7.5).

Second, we consider what Highmore (2004: 314) calls "the life lived aesthetically". This phrase can be interpreted in various ways. For some commentators, the idea of everyday life as a 'lifestyle' or 'project' of identity is a key part of Western consumer culture (Giddens 1991). In fact, one may suggest that the performance of the self through consumer goods such as clothes, mobile technologies and foreign holidays is the kind of performance to which many people most aspire in their daily lives. Highmore (2004: 314) puts it this way:

> As such the aestheticization of daily life in terms of fashion, design, food, music, etc. must be seen as at the heart of the everyday. As soon as we make choices about clothes or the colour of a room, we are part of an aestheticization of daily life.

The implications of such consumption practices are too many and varied to consider here (for a range of examples, see Clarke *et al.* 2003). But consider just one: the cult of celebrity, perhaps most heightened in the UK and USA around sports stars, pop stars and media figures. Simultaneously, the meteoric and more-or-less global rise of 'reality TV' shows like the *Big Brother* franchise has represented a particular fascination with the everyday lives of 'normal' people. The banalities of sleeping, eating, arguing and copulating in 'The Big Brother House' have become the very performances that engross viewers – they constitute 'entertainment'.

Box 7.5
Performance, art and everyday life: cultural geographies of walking/walking cultural geographies

Francis Alÿs is an internationally renowned Belgian artist who lives and works in Mexico City. His work reflects critically on ongoing problems surrounding social segregation, the environment and basic services in Latin American cities (especially Mexico City). Based around a series of walks (*paseos*) in and around the city, his work centres around the mutual production of art and everyday life: "the street becomes a place of invigorating possibility and confluence, a space where the complexity of the popular collides and interacts with the practice of making art" (Schollhammer 2008: 144).

In 1991, Alÿs performed *The Collector*, when he pulled a small magnetic toy around the city on a leash, collecting any metallic objects and debris that were attracted to it.[3] In *Patriotic Tales* (1997) he led a flock of sheep into the city's main plaza and made them circle a flagpole (Schollhammer 2008). Many of his works have been documented by videos.[4] His walks offer a playful but challenging reflection on the social and spatial processes that shape the contrasting everyday lives of Latin American cities. He is particularly critical about the role of 'central' public spaces – places that have been neglected by planners for decades. In Mexico City, the 'centre' has, as a result, lost its importance for the city's inhabitants. In works like *The Collector*, Alÿs "rescue[s] the informal materials of . . . neglected existences in the streets and reinsert[s] them in the symbolic and historical dynamics of the city" (Schollhammer 2008: 148). In other words, Alÿs draws out the tensions between contemporary urban development policies and the many ways in which disadvantaged groups forge a living in and from the city. His walks are designed to draw attention to these pressing urban and social geographies.

[3] http://www.postmedia.net/alys/freematrix.htm

[4] http://francisalys.com/

Box 7.5 continued

Inspired especially by the Surrealists and Situationists, cultural geographers have explored many other ways in which the seemingly banal, everyday act of walking may be an artistic (and academic) performance that can draw out the rhythms and tensions of the city (Pinder 2005a). The 'Mywalks' project in Northumbria encourages members of the public to look again at their local taken-for-granted environment by walking through and making audio or visual recordings of what 'tickles', disgusts or frustrates them (Fuller *et al.* 2008). In London and New York, Butler (2006) creates 'walks of art': creating sound recordings along a walk that enable the often complex cultural and historical geographies of a city to be presented to the public in more accessible ways. Hence walking enables us to reflect on cultural geographies (in these examples of the city) as well as on the ways in which researchers and members of the public actually do cultural geographies as a research practice (Latham 2003).

Meanwhile, in a similar trend, celebrities of all flavours have the intricate and often intimate details of their everyday lives – shopping, going to the beach, having an affair, crashing a car – splashed over the pages of tabloid newspapers and magazines. The odd scandal notwithstanding, aesthetic performances are absolutely key to those celebrities remaining newsworthy. They are expected to model the latest designer fashions, to be seen at particular parties, to be photographed at the latest in-vogue holiday destination. They are, in a sense, the pinnacle (for some the farcical and excessive face) of the 'cult' of the self: an ostensibly never-ending series of performances that have to be *seen* to gain any value. Here, then, everyday life has become a quite different performative project from the radical and critical notions of everyday life put forward by the Dadaists, Surrealists and other artists of everyday life.

7.6 Concluding comments: performing *what*, exactly?

The extent to which different people in different geographical and social contexts can (and want to) participate in 'the life lived aesthetically' varies enormously. Yet the examples at the end of the previous section return us to the question with which we began this chapter on performed geographies: *what,* exactly is being performed? After reading the chapter, you may be able to think of many answers to this question. However, we should like to briefly point to two.

First, the notion of 'the life lived aesthetically' suggests that everyday lives in different contexts are becoming specialist performances in and of themselves. In this way – as Saldanha (2002; Box 7.1) shows in his work on elite young people in Bangalore – music, dance and other 'named' performances are blurred with the more 'mundane', un-named performances that are usually taken for granted. Hence, in aspirational, mass-mediated consumer cultures, *everyday life is blurred with and becomes a set of more specialist performances* to create the notion of 'lifestyle'. In essence, those lifestyles perform and depict places in new ways (as 'modern', 'authentic', 'community-centred', etc.), and may turn places themselves into objects to be consumed or ignored.

Second, in the box at the very beginning of this chapter, we asked you to think about an installation by the performance artist Marina Abramovic. Perhaps what is most striking is that she *seemed,* in essence, to be doing absolutely nothing – not to be performing at all. Alternatively, we could view her vigil as an extreme version of *both* art (a specialist performance) and everyday life. The longevity of her installation (700 hours) concentrates the focus of the audience on the simple banality of 'just sitting there'. But the effect is the reverse: strangely enough, just sitting there becomes *extraordinary* (Kraftl 2010). Michael Gardiner (2004) argues that these extraordinary moments are not separated from everyday life, or, as Highmore (2004: 321) puts it, some "minor subversion" that provides a second's entertainment. Rather, they are moments that allow a critical view of what it is that makes the everyday, everyday. Whether this is 'art' (or can be defined as a 'sport', like parkour), such performances – and performed geographies – are an important way for cultural geographers to critically consider life as it is – and how it might be otherwise (Gardiner 2004; Pinder 2005a).

Summary

- Music does not just 'play in the background'; musical performances are vital to the ways in which spaces are experienced. In particular, music is consumed during the creation of identities – often linking everyday bodily practices to cultural values at the scale of the nation or the globe. Music also plays a key role in cultural regeneration strategies happening in diverse cities around the world.
- Cultural geographers have raised a number of questions about the 'role' of sport in different spaces. Some have focused on sports like parkour and skateboarding, which are at the centre of heated debates about the 'proper' use of streetscapes. Others have explored how sport may constitute the 'glue' that holds communities together – but those same sports may also reinforce the marginalisation of particular groups in those communities.
- Dance has been at the centre of debates in cultural geography about representation and non-representation. Most geographers agree that dance involves both representable cultural values and non-representational bodily practices/emotions. However, there is considerable ongoing debate about their relative significance in understanding how dance matters in different contexts.
- Cultural geographers' work on a range of 'specialist' performances highlights how music, sport and dance are linked to and help produce everyday or 'non'-specialist performances. Equally, though, the performance of everyday life – in public spaces – involves a range of performances. Some of these performances are becoming 'aestheticised' as part of consumer and celebrity cultures; others are aestheticised in different ways as geographers and artists use ordinary practices (like walking) to reflect on the complex processes that make up life in cities.

Some key readings

Anderson, B., Morton, F. and Revill, G. (2005) Editorial: practices of music and sound. *Social and Cultural Geography* **6,** 639–644.

Kong, L. (1995) Popular music in geographical analyses. *Progress in Human Geography* **19,** 183–198.

Carney, G. (1998) Music geography. *Journal of Cultural Geography* **18,** 1–10.

The three papers listed above provide introductory remarks on music as everyday practice (Anderson *et al.*), on the vast possibilities of musical geographies (Kong) and the considerable work undertaken using a 'regional' approach to music (Carney).

Bale, J. (2002) *Sports Geography,* Taylor & Francis, Abingdon.

A fascinating survey of sports geography by one of the original and most well-respected geographers writing on sport. The book covers a number of themes – from globalisation to the location of sport, and – of more relevance to this chapter – the relationships between sport, the senses and imagined geographies.

Leyshon, A., Matless, D. and Revill, G. (eds) *The Place of Music,* Guilford Press, New York.

A collection of essays exploring the social and cultural geographies of diverse forms of music, past and present, ranging from electronica to Victorian brass bands, to forms of American country music. A key resource for considering the ways in which new cultural geographers explored music, and created a space in which geographers could explore music, and other popular cultural phenomena.

McCormack, D. (2008) Geographies for moving bodies: thinking, dancing, spaces. *Geography Compass* **2,** 1822–1836.

An accessible and wide-ranging introduction to conceptual and methodological issues raised by the geographical study of dancing bodies. Extends the debate in this chapter about (non-)representation and points to the potential for future geographical research on dance. Nash's (2000) paper remains a provocative earlier intervention on dance.

Highmore, B. (2004) Homework: routine, social aesthetics and the ambiguity of everyday life. *Cultural Studies* **2/3,** 306–327.

This text explores some of the performances that make up 'everyday life'. Parts of the paper review debates about the 'aestheticisation' of everyday life. Highmore also raises a range of significant political issues about everyday life that provoke further reflection. Part of a, wide-ranging and critical double special issue on everyday life.

Identities

Before you read this chapter . . .

Think about *your* identity. Try to write down a few key words that encapsulate your identity. Use these prompts to help:

- Bodily characteristics
- Social categories (gender, ethnicity, age, religion)
- Where you feel comfortable, out of place, or cannot go
- Personality traits
- Your interests and hobbies
- Beliefs or ideals that you may share with others

Reflect briefly on this task. Was it easy? Why – or why not?

Chapter Map

- Introduction: the complexities of identity
- Piecing together identity
- Adding complexity
- The social construction of identity
- Relational identities
- The performativity of identity
- Concluding points

8.1 Introduction: the complexities of identity

> Identity is a complex and contested term.
>
> (Jackson 2005: 392)

> We . . . use the term 'spatiality' to capture the ways in which the social and spatial are inextricably realized in one another; to conjure up the circumstances in which society and space are simultaneously realized by thinking, feeling, doing individuals and . . . the many different conditions in which such realizations are experienced by thinking, feeling, doing subjects.
>
> (Keith and Pile 1993: 6)

The two quotations above signal two things about identity. First – as Jackson notes – there are no easy, singular ways to understand the term 'identity'. In fact, if we look at them closely, the six prompts in the introductory activity tell us this. Take 'bodily characteristics' and 'social categories', for instance. It is widely assumed that people with physical 'impairments' (such as the loss of a limb, or visual or hearing difficulties) are part of a name-able social category: 'disabled' (Golledge 1993). While this might work on a general level, there are many exceptions. For instance, Skelton and Valentine (2003) show how many deaf young people do *not* consider themselves to be 'disabled'. Instead, they make alternative identities for themselves, centring around their membership of a minority language group – those who use sign language. Thus it is difficult to identify any single definition of the term 'identity', because it can refer to and complicate *all* of the six prompts above, and much more.

The second thing – captured by Keith and Pile's quotation – is that identity is a profoundly *geographical* concept. This is not to say that the study of identity is the exclusive realm of geographers – far from it – but that cultural geographers have a key role to play in examining identity. As Keith and Pile explain, the main reason for this is that the social realm (interactions between individuals) and the spatial realm (landscapes, public spaces, regions) are always interconnected. The two can never be separated. Society creates spaces, which create societies, which create spaces . . . and so on. Keith and Pile note how geographers term this inseparable process **spatiality** (see Chapter 13). Finally, they suggest that individual people ('subjects', with identities) may experience these processes in many different ways. Therefore identity – as we will see – is always a social and a spatial phenomenon. Different spatialities produce different identities, and vice versa.

The rest of this chapter expands on these debates. It explores different explanations of identity that have been developed by cultural geographers. It begins with the relatively simple 'building blocks' of identity – social categories like class and gender. In the following sections, we add layers of complexity. We ask you to think critically about social categories by considering the social constructions, relationships and performances that are so important to the spatiality of identity.

8.2 Piecing together identity: essentialism and time – space-specific identities

> When considering someone's identity, there is necessarily a process of selection, emphasis and consideration of the effect of social dynamics such as class, nation, 'race', ethnicity, gender and religion.
>
> (Sarup 1996: 15)

Think again about the activity that began this chapter. Focus on 'social categories'. The most common include gender, class, age, dis/ability, religion, ethnicity and sexuality. How important would these categories be to your own identity?

For some people, it may be that just one social category is paramount to the way they live their lives. Often (but not always), those people are marginalised or persecuted by mainstream societies. A famous example, although based in historical roots stretching back far further, would be the separation of Black Americans from White Americans during the twentieth century – solely on the basis of their colour. For instance, Black Americans and White Americans were segregated on public transport, being forced to sit on different parts of the same bus. During the 1960s, many Black Americans used their shared experiences of marginalisation, based around a single social category, to promote equal rights for people of all ethnicities.

We shall return to the political uses of identity later in the chapter (when we look at 'subverting the centre'). Before that, we want to consider social categories – what we are terming the 'building blocks' of identity – in two ways. First, we will question *why* it is that social categories – like ethnicity – are of such over-riding importance to particular groups of people. Our explanation will focus on what are known as 'essentialist' understandings of identity. Second, the chapter will highlight what the example of a divided America makes abundantly clear: that social categories take on different meanings depending on the historical and geographical context. In other words – as we will repeatedly stress in this chapter – identities are *always* place- and time-specific.

Essentialism and social categories

Essentialist explanations of identity have a long history, and extend well beyond cultural geography. From the Ancient Greeks onwards, many people thought that having an identity meant having some 'essential' quality (Martin 2005). Essentialist explanations argue that each person has some essential essence that they are probably born with, and which remains unchanged throughout their lives. If one were to take that essence away, they would no longer be the same person. Indeed, for many thinkers – who suggested that a person's essence was their 'soul', or their ability to think logically – to take away that essential quality would be to take away what makes a person human!

Martin (2005: 97) goes on to suggest that "reference to individual identity frequently involves the idea of an interior 'self' or subjectivity that is equal to certain core characteristics that give it [identity] integrity and coherence." Hence the key thing about essentialist understandings of identity is that they see a person as an *individual, bounded* being, with *one* identity that remains more or less the *same*. In a sense, this is the default position that many of us begin with when we think about identity. Most of us (try to) think of something that makes us what we are, and therefore believe that we are unique individuals (i.e. there is something about us that makes us different from others). In fact, essentialist arguments would go as far as to suggest that a person's characteristics are innate or biological: that they are born with them.

Essentialist explanations of identity tend to rely on a range of social categories. Many of these – such as race and gender – are viewed as biological traits, which tend also to refer to specific places (for instance, in Table 8.1, see sweeping reference to places like 'Africa' and 'India'). In fact, social scientists – including many cultural geographers – often use social categories in their research.

Table 8.1 shows an example that you may have come across. It shows ethnicity classifications for the 2011 UK Census (similar classifications exist for many other countries). Every Census includes similar tables for age, gender, religion, socio-economic class, and so on. For each category, when completing the Census, respondents can tick just *one* of these choices. At the national scale, a completed Census gives a neat, quantifiable overview of the proportion of people in each group. It can also display intersectionality – identifying subgroups of the population by collating two or more variables (an example would be Asian British>Indian>males>aged 18–30). Hopkins and Pain (2007: 289–290) explain how geographers "have sought to explore the ways in which various markers of social difference – gender, class, race, (dis)ability, sexuality, age, and so on – intersect and interact." However, their work actually calls for a relational understanding of identity; we think more about relational identities later in the chapter.

Table 8.1 National Statistics standard classifications for presenting ethnic groups of the UK (from ONS 2011)

White	*Black or Black British*
White British	Caribbean
Irish	African
Other White background	Other Black background
Mixed	*Chinese or other ethnic groups*
White and Black Caribbean	Chinese
White and Black African	Other ethnic group
White and Asian	
Other mixed background	
Asian or Asian British	
Indian	
Pakistani	
Bangladeshi	
Other Asian background	

Source: adapted from Office for National Statistics, http://www.ons.gov.uk/about-statistics/classifications/archived/ethnic-interim/presenting-data/index.html, Contains public sector information licensed under the Open Government Licence (OGL) v1.0. http://www.nationalarchives.gov.uk/doc/open-government-licence/open-government licence.htm

Categories like those in Table 8.1 can be useful in three ways. First, they can provide large-scale, quantitative baseline data for national agencies (such as government departments) to better understand, and plan for, large populations. Second, they can reveal basic information about *places* – for instance whether certain ethnic groups or people in the same socio-economic class cluster in particular neighbourhoods. Researchers often use Census data and other similar measures to undertake (and justify) research about particular (usually 'deprived') places. Third, researchers often tend to 'list' different social categories when acknowledging that difference *matters* to the study of cultural geography. When they do this, they are usually evoking the kinds of classification listed in Table 8.1.

Despite the usefulness of Census data – and prescriptive methods of classification like those in Table 8.1 – we can identify a number of problems with essentialist explanations of identity. You may already have identified some of these: try to put yourself into *one* of the categories and consider how well this represents how *you* feel about your ethnic identity. Problems with these kinds of category include the following.

- Many people find it hard to tick just *one* box; many people end up ticking '*other*'. A prescribed list of essential social categories can never capture the diversity of ethnic identities.
- Owing to the incompleteness of the list, significant ethnic groups feel '*excluded*' by the process of filling out a census. In Britain, for example, Cornish pressure groups campaigned for a specific tick-box that allows them to express 'Cornish' ethnicity in the 2011 Census (BBC 2011), although this right was refused (they had to tick 'other' and manually enter 'Cornish' instead). Cornwall is a county in the south-western-most corner of the United Kingdom that has seen an increasingly popular movement for self-rule (it also has its own language and flag). This reminds us once again of the *contestability* of identity (Jackson 2005).
- Simply ticking a box does not express the multiple *meanings* associated with being, feeling and belonging to a particular ethnic group. As we will see in the next part of the chapter, cultural geographers have tried to show how identities are a *process*, rather than a stable thing internal to each of us.
- Just knowing that *n*% of a neighbourhood's residents come from socio-economic class *x* is not enough. It does not tell us much about how class matters to people's everyday lives in *that* neighbourhood, nor does it recognise difference and diversity *within* that group. More problematically still, this knowledge (especially about class and ethnicity) can lead to the broad-brush stigmatising of certain areas, and a reinforcement of processes of social exclusion that may already exist (Sibley 1995).

Shurmer-Smith and Hannam (1994: 90) neatly summarise the problem with social categories thus:

> [i]t should always be remembered that the system of classification is an artifice, there is nothing natural or obvious about it and it invariably serves to facilitate thinking from some perspectives more than others. At the point of contact between categories there are things, places, people and ideas which are hard to classify because they fall between two or more of the available sets . . . [T]hese things will be regarded as anomalous by the sorters, and from the anomaly there emerge ideas of sacredness, wonder, dread or revulsion.

Social categories are place- and time-specific

Thus far, we have suggested that social categories – like class, ethnicity and gender – are the basic building blocks that many geographers have in mind when they begin thinking about identity. But – and this is a huge qualification – the use of essentialist explanations and of 'tick-box' classifications is widely critiqued.

Arguably, very few cultural geographers have used essentialist understandings of identity in a pure sense. Shurmer-Smith and Hannam (1994: 101) suggest that essentialism can be a powerful but problematic tool for analysing geographical processes. For instance, they criticise David Harvey's (1989) reliance on economic, class-based explanations for the rise of postmodern, global cultures. Indeed, for many years, geographers' interest in identity was based on Marxist analyses of socio-economic class. More recently, those studies have been heavily criticised by geographers who have argued that they failed to acknowledge other forms of social difference such as gender (Rose 1993) and ethnicity (Kobayashi 2004).

On the other hand, Shurmer-Smith and Hannam also acknowledge that, in some cases, recourse to particular, essential categories may be a potent political tool. Often, marginalised groups turn their persecution by others (on the basis of an essential category) into a tactic to push for social change. The example of Black Americans in 1960s America used above is a high-profile example of this. Contemporary struggles for land-based rights among indigenous peoples around the world (including those in Australia, North and Latin America) would constitute other examples. However, we must be careful *not* to confuse a *tactical* search for 'common ground' – for temporary agreements that allow political advances to be made – with *actual* essential categories that are unchangeable and universally recognised (Pickerill 2009).

Whether or not we (or you) agree with the above arguments, cultural geographers have made one thing abundantly clear: social categories are *always* place- and time-specific. Geography matters – as does history. And it is from this observation that we can work towards the next section of this chapter and build an increasingly subtle understanding of identity. Quite simply put, by acknowledging that social categories are place- and time-specific, the idea of 'essential' categories begins to unravel. Before you move to the next section of this chapter, you may like to look at Box 8.1, which illustrates a highly place- and time-specific use of social categories in Australia. This is particularly important to geographers, because it highlights how identity categories both enable and constrain the kinds of **spatial** flow and mobility (Chapter 13) that geographers routinely explore.

8.3 Adding complexity: social constructivist, relational and performative explanations of identity

The example of Australia's immigration system (Box 8.1) is fascinating, for many reasons. First, it does not rely on the classic, essentialist categories such as class, race or gender (or, at least, not explicitly). Rather, the system works by classifying and ascribing value to individual traits. Some of these *are* given, structural and often unchangeable – such as family relationships, or the conditions causing a person to become a refugee. Others are not; they may be learnt (like qualifications), acquired or inherited (like having sufficient capital),

Box 8.1
Australian immigration criteria

Like many countries, Australia controls immigration through its borders using a set of published criteria. The Department of Immigration and Citizenship states that it does not "discriminate on the basis of race or religion. This means that anyone from any country, can apply to migrate, regardless of their ethnic origin, gender or colour, provided that they meet the criteria set out in law." There are different criteria for different categories of visas with different criteria:

> *Skill* – most migrants must satisfy a point test, have particular work skills, be nominated by particular employers, have other links to Australia or have successful business or investment skills and sufficient capital to bring to Australia to establish a business or investment of benefit to this country.
>
> *Family* – selected on the basis of the family relationship to a sponsor in Australia – essentially partners, fiancés, dependent children and parents.
>
> *Humanitarian* – refugees and other Humanitarian Program arrivals – must satisfy the criteria concerning refugees or humanitarian cases." (www.immi.gov.au 2009).

Therefore the criteria include "such factors as relationship to an Australian permanent resident or citizen, skills, age, qualifications, capital and business acumen. All applicants must also meet the health and character requirements specified by migration legislation."

The Australian Government uses a 'points system' to test the suitability of potential migrants. This kind of system is also used in many other countries. So the system does, to some extent, reduce people's identities to quantifiable categories. But – whatever the failings of this 'points system' – it represents an attempt to evaluate a person's identity on the basis of more than the classic essentialist categories, such as race or gender. Instead, it focuses on skills, financial acumen and family relationships.

It is, however, worth noting some of the historical antecedents (now officially dismantled) in Australia's immigration policy. The most infamous is the *White Australia Policy*, which privileged (white) British immigrants to the country over other ethnic groups. The policy became passed as an Act in 1901, as one of the first Acts of the National Parliament. In 1975, the *Racial Discrimination Act* was passed, which made it illegal to make decisions about immigration based on race. Rather, the kinds of criterion listed above took absolute precedence.

This brief example highlights how essentialist categories may be deployed in time- and place-specific ways, which may (especially in retrospect) be both contested and problematic.

Source: adapted from Australian Government, Department of Immigration and Citizenship.

or simply part of the way that a person 'performs' or holds themselves (like certain practical skills or medical health). Hence, as we discuss later, identities are performative as well as 'ticked off' from a checklist of categories.

Second, it should by now be clear that there is a significant difference between the categories in Table 8.1 and those in Box 8.1. Each set of categories is designed to do a different thing. And that is precisely the point – they are *both*, somehow, designed or *constructed*, for a particular purpose: hence their difference. Underpinning any suggestion that identities are somehow 'natural' there are always sets of social, cultural, political or economic considerations. Identities are rarely neutral, objective 'things' that people have. Instead, they are socially constructed and then revised. They are *created* and *re-created*. The criteria for immigration into Australia have gradually changed over time (Box 8.1), dictated by changing political, economic and demographic demands, as well as by changing societal attitudes to ethnicity. So if, in the future, there is a need for more medical professionals, then the criteria will shift accordingly. There are many, many other examples: we look at what is called the 'social construction of childhood' later on in this chapter (Box 8.2).

A third, more political, point is that the criteria listed in Box 8.1 are part of the way in which Australians define what is 'inside' and what is 'outside' their nation (Gelder and Jacobs 1998). This is a process that occurs in many spaces – from gated communities to membership of special interest groups (Cresswell 1996). Identity is crucial here: it forms the central part of differentiating between those who belong and those who are 'strangers' (Sarup 1996). This means that identity is relational: individuals gain their identities in comparison with others (those who belong in comparison with those who do not). Likewise, whole nations are said to gain a sense of national identity (Section 8.4) in distinction to what is 'outside' their nation. Thus, like many countries, Australia has a 'citizenship test' that assesses the extent to which a person has the capacity to be (or become) 'Australian' (Löwenheim and Gazit 2009). The test – first introduced in 1901 and most recently revised in 2007 – includes 20 multiple-choice questions on Australian citizenship, history, values and traditions. Like similar tests in other countries – indeed, like the whole topic of immigration and its relationship with ideas about multiculturalism in many countries – the process is rife with controversy, and is intensely divisive (Löwenheim and Gazit 2009).

The example of Australian immigration policy is a useful introduction to this section of the chapter. The key point is that we are building away from an assumption that identity is natural or essential. Rather, the majority of work in cultural geography – certainly over the last 20 years – has explored the manifold complications of identity. Despite this complexity, we can say that – broadly speaking – cultural geographers have taken a threefold approach to identity, focusing on the social construction of identity, relational identities, and the performativity of identity. To consider each of these ideas in more detail, read on.

8.4 The social construction of identity

We have argued that identities are rarely (if ever) truly 'natural', fixed or essential. Therefore the idea that identities are socially constructed is an important one. In fact, this argument is central to a range of work in cultural geography that, over the last 30 years or so, has explored diverse kinds of identity.

The idea of social construction emphasises that identities are only ever provisional, and are constantly being (re)created, worked on, negotiated and contested. For instance, while a person might have a biological age of 18, that age is imbued with particular values in particular societies (i.e. its meaning varies over space and time). In many European countries this is the age at which a person becomes an 'adult' – a sharp transition stipulated by laws governing voting, alcohol consumption, marriage and so on. However, these laws – and many like them – have taken centuries to evolve and (like the age at which a homosexual couple may marry or enter into a civil partnership) are often still contested (Valentine 2003).

Identities may be socially constructed in a variety of ways. This section summarises four important ways in which identities can be constructed (see also

Chapter 3). We focus on *language, narratives,* the role of *institutions,* and the significance of *material spaces.* Bear in mind that, in practice, the construction of identity may involve some or all of these aspects.

First, as Martin (2005: 99) argues, "[t]he notion of identity [is] a 'discursive' construction – that is, one constructed in and through language." *Language* is important to identity, because the words we use to describe identities are often loaded with political or cultural assumptions. Those assumptions vary over space. In the UK, a phrase such as 'working class' has moral connotations associated with respectability, endeavour, 'roughness' and/or the struggle to make ends meet. In Poland, however, the working classes are often labelled either as '*dresiarze*' (those wearing tracksuits) or '*blokersi*' (those living in tower blocks) (Stenning 2005). At its simplest, identity is commonly a process of labelling – of using language to refer to and generalise about particular groups – and therefore to construct identities through particular words and phrases.

Second, Sarup (1996) argues that the process of labelling is not the only way in which identities are constructed through language. Instead, he thinks that "we all link these dynamics [labels, categories] and organise them into a *narrative*: if you ask someone about their identity, a story soon appears" (Sarup 1996: 15). That story might well be an individual, personal story or a collective 'history'. Cultural geographers have shown how various forms of 'life history' – autobiographies, travel writings, diaries, educational texts, and the stories we tell to others – are important to the making of identities (Daniels and Nash 2004). Many geographers have focused on the construction of migrants' narratives of identity because of the importance of space and flow (the journey) to their individual or collective histories (e.g. Hopkins and Hill 2008; on mobility, see Chapter 13).

Third, geographers have shown how *institutions* have a key role to play in constructing and structuring identities (we also look at their role in producing cultural spaces in Chapter 2, and in the production of **space** in Chapter 13). Place is fundamental here, as are structural explanations that suggest that individual people do not have complete control over their own identities (or their own destinies). As Box 8.2 shows, institutions like governments, schools and academia set out expectations for how people should behave, design spaces to monitor and regulate their behaviour

Box 8.2

Colonial constructions of childhood in British Columbia, Canada

Geographers have shown how many 'essential' identities are social constructions. Recently, a large body of work has explored how childhood – which until recently has seemed a 'natural' identity category – has a long history of creation and reinvention. For instance, James and James (2004) make a compelling case that adults have constructed childhoods since medieval times. Much of this construction has involved the creation of special laws and places dedicated to children. For instance, during the nineteenth century, the UK introduced a series of *Education Acts* (e.g. in 1870), which gradually increased the school-leaving age. These Acts removed children from the dangerous, dirty, 'adult' spaces of the Victorian factory to the 'safe', 'childhood' spaces of the school. Since then, geographers have explored precisely how school spaces construct childhood – focusing on behavioural ideals (Pike 2008), the kinds of 'desirable' knowledge children should learn (Ploszajska 1996), and the notion that school should, somehow, be a home-like space (Kraftl 2006a).

Over time, the social construction of childhood has involved the legal, cultural and especially spatial separation of children from adults. Problematically, this has been done almost exclusively by adults, with little say from children. Moreover, it has been done by powerful adults who tend to treat all children the same way, ignoring difference and diversity.

This is poignantly and starkly illustrated in a study by de Leeuw (2009) on aboriginal children in British Columbia. De Leeuw shows how, during the nineteenth century, European settlers attempted to 'integrate' aboriginal groups into what settlers saw as their more superior, modern society. Settlers sought to 'break' aboriginal

Box 8.2 continued

cultural identities by isolating children from their native background. Children were sent to 'residential schools' for lengthy periods (often years). The schools were imposing places that tried to instil 'European' skills, knowledge and values. Often, however, methods for doing so were brutal. Children with darker skin were frequently scrubbed because they looked 'unclean' (a colonial ideal of whiteness was imposed). Many children were mentally or physically scarred by their schooling, and some died at or upon leaving school. This is an extreme example, but it highlights how the construction of identities is a geographical process that often involves spatial exclusion and the literal construction of places for the creation and control of identities.

and, crucially, set the parameters for what kinds of identity are *acceptable* (Rose 1993; on policies for schools, see Chapter 6).

A stark example of the role of institutions in constructing identities has been provided by geographers studying disability (Holloway 2005). Imrie (2001) argues that disability is not a natural state; rather, it is constructed and perpetuated by mainstream (able-bodied) society. While some people have impairments (such as 'visual impairments'), Imrie shows how those impairments are exacerbated – how they are turned into disabilities – by unthinking, inaccessible and often silently prejudiced institutions. The argument is that society *dis-ables* individuals through particular features of urban design – from inaccessible automated teller machines (ATMs) provided by banks to high kerbs, and from narrow supermarket aisles to demeaning signs asking wheelchair users to use a 'back entrance'.

Fourth, cultural geographers have shown how *material spaces* are complicit in the social construction of identities. Once again, we return to the notion of spatiality, but in this case to see how the very physical fabric of a place is entwined with a particular social process – the social construction of identity. This is particularly – if not blatantly – evident in the construction of national identities. National identity has been a long-standing concern for cultural geographers (e.g. Gruffudd 1995; Matless 1998). For instance, Matless (1998) showed how geographical processes have been central to the construction of English national identity – from images of England's 'green and pleasant' landscapes to the ways in which 'Englishness' has been produced in films, topographical guides, photography, poetry, statues and more. To read about an example of how a nation's landscapes are portrayed in texts, turn to the discussion of Yasunari Kawabata's (1956) *Yukiguni (Snow Country)* in Chapter 6.

But our focus here is on *material spaces*. The ways in which national identities are constructed through material spaces can vary enormously, even if they appear similar on the face of it. We focus here on two statues, both in prominent public places in capital cities. Compare Photographs 8.1 and 8.2. Photograph 8.1 shows Winston Churchill, who was the Prime Minister of the UK during the later years of the Second World War. He is a celebrated figure in English history for his part in securing the victory of the allied troops during that war. While the statue has, like Trafalgar Square and Buckingham Palace, been subject to countless disfigurements and satirical manipulations in the media, it is nevertheless one of many material (i.e. physical) symbols of Britain's national identity. In this case, the statue is a material symbol of a relatively recent event (the Second World War) that, to some extent, binds together the collective identities of a country's populace.

Photograph 8.2, on the other hand, shows a monument erected by the Prime Minister of Turkmenistan – Saparmurat Niyazov – who was in power from 1990 to 2006. He was criticised outside the country for being one of the most repressive, totalitarian and, famously, egotistical leaders in modern world history. For instance, during his rule, he changed words in the national anthem to include his own name, and renamed schools, hospitals and even a meteorite after himself. He also engaged in some spectacular building projects – such as the 75-metre-tall 'Neutrality Arch', which featured a 12-metre gold statue of Niyazov that rotated to track the sun's path (Photograph 8.2), and

Photograph 8.1 Statue of Winston Churchill in London.
Source: Alamy Images/Peter Greenhalgh.

which was placed in a prominent public square in the Turkmen capital, Ashgabat. Through these projects, and a set of related innovations (such as a new alphabet and public holidays), Niyazov constructed a particularly top-down kind of national identity. But his statue was viewed by many as the ultimate material symbol of a dictatorial notion of national identity, not based upon popular or collective accord (in distinction to the British example). This was, indeed, perhaps the most extreme example of the social construction of national identity in living memory, and material spaces played a key role therein.

Of course, the relative moral worth of any national identity – and its remnants – is always up for debate. We are not making the point that the English version of national identity is somehow 'better' or less problematic than the Turkmen version (remember that England has a long history of colonialism and exploitation, which also forms an inextricable part of what it means to be English). Rather, we are observing that a seemingly simple material object (a statue) positioned in a prominent public place (a capital city) can be embroiled in some very *different* constructions of national identity, depending on the time and place (see 'Social categories are place- and time-specific' in Section 8.2 above). Had the allies not won the Second World War, then the world might not look upon statues of Churchill in the same way (indeed, the statue might never have been erected). Indeed, it may be that English national identity, and its material remnants, would have been viewed with the same suspicion as that of the Nazi party and its architectural projects

Photograph 8.2 The Neutrality Arch in Ashgabat, Turkmenistan.
Source: Alamy Images/Guillem Lopez.

following the Second World War (for more on the latter, see Chapter 2). The construction of national identity is always something that is contested (after all, it involves so many people!) and contextualised by both history *and* geography. Incidentally, as if to reinforce this point, the statue in Photograph 8.2 was removed in 2010, as the new president sought to remove all traces of Niyazov's 'cult of personality' from Turkmenistan's urban fabric, in the search for a more benign version of national identity. To read about a very different – and less obviously material – expression of national identity, please turn to Box 8.4.

The last examples show how identities are constructed not simply via language or narrative but *also* through the (sometimes unthinkingly prejudiced) ways in which institutions, governments and societies all over the world design, use and control everyday spaces. This reinforces the point that the social construction of identity is a multifaceted process. Moreover, this reminds us again of the *spatiality* of identity – as Keith and Pile (1993) suggest, the many ways in which people both express and perceive their identities as a result of interaction with different places.

8.5 Relational identities

The preceding discussion about social construction focused on the *construction* side of the equation. On the other side – but intimately linked – is the *social*. By this, we mean that, in the vast majority of cases, identities are far from free-floating, bounded things. Instead, they are relational (Jackson 2005). In this section, we begin by defining what is meant by the term 'relational identities'. We then focus on two very different areas of cultural geography where relational identities matter: feminist geographies and geographies of the Internet.

To understand what we mean by relational identities, we have to understand the notion of the 'self–other dualism'. The argument goes that, when humans construct their identities, they are always constructing their 'self'. The notion of identity – at least in contemporary cultures – implies an image, narrative or legal definition of the self. But the self is not a free-floating entity. You do not – we assume – live in an impenetrable bubble. We assume that you interact with people on a daily basis. One of the most important elements of this interaction is a constant process of comparison. As Crang (1998: 61) argues, "identity can be defined as much by what we are *not* as by who we *are*." People constantly compare themselves with others. People look at other people – both those who are like them, and those who are different – and, often subconsciously, ask questions. They might consider whether they want to emulate a person's hairstyle, music choice, political affiliation or moral standing on a subject. Or they might consider whether they want to deliberately distance themselves – literally or metaphorically – from someone they perceive as 'not like me'. We return to this process of 'othering' a little later on in this section.

Many theorists argue that the relational process of identity formation begins when a child is very young. Many of these ideas are influenced by the famous psychoanalyst Sigmund Freud, whose ideas have, more latterly, been revised by Jacques Lacan. For Lacan, the "individual subject [the young child] establishes a sense of self through visual identification with its image in a mirror" (Elliott 2001: 53). For the first time, Lacan argues, this provides a child with a sense of *self* and, moreover, what Doel (1994) argues is an illusory sense that their body has a unique wholeness or unity. This means that the image of the self is just that – a delusion. For Lacan, this delusion becomes the very basis for all our social relationships and, therefore, our perception of ourselves in comparison with others (Elliott 2001).

We shall not explore psychoanalytical ideas in any more depth here (but if you want to know more, see Elliott 2001 and Pile 1996). This is because the majority of geographers working on relational notions of identity tend to emphasise something else: the importance of 'difference'. We turn to this idea next, through an examination of feminist geographies.

Feminist geographies and geographies of difference

Feminist geographers (Box 8.3) have been instrumental in informing cultural geographies of difference. Rose (1993), for instance, argues that until the 1990s the vast majority of academic geographers were male, white, probably heterosexual, and almost certainly from middle-class, educated backgrounds. For Rose, this brought two problems. First, that this was the assumed 'norm' from which academic geographers were writing. Therefore the voice of the geographer was predominantly male, white, etc. – it was remarkably narrow and homogeneous. Second, this also meant that the kinds of research being done in human geography (with relatively few exceptions) made little attempt to consider *other* social groups – especially the varied experiences of place felt by women, minority ethnic groups, people with diverse disabilities, children and gays, lesbians or bisexuals. During the 1990s there were several calls to acknowledge social difference in geography, and today geographers continue to press for an acknowledgement of diversity between and *among* identity groups.

An important implication of work on relational identities and, especially, difference is that interactions between people can lead to forms of exclusion and marginalisation. If you look again at some of the examples above (especially Boxes 8.2 and 8.3), you will see that social and spatial exclusion often occur when different identity groups interact. This is because dominant groups often engage in processes of marginalisation in order to continuously reassert their dominance. This process – known as 'othering' – happens when people from a relatively similar group encounter someone who falls outside their assumed 'norm'. That outsider or 'other', as they are often called (Crang 1998), poses a challenge to people who consider themselves 'insiders'. That challenge – that 'otherness' – might come from having a different appearance (skin colour or dress), behaviour (bodily gestures or language) or beliefs (religious or moral standpoints).

Box 8.3

Feminist geographies; gender in public and domestic spaces

Feminist geographies have been instrumental in foregrounding different ways of knowing the world that go beyond male, white, middle-class forms of knowledge. This is because some of the most important work in cultural geography - indeed in geography as a whole - has explored how gendered expectations permeate social spaces (Rose 1993; Massey 1994; WGSG 1997). Feminist research takes a relational understanding of identity because it seeks to illuminate how men have constructed (and thereby often dominated) women. It identifies gender as a social construction, but suggests that men make assumptions about women's roles, sexuality, skills and knowledge - typically in order to reaffirm their own identities. Once again this is a process of 'othering' that is designed to provide a more powerful group (in this case men) with an inflated image of the self.

Here, we take a brief look at some examples of feminist research (and research about gender) in geography. Many feminists have shown how women's 'natural' sphere is assumed to be the home, in distinction to the public sphere, which has been dominated by (and thereby is assumed to be 'for') men. This is a myth that is perpetuated in many ways. Domosh (1989) shows how, in late nineteenth-century New York City, particular streets were 'feminised' - shops displayed clothes and jewellery. In this way, New Yorkers (read 'men') were bringing the 'private' into the public realm, and making the presence of women more acceptable. Elsewhere, a range of work on India (Blunt 1999; Gowans 2003) has explored how the home is a key locus for debates about femininity and nationalism in a postcolonial context. For instance, Legg (2003) argues that, at the same time as being confined to their homes by cultural expectations, women have managed to participate in Indian nationalist movements by supporting public processions from their homes. Finally, geographers have studied women's fear of public spaces, arguing that assumptions about public space as 'masculine' (while private, domestic spaces are 'feminine') underlie the many anxieties that women have about going out alone or at night (Valentine 1989).

Arguably, all this work is relational, because it focuses on the forms of exclusion, domination and resistance that are at the heart of relationships between women and men in many geographical contexts. We shall revisit this argument in the last section of this chapter, when we discuss performativity.

Cultural geographers have shown a great deal of interest in processes that cause the exclusion or marginalisation of 'others'. Most importantly, geographers have considered the importance of place in helping societies define what is 'appropriate' and what is 'inappropriate'. In Cresswell's (1996) seminal text, it is argued that certain identities and their related behaviours are understood to be 'out of place'. This is a term we often use to describe something a little odd or uncomfortable to our ordinary ways of doing things. Cresswell's study is important, though, because it highlights a range of examples of identity groups whose actions transgress (undermine or question) the norms of those who are 'in place'. Other work on exclusion has, for instance, looked at how planning laws and societal prejudice have reinforced the position of Gypsy, Romany and Traveller communities as 'out of place' in British society (Sibley 1995; Holloway 2003). Again, take a look back at some examples in this chapter and consider the ways in which 'other' groups are defined as (and sometimes *forced* to be) 'out of place' (see also the example of young people in shopping malls in Chapter 3). We look at the transgressive or subversive acts that 'out of place' groups use to (try to) assert their identities, and to deal with exclusion, in the last part of this chapter.

Geographies of the Internet

Since the late 1990s, many cultural geographers have engaged (albeit sporadically) in research on the Internet. It has to be observed from the outset that this work has by no means kept up with the incredible pace, complexity and (increasingly) geographical reach of the World Wide Web. Curiously, when compared with other areas of cultural geography, cultural geographies of the Internet remain quite small in number. An early collection of essays (Crang *et al.* 1999) sought to explore how various communication technologies (including telephones and the Internet) have recast social expectations, changed perceptions of distance ('compressing' time and space, as Harvey 1989 has it), and provided new opportunities for the kinds of social relations that are, as we have argued in this section, so critical to identities.

Since then, a range of studies has sought to cut through some of the more sensationalist predictions surrounding cyberspace (Pickerill 2003) to explore whether and how the Internet *is* changing the ways in which social relations, social spaces and identities are produced. Notably, geographers' research about people's actual *use* of the Internet has tended to show that technology does *not* always radically alter the ways in which people relate to one another. A great example of this (by a non-geographer) is Hodkinson's (2007) research on the online social forums used by members of the 'goth' **subculture**. In it, he argues that goths simply use forums to prop up their existing social relationships, and as more-or-less 'enclosed' spaces in which they can express the identities that they would also display in face-to-face group settings. Similarly, important early research on gender identities in the school classroom showed that pervasive (heterosexual) gender norms tended to be reinforced in girls' and boys' usage of the Internet. In particular (at least in the early 2000s), the teaching of information technology (IT) in UK schools tended to favour boys (Holloway *et al.* 2000). Thus, despite attempts to equalise access to IT through targeted schools policies, such policies actually tended to reinforce gendered assumptions about boys' aptitude with and interest in computers, programming and software (especially games).

Sticking with young people – this time, undergraduate students – it was found that, in 2009, around 95% of UK students used social networking sites (such as, at the time, Facebook and BeBo; Madge *et al.* 2009). This will probably come as little surprise to you. They also found that students tended to use sites like Facebook more to bolster their social networks than to bolster their learning, although more recent studies have shown that social networking sites can be important resources for students' learning, too. However, a significant part of this study – which differs from Hodkinson's (2007) work – was that Madge *et al.* (2009) talked with students about their use of Facebook *before* they arrived at university. In other words, they explored the impact of social networking sites on subsequent friendships and identities (rather than looking at how such sites supported already-existing relationships). Over three-quarters of students in their study were Facebook users before coming to university, with the most frequent uses being to make new friends before coming to university, and to get in touch with students who would be living in the same halls of residence. At the same time, students also used Facebook for keeping in touch with existing friends from home.

Thus, as Madge *et al.* (2009) argue, this is a complex picture of Internet usage and social relationships. On the one hand, the traditional experience of arriving at university and trying to make friends in the first few weeks has shifted earlier and, in part, online. On the other hand, Facebook means that students keep in touch with existing friends so that the transition to university (and a 'student' identity) is, arguably, *less* of a break for people's social relationships (and identities) than it was before. Of course, as you may know yourself, uses of Facebook and other social networking sites vary widely, so you may like to think about how your experiences relate to the findings of this particular study. To read a very different case study about young people and the Internet – focused on Chinese national identities – please turn to Box 8.4.

It has also been argued that the Internet may herald new possibilities for social movements and activist identities (for more on activism, see Chapter 3). Manuel Castells (2000) suggested that the Internet would do three things for social movements and

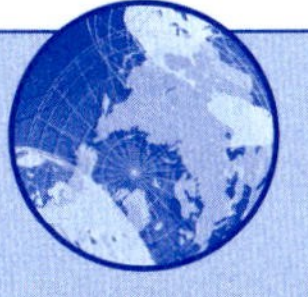

Box 8.4

Young people, Internet usage and Chinese national identities

There have been many debates about young people's uses of technology, and especially the Internet. Some arguments – based on robust research – have picked out some of the nuances in how young people actually use the Internet in the course of their lives and identity construction (e.g. Holloway *et al.* 2000; Hodkinson 2007; Madge *et al.* 2009). Other commentators are a little more sensationalist – for instance, Sue Palmer's (2007) book blames, in part, Internet technologies for creating *toxic childhoods* that are literally and metaphorically 'poisoning' young people. Similarly, you may remember that during 2011 there were riots and uprisings in several countries – including the so-called 'Arab Spring' and the summer riots in the UK. Again, young people – and their use of social networking sites – were blamed for instigating these riots, even though the reasons behind them were often complex.

Notwithstanding these differences, it is also notable that many of these debates are located in and around Anglophone contexts, such as the UK and USA. Inevitably, this means that we hear about quite specific kinds of identity and social relationship among young people. This raises the question of how people – especially young people – use and create their identities online in other contexts.

Research by Liu (2012) explored Chinese university students' use of the Internet. The article begins by noting that – as in many other geographical contexts – Chinese youth are characterised as politically indifferent. Yet, looking at their social networks online, this myth was exploded. Rather, it transpired that university students were engaged in a "new politics" of "nationalist passion" (Liu 2012: 53). For instance, following anti-Chinese comments from outside China, surrounding the Olympic Games and China's policy on Tibet in 2008, there was an upsurge of nationalism among the populace. Significantly, because the Chinese government tends to quash displays of political passion, the Internet became a key outlet for such revitalised expressions of Chinese identity. This was all the more significant because a large proportion of users of the Internet in 2008 were educated, young Chinese (often students).

In this way, following negative representations of China by the news network CNN, many students were motivated to go online. As two students put it:

> I was so angry. The national dignity was insulted by CNN, Dalai and the West. The West represented by the US is always afraid of China becoming strong. They would grasp every chance to disturb us . . . We Chinese people have suffered too much in the past. It is high time that China became powerful. I sent out many postings in the bulletin boards to cheer up the Chinese people.
>
> (Student, quoted in Liu 2012: 60)

> Chinese people have not been so united [*tuanjie*] for many years. People have been divided since the reform started. All have been bent on their private interests in the competitive market economy . . . We should thank the anti-China forces for making us so united. The more they are against us, the more united we are. This has greatly encouraged us

Box 8.4 continued

Chinese. They should study Chinese history before claiming that Tibet is a country. How ridiculous!.

(Student, quoted in Liu 2012: 61)

These examples go to show that the Internet can be a key element in identity construction, in some very particular ways. Unlike in other studies, it is particularly important to point out that the Internet is being used here in relation to *national* identities, rather than those that have been the more frequent interest of Anglophone scholars (**subcultures**; see Chapter 3). Arguably, the mobilisations of such new political passions would have been difficult in the Chinese context *without* social networking sites. Yet, at the same time, the Internet also acted as a mere outlet for expressions of anger (and nationalist identity) that many young people already felt. Therefore we can at best say that the relationship between identity and the Internet is both complex and recursive. In other words, the Internet has no causal power over the kinds of identity we create or the social *relationships* that support them (Pickerill 2003) – but it can be an important tool or cipher for those processes.

identities: it would help diverse groups of people, previously separated by time or space, to mobilise around particular cultural values (say, an alternative religion); it would provide, unlike 'real' spaces, the opportunity for genuinely non-hierarchical modes of social organisation, which are an important feature of autonomous social movements; and it would allow people to remain rooted in their local (or national) culture, but at the same time to operate at the global level. In this way, social scientists (such as Featherstone 2008) show how social movements and protest networks can be translocal, taking place across and linking people at different spatial scales (see Chapter 13 for more on scale), making them all the more effective.

However, Pickerill's (2003) important study of environmental activists' use of the Internet showed that new technologies were making less of a difference to activists than Castells had predicted. In particular, she argues that many environmental activist groups (such as Friends of the Earth and Earth First!) were *already* organised non-hierarchically, and *already* invested in the idea of 'thinking globally but acting locally'. In this way, Internet technologies and social media simply enabled those already-existing social relationships to take place in more, denser, or other ways. From this observation, she also argues that the implications for activists' *identities* were far from straightforward. On the one hand, she found that, among some activist groups, online forums were key places for the expression of solidarity, and the working-out of collective identities around particular environmental-cultural values (as per Castells' argument). However, for larger, more established groups (like Friends of the Earth), the impact on members' identities was minimal – instead, the Internet was a simple conduit for making their work more efficient.

As a way of summing up this part of the chapter, it is helpful to look at a recent paper on the geographies of age. In it, Hopkins and Pain (2007) clarify some of the gains to be made from relational geographies of identity and difference. Two important concerns stand out. As you will see, geographers are a long way from providing definitive statements about either of these concerns!

- Geographical work on difference has tended to focus on social groups at the *margins* of society. The argument is that while, in the past, geographers tended to *assume* the position of the white, middle-class, middle-aged, heterosexual male, today, research critically explores *other* groups in order to foreground their experiences. This is a good thing, as this chapter has tried to argue. But the problem is that the 'centre' remains immune from critical geographical study, apart from a few exceptions – such as work on masculinity (McDowell 2003). Thus "geographers have still to break out of the tradition of fetishising the margins and ignoring the centre" (Hopkins and Pain 2007: 287). Therefore the centre remains just that, as it escapes detailed investigation. Thus a relational approach would involve studying the centre *and* the margins of society – but there remains much work to do in terms of this study.
- One way round the problem of the 'centre' is to undertake *intergenerational* and *intersectional* studies

(Hopkins and Pain 2007: 288–289). Such work "refers to the relations and interactions between . . . groups" (Hopkins and Pain 2007: 288). For instance, age-based identities are not simply a result of an individual person's 'youthfulness' (or not). Rather, our understandings of youthfulness are always comparative. Furthermore, interaction *between* people of different generations tends to highlight differences (and similarities) in terms of expectations, behaviour, dress and taste. Those differences – those characteristics that really *make* an identity – are only ever apparent in those interactions or relations. Yet – as with the previous point – there remains much work to be done on the kinds of relations that constitute identities – whether in terms of age, sexuality, class or ethnicity.

8.6 The performativity of identity

If you return briefly to the Australia case study (Box 8.1), you will see that we suggested that one interpretation of Australia's immigration policy centred upon the performativity of identity. In this final section of the chapter, we want to expand on this idea. In so doing, we want to put forward four arguments about identity that may seem quite obvious to you if you reflect on your *own* identity. First, that the way we *present ourselves* (how we dress, our manners and habits) is a key part of how we construct our identity in relation to others – and in relation to common social norms (see also Chapter 12 on **bodies**). Second, that the social construction of identity (which we discussed earlier) is achieved precisely through these kinds of *performative* act. Third, that acting and experimenting may be important ways in which (especially) marginalised groups can make spaces for their identities in relation to those more accepted by dominant groups. Fourth, that assumptions about **spaces** – and how we act in them – help to 'code' what is generally deemed 'acceptable' (or not).

First, then, we will reflect on how the way we *act* matters to our identities. As noted in Chapter 7, the sociologist Erving Goffman provided an insightful, explanation of this process, focusing on the roles that people play in different contexts, and arguing that in different places people act differently in order to create different impressions upon others. This is, then, the first part of an explanation of performativity (although Goffman did not actually call it this).

Second, we want to underline how the social construction of identity is achieved through the kinds of performance and gesture hinted at above. Although feminist philosopher Judith Butler is highly critical of Goffman's use of essentialist categories, she is particularly noteworthy because she developed the notion of performativity. Her work develops a particular way of understanding the social construction of identity, focusing on gender. As Gregson and Rose (2000: 437) put it, "Butler's project is to disrupt dominant understandings of sex, gender and sexuality, which assume that there are two bodies [male and female], two genders, and that heterosexuality is the inevitable relation between them." Rather than being essential, Butler's understanding is that gender identities are fabrications that are only kept alive through discourse (to read more on **discourse**, see Chapters 2 and 6). By discourse, Butler is referring to the language and moral codes that most commonly – but *not universally or naturally* – make up expectations about what each gender should consist of, and especially what their bodies should look like (to read more about Judith Butler on **bodies**, please turn to Chapter 12). These fabrications may be kept alive for many reasons, and may preserve or subvert the dominant make-up of society and the position of power men hold over women (Photograph 8.3). This should now be a familiar argument about social construction (Box 8.3), although it is a specific version of this idea, which focuses on what bodies do, how they look, and, especially, how they are created by discourses.

Butler suggests that the example of 'drag' (a form of cross-dressing), destabilises or exposes gender discourses as mere fabrications. In effect, any act, gesture or talk is gendered (which may be most of the kinds of act Goffman writes of) – and therefore is stylised. It is almost as if *all* one is doing when one is acting along gender 'norms' is *actually* performing a caricature *of* those gender norms. Butler's contribution, then, is, first, to expose gender as a social construction, and, second, to show how particular kinds of performance may reinforce or undermine the *seemingly* stable

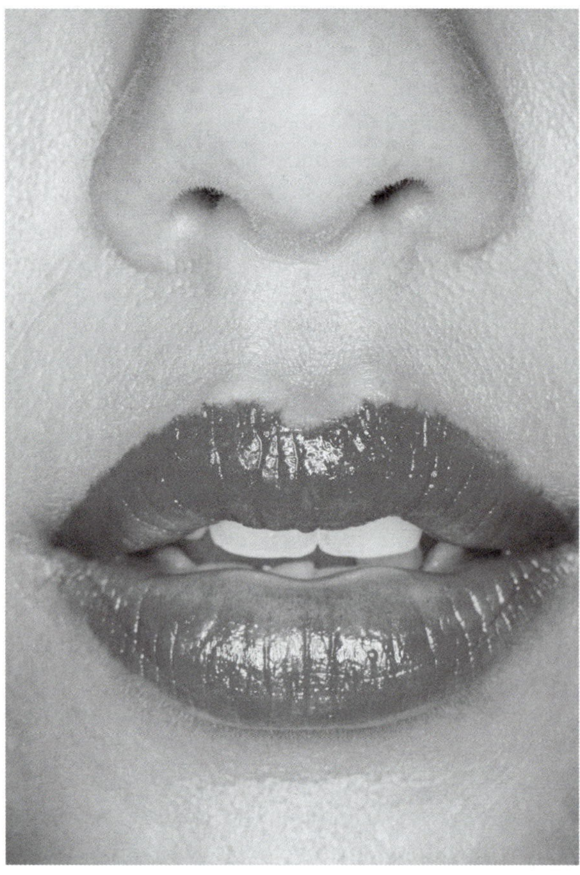

Photograph 8.3 A woman wearing make-up. This image could be read as: a choice to look more 'beautiful' (but who defines beauty?); wearing make-up to appeal to male expectations about femininity; or, a 'deviant art' self-portrait of a self-defined 'lipstick lesbian' who exhibits excessively feminine attributes as part of her lesbian identity (and may in fact be parodying male expectations about a woman's sexuality).

Source: Alamy Images/Corbis Super RF.

foundations of those constructions. Butler's work has been criticised for using few real-world examples, and for its apparent disregard for the importance of space (Thrift and Dewsbury 2000). Nevertheless, several cultural geographers have developed Butler's work in empirical studies: please take a look at Box 8.5 to see some examples.

Third, as Box 8.5 also highlights, it has been the project of many cultural geographers to show how (often marginalised) groups may cultivate particular performances in order to subvert or challenge dominant ideas. The range of work in this area is enormous, as many cultural geographers have a keen interest in criticising the kinds of marginalisation and exclusion that dominant identity discourses (like gender) can produce. In Chapter 3 we highlighted the performative **consumption** practices of **subcultures**: i.e. "groups of people that are in some way represented as non-normative and/or marginal through their particular interests and practices, through what they are, what they do, and where they do it" (Gelder 2005: 1). We also began to explore the performance of activism. Activism can take many forms and involve demonstrations, rallies, forms of direct action (like sabotage) and events such as festivals or temporary camps (Brown and Pickerill 2009). For instance, Brown (2007) explores radical forms of queer activism in marches and activist networks. He shows how 'mainstream' gay rights activism has become infused with commercial interests and those of the professional middle class. In order to parody the assumed performativity of class in London's (now mainstream) 'Gay Pride' march, which is now held in a ticket-only, fenced-off space, activists

Box 8.5
Performativity in cultural geography

The concept of performativity highlights how identity – including gender and sexuality – is created through practices. Often, those performances are stylised acts that are repeated to reinforce norms and social codes (Butler 1990b). Butler's work has been taken up by geographers in a number of ways; we want to highlight just two in this box.

First, several geographers have developed her work to highlight the role of space in performing gender. For instance, Longhurst (2000) explores a bikini contest held for pregnant women in a public space in Wellington, New Zealand. She suggests that society expects pregnant women to act in certain ways: to be demure, to be modest, and to 'cover up'. As Butler argues, these kinds of act get reinforced over time so that they appear natural. Longhurst shows how the bikini contest momentarily involved the near-naked exposure of pregnant women's bodies in public spaces around Wellington. By behaving 'with attitude' – flaunting their bodies, without 'shame', in public spaces – these pregnant women momentarily destabilised the expectations that society has of pregnant women's behaviour.

A second line of research has focused on the performative geographies of work. Turner and Manderson (2007), for instance, show how elite groups of Canadian law students 'play at' being lawyers during social events organised by big law firms to promote their brands. Hence, although not actually 'at work', these students begin to learn the rituals, talk and social networking skills necessary to become high-flying lawyers – and to gain acceptability in the profession when they do enter a firm.

In a very different context, Blumen (2007) explores the routine of commuting to work in Israel. She explores the distinction between those who go to work and those who do not. In terms of the former (those who work), she shows how the routine of going to work is performed through urban landscapes – through public transport systems, routes, routines, and the 'proper' clothing that marks people out as ready for work. In terms of the latter (those who do not work), she focuses on ultra-Orthodox Jewish men who undertake unpaid, religious study. Their striking appearance (wearing old-fashioned dark clothes) and the very fact that they are not commuting for paid work make them appear as 'others' in Israeli public spaces, especially during morning and evening commutes. Emphasising the juxtaposition of the two groups in public space, Blumen shows how two contrasting notions of 'Jewish' identity ('modern' and 'ultra-Orthodox') are in play at the same time. Hence landscapes of work and alternative performances within them are key sites at which struggles over Jewish identity are played out.

created a spoof 'VIP enclosure' behind a small, rickety fence with fake 'security' guards. Elsewhere, cultural geographers have explored creative artistic practices in the city – from walking tours to performance art (see Chapter 2 on producing cultural spaces and Chapter 7 on performance art and walking in everyday spaces). It is argued that such practices may allow their participants to imagine alternative 'visions' to the dominant (often capitalist) way of organising urban spaces.

Fourth, many geographers have argued that spaces tend to be 'coded' in certain ways, so as to make particular kinds of performances acceptable, while others are not. 'Queer theorists' – writing predominantly about sexuality – have been at the forefront of these debates in cultural geography. Early geographical work on sexuality sought to understand the experiences of people who identified themselves as gay or lesbian in spaces that tend to be coded as 'heterosexual' (Oswin 2008). In Cresswell's (1996) terms, they were made to feel like bodies 'out of place'. Thus geographers focused on how spaces are never essentially heterosexual in nature – there is no such thing as a pre-given or natural set of assumptions about sexuality that is written into space before it is inhabited. These geographers thus critiqued the idea that spaces – just like identities – are pre-given and natural. See the critique of the essentialist approach earlier in this chapter.

Instead, as Bell and Valentine (1995: 18, original emphasis) argue, many contemporary social spaces "have been *produced* as ambiently heterosexual" – through, for instance, the policing of 'appropriate' acts of affection in public spaces, through adverts that are predominantly heterosexual in nature, and so on. This is an example of the production of **space**, which we also look at in detail in Chapter 13. These insights have led cultural geographers to explore how diverse subcultural groups try to carve out spaces in which they can (more) comfortably express their identities. Some, like Brown (2007), in the example mentioned previously, explore how temporary activist events provide one way to occupy, resist and rupture heterosexual spaces (and broad, societal expectations about normative kinds of homosexuality) by occupying them or marching through them. Others explore the experiences of gay, lesbian or bisexual people in 'separatist' spaces – such as clubs, bars and private venues.

For instance, Browne's (2009) study of the Michigan Womyn's Music Festival (in the USA) looks at how its location in a rural environment afforded an almost utopian sense of possibility for its participants. Specifically, owing to its *rural* location, the festival allowed lesbian women the opportunity to experiment with expressions of identity that they could not in 'everyday' (ambiently heterosexual) *urban* spaces, such as a city street. However, Browne argues, the rural location of the festival did not allow some women to escape normative notions of womanhood completely, because its entry policy allows only 'womyn-born-womyn' (i.e. not transgendered women, born as 'men') to attend.

Browne's final point speaks again of the importance of attending to the complex, dynamic and intersectional nature of identities as they are performed. This is something that we considered at the end of Section 8.5. In the example of the Michigan Womyn's Music Festival this is important, because seemingly singular identity categories (like 'woman') are in fact complex and contested, and often mean very different things to different people – not least because they are ongoing social constructions. This is perhaps all the more striking in Browne's study (and that is her point), because she looks at what is meant to be a separatist space for one identity category – 'women' – but which is actually characterised by considerable uncertainty about what a 'woman' is or should be, especially in ongoing debates about the Festival's entry policy.

Considering intersectionality more broadly, cultural geographers have tended to focus on particular *kinds* of sexual identity. As Brown (2009) points out, most studies of gay men have tended to focus on relatively normative gay identities, to the detriment of a series of diverse gay spaces and economies (which his work charts). And, as Brown (2012) notes, cultural geographers have explored intersections between sexuality and race in far more detail than they have between sexuality and socio-economic class.

8.7 Concluding points

There remains, then, much to be done in exploring how identities are performed – not only in research on sexualities, which we discussed at the end of the last section, but also in respect of many kinds of identity. Thus, the critiques of Brown (2009, 2012) are actually valid for most of the identity groups whom cultural geographers choose to study.

This should be a prompt for your own reflection about the cultural geographies of identity. Perhaps you can think of different examples of where people use their identities to *resist* dominant laws or moral codes in particular spaces. Perhaps you have been *involved* in or witnessed a rally to promote the rights of a particular social group; or perhaps you or someone you know listen to a style of music that promotes an *alternative* political message. Perhaps you try to *avoid* any attempts to challenge or subvert globalisation, consumer capitalism, or any other dominant cultural-economic forces. Or perhaps, in thinking about your own identity (as we encouraged you to do at the beginning of this chapter), you are caught up in *concerns* about how different forms and performances of identity intersect with one another in ways that we have not even touched upon in this chapter.

Yet the key point here is that – even if we feel we are simply going along with the flow, that our identity mirrors that of the majority – our identity is never fixed, and requires various forms of performance to keep it (and us) alive. Once again, as Keith and Pile (1993) suggest, space and spatiality matter, because, without them – at least for some phenomenological philosophers – human action and therefore human identity would be impossible (see Chapter 13). This point might sound like a bit of a truism, but we have outlined in this chapter that – through essentialist, social constructivist, relational and performative lenses – cultural geographers have shown how identity performances always take place *in spaces*, and very often help to *make places*. Perhaps dictators and national governments, subcultures and activist groups provide more overt examples of place-making as they create knowingly 'mainstream' or 'alternative' spaces. Yet, as Laurier and Philo (2006a) show, even the 'smallest' expressions of identity are the very stuff of social and spatial relationships (see also **Chapter 7, on performances of everyday life**). Those expressions of identity can be the very things that keep spaces like cafes, commuter routes and public spaces ticking over – or, indeed, that can disrupt them. Identity is not the be-all and end-all of social spaces. But cultural geographers have shown how the very assumptions we make about who or what is 'in' or 'out' of place tend to rely on what *seem* natural but are usually socially constructed, relational, complex, dynamic and performed expressions of identity.

Summary

- Identities are complex and contested. But identity can be properly understood only by looking at the *spatiality* of identity – how society and space are always fused together by people as they create and experience their individual or group identities.
- Essentialist categories (like gender and ethnicity) can be understood as the building blocks of identity. But it is important to remember that they are not 'natural' labels. Cultural geographers are critical of essentialist categories – a simple reason being that use of these categories varies over time and space.
- Rather than being essential, identities are nearly always socially constructed. This means that social categories – such as childhood – are predominantly inventions that fulfil particular social roles. Often, identities are constructed *for* marginal groups *by* dominant groups in order to control, assimilate or even exclude them. Often, exclusion takes spatial forms (e.g. in debates about who is 'in' and who is 'out' of place in public).
- Identities are relational. It is (nearly) impossible to think about an individual's or group's identity without comparing them with another individual or group. Often, experiences and representations of identity are most obvious when groups interact with each other.
- Identities are performative. Identities are made by bodies that talk, move, gesture, commute, wear clothes and play. Often, people's identity performances simply reinforce dominant assumptions about public spaces or gender roles in those spaces.
- However, geographers have become increasingly interested in the ways that people use creative performances to make alternative spaces where they can resist or subvert dominant political forces.

Some key readings

Cloke, P., Crang, P. and Goodwin, M. (2005) *Introducing Human Geographies*, Hodder Arnold, London.

The chapters by Peter Jackson and Sarah Holloway provide accessible introductions to identity. This book also expands on some of the key terms used above – for instance, there are chapters on 'self and other' and 'masculinity and femininity'.

Duncan, J., Johnson, N. and Schein, R. (eds) (2004) *A Companion to Cultural Geography*, Blackwell, Oxford.

The chapters in Part IV (on culture and identity) provide more detailed and theoretically complex discussions of different kinds of identity, including nationalism, race, class and sexuality. They all have detailed bibliographies that will help you to direct your reading if you want to explore any of these identities in greater depth.

Sarup, M. (1996) *Identity, Culture and the Postmodern World*, Edinburgh University Press, Edinburgh.

Not strictly a cultural geography book – but a readable introduction to many of the key terms used in this chapter. Moreover, Sarup is especially interested in the ways in which identities are made and experienced geographically (especially in Chapters 1, 2, 10 and 12). Sarup also introduces notions of identity not explored in this chapter.

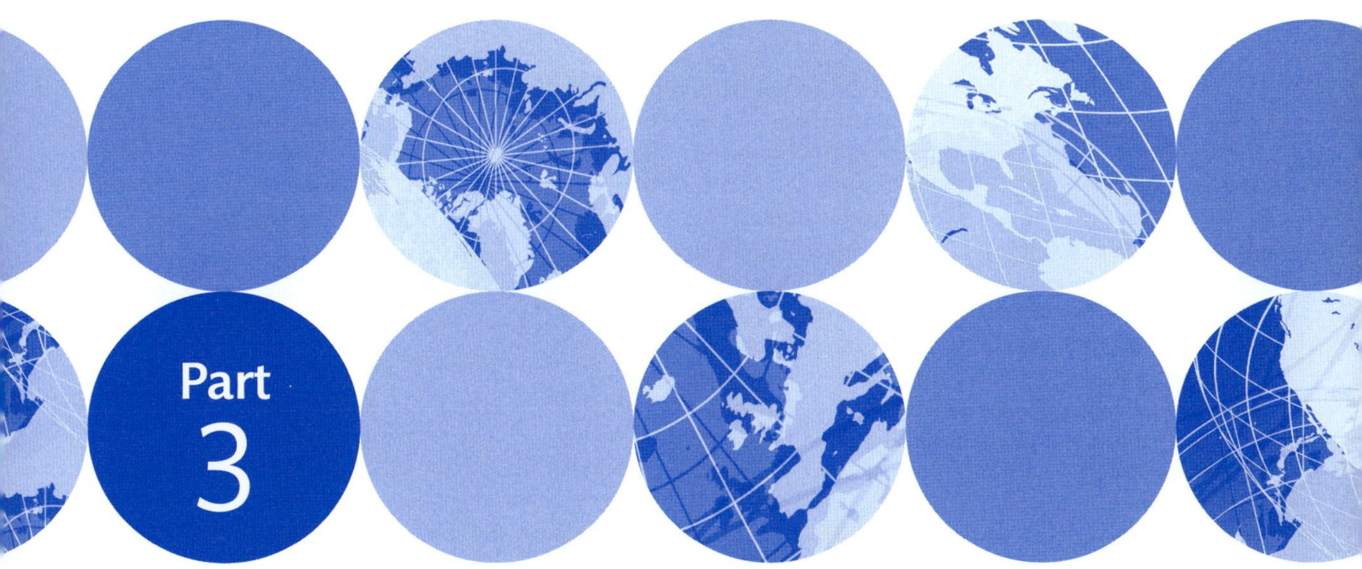

Key concepts for cultural geographers

Part map

In Chapter 1 we identified three major ways in which the work of cultural geographers is important:

- researching cultural processes and their complex geographies and politics;
- exploring the importance of cultural materials, media, texts and representations in all kinds of space and geographical context;
- engaging with new lines of social and cultural theory, and reflecting upon their importance for human geographers.

Part 3 deals with the third of these themes. As we noted in Chapter 1, cultural geographers have often produced some of the most conceptually adventurous work within the discipline of human geography. Their work has often involved exploration of, and engagement with, the contemporary avant-garde of social and cultural theory. Chapters 9–13 introduce some key groups of concepts that have received considerable attention from many cultural geographers, and which we argue should be considered deeply important for understanding the kinds of cultural process, politics and phenomenon that we have dealt with in Parts 1 and 2. The following chapters can be read as stand-alone introductions to each concept: certainly, we hope that these chapters will give you some 'ways in' to understanding each concept. More than this, however, the chapters could be understood, collectively, as a kind of conceptual terrain: a series of interrelated concepts that require us to pay serious attention to the complex, everyday, material, emotional-affective, bodily and spatial characteristics of *all* cultural geographies. It will probably become clear that a particular group of concepts – derived from a selection of contemporary non-representational, feminist and actor-network theories – have been deeply influential to our own work. However, it is not our intention to drum up support for these concepts; instead, we invite you to reflect carefully and critically upon each group of concepts, and work out where you stand in relation to each of the ideas under discussion in the following chapters.

Chapter 9 explores the importance of **everyday** spaces, practices and events for the work of cultural geographers.

We particularly highlight the challenges of representing and writing about **everydayness**, and introduce a body of geographical work, known as **non-representational theories**, that has reflected upon this challenge. Chapter 10 recognises the importance of **material objects** for cultural geographers. We introduce several influential ways in which material objects have been approached, via concepts of **Marxian materialism**, **material culture** and **actor-network theory**. Here, we also recommend the work of geographers who have 'followed the thing' as important for any geographers interested in cultural **production** (see Chapter 2), **consumption** (see Chapter 3), and especially the connections between cultural processes (see Section 3.5). Chapters 11 and 12 explore some important geographical work that has recognised that all human geographies are **emotional-affective** and **bodily**. While it can seem obvious that human beings are emotional and have bodies, we note that, historically, human geographers overwhelmingly tended to overlook these characteristics of human life. We then explore work by geographers and social scientists which demands that we acknowledge the bodily and emotional-affective aspects of all cultural processes and geographies. Finally, Chapter 13 explores notions of **space** and **place**. These terms are central to the discipline of human geography, and here we summarise why it is important to explore the **spatiality** of cultural processes and geographies. In effect, then, we reiterate the value of taking a geographical approach to the kinds of cultural issue, process and phenomenon we discussed in Parts 1 and 2. We hope you will find something that sparks your interest and prompts you to explore some of these concepts in more detail. It should become clear that we have a particular enthusiasm and commitment for these ideas; indeed, the conceptually adventurous and experimental spirit of cultural geography is one of the things we love about the subdiscipline.

After reading Part 3, there are a number of activities that could help you to extend and develop your understanding of the key concepts under discussion.

- It is worth noting that many of the concepts discussed in this section are rather controversial. While they are valued by many cultural geographers, other geographers may have very different ideas about their value and usefulness. It would be interesting to ask a range of different geography tutors what they think about, say, emotional and bodily geographies. They are likely to have a strong opinion, one way or another.
- Back in this book's introduction (Box 1.5) we outlined some key principles for approaching cultural geography. After reading the chapters in Part 3 it would be particularly useful to return to two points from Box 1.5. First, work out what you think. Ask yourself: where do I stand in relation to this concept? Develop critiques of your own. But remain open-minded to anything that you feel negative about. Second, reflect on the 'point', the usefulness and the application of these concepts by cultural geographers. Ask yourself: how, exactly, do these concepts extend and enhance our understanding of key issues?

Everyday geographies

Before you read this chapter . . .

- Identify some things that you do every day.
- Identify a place that seems totally unremarkable to you: perhaps somewhere that you pass every day, without really noticing it.
- Think about something that matters a great deal to you, in your everyday life. Try to write down how and why it matters so much.

Chapter Map

9.1 Introduction: waiting . . .

Picture the scene: a bus stop on the edge of town. Seven people and a dog, waiting for a bus that is already 15 minutes late. A windy, rainy evening. The bus stop roof provides some shelter. A man wearing headphones and a Bob Marley t-shirt eats a hot dog. A man with a walking stick checks his watch, and tuts. His dog sits obediently at heel. I check my mobile phone. A woman loses patience with her three children: "I won't tell you again – just *sit down*!" We all have heavy shopping bags, and look weary. Time passes. The wind blows. No bus comes. No-one speaks. Nothing much happens . . .

For two years, one of the authors waited, almost every day, at the bus stop pictured in Photograph 9.1. Nothing much happened there during this time. The many hours spent there were characterised by absent-mindedness and boredom (waiting for a bus that never came), tiredness (*another* day of work), and passing, avoidant social interactions. The wait at the bus stop proceeded in this dull, largely unmemorable manner, day after day after day after day. We can therefore think of this bus stop as an *everyday* **space** (see Chapter 13), characterised by routines, **practices** (see Chapter 12), **emotions** and **affects** (see Chapter 11) that seem mundane, normal in this particular social-historical context. Of course, waiting at a bus stop is just one example of an everyday practice. No doubt you can think of examples from your own experience: perhaps preparing breakfast, getting dressed, cleaning, waiting in a traffic jam, preparing for sleep.

At first glance, everyday spaces and practices may seem somewhat banal, unremarkable and unimportant. They may feel out of place in an academic textbook.

Photograph 9.1 A bus stop in the West Midlands, England. Perhaps you too wait every day, in the same place, for public transport.
Source: Dr Faith Tucker.

However, in this chapter we introduce a turn within cultural geography wherein everyday geographies have become an important focus for much research and theory. We shall explain how one important consequence of the **cultural turn** (see Chapter 1) has been a growing awareness of the significance of everyday spaces for understanding geographical issues – and a growing range of concepts that seek to articulate and make sense of everyday geographies. Indeed, in the remainder of this book we shall argue that cultural geography has been revitalised, in many ways, by geographers' engagements with concepts dealing with **everyday life** (this chapter), everyday **material objects** (Chapter 10), everyday **emotions and affects** (Chapter 11), everyday **bodily practices** (Chapter 12), and everyday **spaces and places** (Chapter 13).

9.2 Acknowledging everyday geographies

The bus stop shown in Photograph 9.1 is near West Bromwich, an industrial town in the English Midlands. In 1994, the geographer Doreen Massey (1994: 163) wrote:

> Amid the Ridley Scott images of world cities, the writing about skyscraper fortresses, the Baudrillard visions of hyperspace . . . most people still live in places like Harlesden or West Brom[wich]. Much of life for many people . . . still consists of waiting in a bus shelter with your shopping for a bus that never comes.

Massey's first sentence caricatures some issues that excited contemporary Anglo-American cultural geographers in the early 1990s: skyscrapers, cinematic representations, postmodern theories of spectacular urbanity (see Chapters 2 and 4). By contrast, she reminds us that 'most people still live in places like Harlesden or West Brom'. In English popular culture, West Bromwich is a kind of by-word for an unglamorous, provincial, 'blue collar' industrial community in decline. Similarly, Harlesden, to the west of London, exemplifies the many tracts of down-at-heel, suburban sprawl that encircle most major British cities. Massey was therefore making the point that, although cultural geographers had paid considerable attention to exciting theorists and spectacular examples of urban geography (see, for example, the discussion of Las Vegas in Chapter 2), they had overwhelmingly neglected places 'like Harlesden or West Brom'. That is, places that were definitely *not* vibrant world cities, that did *not* regularly feature in the work of exciting theorists, and which were *not* noted for post-modern architecture, avant-garde culture, or radical social movements. Moreover, Massey draws attention to what appear to be *especially* mundane aspects of life in such places: people sitting in a bus shelter, shopping bags in hand, waiting for a bus that never comes. In short, it was Massey's contention that these kinds of people and this kind of mundanity, in these kinds of contexts, had often gone unchronicled by cultural geographers.

Massey's reminder that 'most people still live in places like Harlesden or West Brom' should be understood in the context of a broader turn towards 'the everyday' within cultural geography, as well as other disciplines such as sociology, history and cultural studies. As we noted in Chapter 1, this shift in focus was a major characteristic and consequence of the so-called **cultural turn** in human geography and other social scientific disciplines. Within each of these areas of research and theory, it has increasingly been argued that more attention should be paid to 'everyday' places, practices, people and concerns.

Despite widespread talk of a turn towards the everyday within cultural geography, 'the everyday' remains a rather difficult term to define. As a starting point, we can say that cultural geographers' turn towards the everyday has entailed paying closer attention to one or more of the following topics:

- experiences, issues and 'lay discourses' (Jones 1995) of "all manner of everyday (dare I say non-academic?) people in all sorts of everyday circumstances" (Murdoch and Pratt 1993: 434);
- spaces and landscapes that appear to be banal, unspectacular and un-noteworthy;
- 'ordinary' daily practices and routines: "what we all do, what we take part in . . . our everyday processes of making meanings and making sense of the world . . . the routine activities and control of ordinary people as they go about their day-to-day lives" (MacKay 1997: 7);
- commonplace, 'low key' experiences and events: the "milieu where, commuting to work, doing one's job in the home or the factory or the office or the mall, going to the movies, buying groceries, buying records, watching television, making love, having conversations, not having conversations, or making lists of what to do next, people actually [live]" (Marcus 1989: 3);
- all "that which does not get recorded in the History made up of important 'events'" (Attfield 2000: 50), where 'History' (with a capital 'H') denotes the 'official', 'academic' or **hegemonic discourse** (see Chapter 2);
- that which is taken for granted or unsaid within contemporary social and cultural life: "the forms of life we routinely consider unremarkable and thus take for granted" (Chaney 2002: 10);
- the very 'normality', 'taken-for-grantedness', 'all-around-us-ness' of all of the above, which makes 'the everyday' "elusive to pin down, to take stock of, and think through" (Valentine 1999: 348);
- the challenges of all of the above for academic research and theories.

Many geographers have found the work of the French sociologist Henri Lefebvre especially useful in reflecting upon 'the everyday'. In an oft-cited footnote, Lefebvre (1988: 87) suggested that three, related ideas were at work within the word 'everyday':

- *la vie quotidienne* (often translated as 'everyday life');
- *le quotidien* ('the everyday');
- *quotidienneté* ('everydayness').

These ideas are somewhat enigmatic, and have been variously interpreted. The cultural theorist Gregory Seigworth provides one helpful interpretation, which we shall use throughout this chapter. For Seigworth (2000: 245), 'everyday life' refers to "life in its most concrete materiality": that is, everyday happenings, situations and events as perceived and subjectively experienced by us as individuals every day. 'The everyday', by contrast, refers to the ways in which 'everyday life' is depicted and transformed into a concept via contemporary systems of **representation** and **discourse** (see Chapters 2 and 6). So whereas 'everyday life' refers to life as it is lived and experienced, 'the everyday' might be understood as the form that life takes after it is captured and transformed by a system of representation (e.g. language, photography or media text). (Note that, for Lefebvre, this distinction was the starting point for a much broader critique of the transformation of everyday lives under modernity and capitalism.)

Finally, 'everydayness' can be understood as "the processual excess of the first terms" (Seigworth 2000: 245) in two senses. First, everydayness can refer to all of those aspects of 'everyday life' that are lost or overlooked in representations of 'the everyday' (we shall return to this theme in the discussion of the **crisis of representation** later in this chapter, and also in Chapter 11 when we distinguish between **emotion and affect**). Second, simultaneously, everydayness can refer to all of those aspects of lived experience that are overlooked by us in *both* our 'everyday life' and 'the everyday'. This idea of 'excess' is explored in more detail in Section 9.4.

Using these terms, we can describe the theoretical development underlying much recent cultural geography in the following way.

- In the aftermath of the cultural turn, it was increasingly recognised that cultural geography had too often neglected many forms, experiences and aspects of 'everyday life'.
- In the process, as geographers have attended to 'everyday life', it has been increasingly argued that geographers' ways of representing and understanding the world have failed to appreciate the implications of 'everydayness' for academic research and writing.

This chapter explores each of these points in turn. Section 9.3 introduces the breadth of geographers' research about everyday life. In so doing, the argument that everyday life should be considered fundamentally important for cultural geography is outlined. Section 9.4 focuses on the 'excessive' nature of everydayness, and explores some of its implications for research, theory and writing in cultural geography. This latter set of arguments sets the scene for Chapters 10–13.

9.3 Why everyday life matters

Perhaps, at this point, you are wondering why inconsequential details of everyday life (like standing at a bus stop) belong in an academic textbook. After all:

> At first sight, everyday life may not appear to be a subject worthy of academic investigation. When we think about what we do in everyday life, we tend to think of mundane, routine activities (often seen as chores) which we all appear to do. Indeed, because of its routine nature, not many of us think about our everyday lives very much at all.
>
> (Eyles 1989: 102)

The preceding quotation was written by Eyles (1989) in an early, explicit call for human geographers to take more notice of the everyday, unquestioned background to our lives. Many cultural geographers have subsequently attempted to do precisely this, and have thus reiterated Eyles' argument. This section summarises some key examples of cultural geographical work to this end, and introduces some broader social scientific concepts that have been influential in this context. In so doing, the chapter outlines seven reasons why, it is argued, everyday life matters for cultural geographers.

Reason 1: Everyday life is too often overlooked by human geographers

Fundamental to cultural geographers' turn to everyday life is the claim that much of the world has been habitually overlooked in academic research, discourse and understandings. For example, it is now widely argued that approaches and methods conventionally used by many human geographers, and other social scientists, have "neglected the everyday in their enthusiasm to document the exceptional, the new and the exotic" (Holloway and Hubbard 2001: 36). To be sure, human geographers have always concerned themselves with topics that encompass everyday lives: landscape, transport, society, culture, etc. But, crucially, it is now apparent that such work often neglected the everyday lives at the heart of these topics: people dwelling within landscapes at particular moments; the everyday users of transport networks; the everyday lives, stories, emotions of people inhabiting societies and cultures; and so on.

This critique of an earlier tendency to emphasise the exotic or dramatic recurs in most of this book's chapters (recall, for example, Massey's quotation at the beginning of this chapter, or the critique of the **Berkeley School** by **new cultural geographers** in Chapter 1). The following extract provides two further examples of this kind of critique.

> For instance, while there is an extensive geographical literature on shopping as an activity central to people's lives, much of this currently concentrates on a few mega-malls (such as the West Edmonton Mall in Canada), overlooking routine acts such as the weekly grocery trip to Wal-Mart or Safeway, or popping out to a local corner shop for a pint of milk and a packet of cigarettes. Equally, much writing on the geography of leisure and recreation examines the types of activities occurring in spectacular settings and themes parks such as Disneyland . . . rather than the leisure time which people spend watching television or pottering in their garden.
>
> (Holloway and Hubbard 2001: 36)

Thus, as in Massey's quotation at the beginning of this chapter, it is argued that much human geography has been characterised by a multiple, recurrent 'overlooking' of everyday life. In particular, cultural geographers increasingly note that this 'overlooking' has resulted in the neglect of three key aspects of everyday life: everyday spaces, everyday practices and everyday events. Boxes 9.1, 9.2 and 9.3 provide some illustrative examples. Let us consider each example in turn.

In the quotation in Box 9.1, Augé uses the example of waiting to evoke the kinds of space where (whether at an airport terminal, a bus stop or wherever one happens to be) so much of everyday life is spent. Augé (1995: 77–78) suggests that much of life is spent in spaces "which cannot be defined as relational,

Box 9.1
An everyday space: waiting, again

Photograph 9.2 An airport departure lounge.
Source: Shutterstock.com/Claudio Zaccherini.

In *Non-Places* (1995: 2), the anthropologist Marc Augé evokes the range of spaces and activities that might be encountered when one "ha[s] nothing to do but wait". As an example, he imagines one traveller's wait for a flight at Paris Charles de Gaulle airport. As you read the following extract, think about any similar experiences of waiting and boredom from your own life: in particular, think about the characteristics of spaces designed for this kind of waiting.

> He parked in row J of underground level 2, slid his parking ticket into his wallet and hurried to the Air France check-in desks. With some relief he deposited his suitcase (exactly 20 kilos) and handed his flight ticket to the hostess . . . He went through Passport Control to do a little duty-free shopping. He bought a bottle of cognac . . . and a box of cigars . . . he strolled past the window-displays of luxury goods, glancing briefly at their jewellery, clothing and scent bottles, then called at the bookshop where he leafed through a couple of magazines.
>
> (Augé 1995: 2)

Box 9.2

An everyday practice: driving to and from work

Photograph 9.3 A driver stuck in rush hour traffic.
Source: Dr Faith Tucker.

Increasingly, social scientists have begun to explore the cultures, economies and geographies associated with the automobile (see Featherstone *et al.* 2005). Attention is thus drawn to the routine, daily driving practices that are fundamentally part of millions of contemporary lives. For example, Laurier (2003: 4) notes the taken-for-granted bodily driving practices that characterise an experienced commuter's drive to/from work:

> The experienced driver dwelling in their vehicle no longer notices the techniques they have to use to control clutch and accelerator, they no longer look at gaps without sensing whether they are tight or wide for *this* vehicle. Driving . . . is a thoroughly learned and living inhabitation of vehicular equipmental complexes situated in traffic.

The skills involved in the commute, and the taken-for-grantedness of those skills, become evident when one observes less experienced drivers attempting to navigate the same city streets:

> struggling with foot pedals, forgetting to check in both directions before pulling out of a junction, missing the gates of gears and being honked at for being in the wrong filter lane; they are well aware of the array of skills of handlings of controls and lookings at traffic required to drive a smooth journey around the city.
>
> (Laurier 2003: 4)

Try to identify some practices that you do every day (without really thinking about it), and try to remember how you learnt these skills.

or historical, or concerned with identity". That is, in spaces:

- with which one has no particularly significant relationship or attachment;
- that appear to have no particularly significant history of their own;
- that are not especially significant in the shaping of one's identity or cultural life, or which have no particularly compelling identity of their own.

Augé calls this kind of space a 'non-place'. He argues that, perhaps increasingly, everyday lives are full of this kind of space, citing examples such as ATMs,

Box 9.3
An everyday event: buying a coffee

Photograph 9.4 Decaffeinated cappuccino, no sugar or chocolate sprinkles please.
Source: Dr Faith Tucker.

Laurier and Philo (2006a: 198–199) reflect upon the act of buying coffee and sitting in a café. They show how this seemingly simple transaction involves a complex set of unspoken rules, norms and associations:

> When we buy a (non take-away) cup of coffee (in a café) we are well aware that we are also buying rights to a seat at a table to drink it. At the very least, the café is a device for allocating temporary possession of seats, tables and some shelter in the city. For many of us, a seat and a table to read, write, use a laptop, or whatever while being 'undisturbed' is a scarce resource. The café, while carrying expectations of conviviality provides this form of temporary dwelling for its customers and, with it, some rights to privacy in public. It is also a place where we can be left to our thoughts away from the pressing matters of the workplace or the home.

Think about the unspoken rules and norms that characterise a space which is familiar to you (see also Chapter 13). How are these rules and norms learnt and maintained?

hotel lobbies, motorways and bus stops. In so doing, Augé directs attention to the everyday **landscapes** (see Chapter 5) that, although often unloved and unnoticed, are fundamentally important components of contemporary geographies. Exploring similar territory, fellow travellers and cultural geographers have begun to explore the importance of everyday spaces such as airports (Adey 2007), motorways (Merriman 2007), public transportation (Bissell 2008), infrastructure (Hayes 2005), civic spaces (Clay 1994), and other spaces of waiting and boredom (Anderson 2004).

In Box 9.2, Laurier uses the example of driving (whether confidently or badly) to draw attention to many, complex practices that each of us does everyday. For example, driving to work involves the accumulated, "occasioned use of . . . [knowledge] from driving these roads many, many times and at this moment, being on this road, at this junction, waiting to take this exit, or this third turn on the left" (Laurier and Philo 2003: 99). Typically, these practices are so well rehearsed, and so taken for granted, that we may not be aware of the skills involved. Skills such as driving are simply done, often in a distracted manner, and often while doing other things simultaneously. The skills and knowledge involved in driving exist "less as an abstract concept in [the] head, and more a series of embodied engagements with hands and feet, upon wheels, buttons and levers in response to . . . looking ahead at the [road]" (Laurier and Philo 2003: 99).

In Box 9.3, Laurier and Philo draw attention to many, small, often-fleeting events and encounters that characterise everyday lives. Buying a coffee may feel like a small, inconsequential event, but, as they show, each cup involves unspoken rules and norms. Drinking a cup of coffee also involves the intersections of different people's everyday lives, at a particular point in time, in the everyday space of the café. Typically, such encounters will be unmemorable and insignificant. But such everyday events and encounters *could* be important and affirmative (one might meet a future spouse for the first time over coffee), or disappointing or problematic (an argument over coffee could lead to divorce!) Thus the nature of everyday events can never fully be anticipated:

> In . . . [a] first encounter in a café we cannot know at the time whether it will be the beginning of an acquaintanceship, based perhaps on both being regulars at the café. Nor, relatedly, can we be sure whether it will be a likely one-off meeting in the way that characterises so many encounters in the city.
>
> (Laurier and Philo, 2006b: 356)

Reason 2: Understanding everyday workings of social/cultural geographies

It follows, from Reason 1, that attention to everyday lives should provide fresh perspectives on the issues and questions that have long preoccupied human geographers. In this spirit, it is increasingly recognised by geographers working in diverse social-cultural contexts that:

> all manner of insights about the workings of human society can best be found, not from the most obvious sources comprising or recording the seeming economic, political or social core of a given people's world, but from sources which on first glance may seem quite marginal, peripheral, insignificant and even esoteric . . . [the] 'bits and pieces' of human life.
>
> (Philo 1994: 1–2)

Consider, for instance, some 'big' topics that have been central to cultural geographers' research and enquiry (see the chapters in Parts 1 and 2): economy, **landscape**, **identity**, **cultural production** and **consumption**. Or consider some major societal issues and inequalities that many geographers have tackled in their work: poverty, racism, sexism, social class. In relation to each of these terms, there exists a significant body of geographical work. And yet in relation to each of them it is increasingly argued that an overlooking of everyday life has produced limited understandings of how such issues 'work', happen and matter, *in situ* and in practice. Therefore sustained attention to the everyday aspects of these issues might reveal some of their fundamental, but hitherto hidden, details and workings. Take the example of social class: during the twentieth century, many social scientists, including human geographers, explored and theorised class-based inequalities in diverse contexts (see Blunt and Wills 2004). However, this work typically neglected the 'hidden injuries of class' (Sennett and Cobb 1977): the small, everyday indignities that actually constitute class-based inequality in practice, and the associated **emotions** (anger, inadequacy, shame) through which 'class' is actually experienced in practice, on a day-to-day basis. The sociologist John Shotter (1993: xi) provides some examples of this kind of 'hidden injury', recalling his first job as an engineering apprentice in an aircraft factory:

> We thousand or so workers trooped in at 7.30 a.m., through a single little door at the back of the factory, jostling and pushing each other to make sure we clocked in on time, as every minute late cost us 15 minutes' pay. The . . . management . . . and the Royal Air Force officer customers came in ('strolled in', we thought) through the big double doors at the front, up imposing steps at 9.00 a.m. But more than that, 'they' had their lunch on a mezzanine floor raised five feet above 'us' in the lunch room; 'they' had waitress service and white tablecloths, 'we' buttered sliced bread straight from the paper packet on the formica top of the table.

Shotter (1993: xii) describes the everyday emotional geographies of class in the factory, whereby 'workers' were constantly angered and aware of the kinds of petty indignity detailed above, but the 'staff' "seemed impervious to the fact that our anger was occasioned by their behaviour, their perks":

> When workers . . . returned to floor, seething after a brush with management, and everyone . . . said 'Oh, you've just got to complain about that', no-one ever did. In the end, it seemed too trivial, and one knew it would be useless. To complain, for instance, about the windows in the men's toilets – put there so the foreman could see that what was being done there was being done properly, and not wasting time – to complain just by saying 'Well I don't like being watched at those times' seemed both inadequate to the anger and unlikely to be effective . . . There seemed to be no adequate language with which to express why we had become so angry, to explain why these little degradations mattered so much to us. And this, I suspect, made us even more angry, for we also became angry at ourselves,

for trivialising ourselves for being so bothered by such trivial things.

Shotter's observations of factory life provide one example of how an attentiveness to everyday life can reveal the detailed workings that lie behind somewhat taken-for-granted terms such as 'social class'. Feminist geographers' work on the cultural production of the taken-for-granted notion of gender (see Chapter 12, and Reason 3 below) has also been especially important in this context.

The work of the social theorist Bruno Latour has often been important for geographers working in this spirit. Latour's method is, essentially, to reveal the small, everyday details, lives and **actor-networks** (see Chapter 10) that constitute major contemporary discourses and social constructs, such as 'science' (see Box 9.4). In so doing, such constructs are revealed as less formidable, powerful and inevitable, and also as more complex, than they are commonly imagined. We shall return to Latour's work – and the broader concepts of **actor-network theory** (ANT) – in Chapter 10. For now, the key point is that everyday spaces, practices and events must be understood as central to large processes and issues that matter for human geographers (think, for example, about the many everyday spaces, practices and events involved in processes of **cultural production** and **consumption**, as discussed in Chapters 2 and 3).

Reason 3: Everyday spaces as exclusionary and problematic

As a great deal of work by social geographers has made clear, everyday spaces can often be deeply exclusionary or problematic for particular social groups (see Valentine 2001). As we shall explore when discussing **emotional and affective geographies** (Chapter 11), spaces that seem taken for granted, normal and unproblematic from one person's perspective may often feel fearful, distressing or uncomfortable for others. For cultural geographers, then, it is important to recognise that **cultural spaces** (see Chapter 2) can be

Box 9.4
Laboratory life

In *Laboratory Life* (1979), Bruno Latour and Stephen Woolgar observe everyday goings-on in a prestigious Californian neuroendocrinology laboratory. The book begins with two pages of field notes, detailing one morning in the laboratory. Consider the following extract.

> 6 mins. 15 secs. Wilson enters and looks into a number of offices, trying to gather people together for a staff meeting. He receives vague promises . . . He leaves for the lobby.
>
> 6 mins. 20 secs. Bill comes from the chemistry section and gives Spencer a thin vial: 'here are your two hundred micrograms, remember to put this code number on the book' and he points to the label . . .
>
> Long silence. The library is empty. Some write in their offices, some work by windows in the brightly lit bench space. The staccato noise of typewriting can be heard . . .
>
> 9 mins. Julius comes in eating an apple and perusing a copy of *Nature*.
>
> (Latour and Woolgar 1979: 15)

When we read a 'scientific fact' we may rarely think of the everyday lives of scientists, as they label vials, eat apples, become irritated with colleagues, or peruse magazines. But with attention to such everyday details, Latour and Woolgar (1979: 40) attempt to chronicle "the ways in which the daily activities of working scientists lead to the construction of facts". They suggest that, in contemporary societies, 'science' is widely perceived as an especially prestigious form of activity and knowledge. For example, 'scientists' are often perceived as formidably intelligent discoverers of important knowledge, and 'scientific facts' are often understood to be indisputable. *Laboratory Life* contrasts this perception of 'science' with the everyday realities and uncertainties of 'science in the making'. The book demonstrates that prestigious scientific discoveries made by highly trained academics involve all manner of mundane, fallible work and events, such as those in the above extract. A complex, messy heterogeneous assemblage of **materials**, technologies and associated practices is also central to this work (see Chapter 10). This work is fundamentally what science *is*. And yet, Latour and Woolgar argue, most key cultural representations of science (e.g. scientific journal articles and media reports) hide this everyday work of 'science in the making'.

experienced very differently by people with different **identities** (see Chapters 3 and 8). Indeed, as we discussed in relation to the regulation of cultural spaces in Chapter 2, such spaces can sometimes be intentionally designed to exclude certain social groups while serving the interests of others.

The work of feminist geographers has been especially significant in this context, in mapping, and raising awareness of, the exclusionary and problematic nature of many everyday spaces for diverse women (see Rose 1993; Valentine 1989, 2001). Feminist critiques of everyday spaces have drawn attention to multiple forms of gendered spatial inequality in diverse everyday contexts (see Smith 1989). Let us broadly summarise four problematic forms of spatial inequality and exclusion recognised by feminist geographers.

First, it is clear from the work of many feminist geographers that experiences of everyday spaces are highly gendered. In particular, it is often the case that certain spaces (such as parks, streets, civic spaces, workplaces or cultural events) are frequently experienced as sites of anxiety and fear by many women, whereas male contemporaries may often consider these spaces unremarkable, or even relaxing and enjoyable. It is now clear that the relative absence of female voices and perspectives from human geography prior to the **cultural turn** constituted a substantial overlooking of these everyday geographies of fear and anxiety (see Valentine 1989).

Second, as we discuss in Chapters 8 and 12, it is in and through everyday spaces that powerful cultural representations of 'ideal' femininity are coded, circulated and encountered. For example, feminist critics have detailed how norms, representations and **discourses** about body size, beauty, lifestyle, heteronormative relationships and motherhood are encountered in all manner of everyday spaces, to the extent that they become taken for granted. Again, it is argued that the ubiquity and power of these normative and hegemonic cultural discourses went unnoticed by earlier (overwhelmingly male) cultural geographers.

Third, moreover, it is noted that the regulation and material construction of many everyday cultural spaces are themselves instrumental in the production of gendered difference and inequalities. For example, think about how particular cultural spaces and consumption opportunities are specifically designed and marketed 'for men' or 'for women' (we shall return to this topic in Chapter 12). More generally, consider the ways in which gendered difference is literally built into many spaces: via gendered toilets and changing rooms, as well as via gendered expectations about behaviour, for example.

Fourth, Rose (1993) highlights how everyday spatial practices and time–space routines are often profoundly gendered. In particular, she notes how gendered norms about childcare, domestic labour and meal preparation continue to limit the everyday movements and lifestyles of many women in diverse contexts.

Reason 4: Revealing everyday consequences of geographical issues

If social/cultural issues are understood as having tangible, everyday causes and workings (see Reason 2), they surely have tangible, everyday, lived consequences. However, it is increasingly apparent that many chief social scientific accounts have overlooked precisely this kind of consequence. This realisation is especially important for human geographers: for relationships between societal structures and individuals' biographies are enmeshed in, and happen through, everyday spaces, lives and events. For Eyles (1989: 103), this relationship, between social-cultural, political, economic and historical issues and their everyday consequences, ought to be fundamental to any version of cultural geography:

> Through our actions in everyday life we build, maintain and reconstruct the very definitions, roles and motivations that shape our actions. We continuously maintain the culture and society that are the unquestioned background to our experience. But our actions do not occur within a vacuum. The patterns and meanings of, and reasons for, human actions are structured into and by the societies into which we are born. We both create and are created by society, and these processes are played out within the context of everyday life.

Boxes 9.5 and 9.6 provide a brief summary of some everyday consequences of two ongoing global events: neoliberal economic development and the HIV/AIDS pandemic. These case studies exemplify the kinds of

Box 9.5

Neoliberal economic development and young people in rural Mexico.

In 1994, the governments of Canada, the USA and Mexico signed the North American Free Trade Agreement (NAFTA). The Agreement created the world's largest free trade area, with unprecedented opportunities for transnational economic exchanges between the three member states. The consequences of this major trade liberalisation project are complex and multiple. Research by Fina Carpena-Méndez (2007) illustrates some impacts of this transnational development project for everyday lives in one central Mexican village. Since NAFTA's implementation, many rural Mexican communities have experienced substantial out-migration of young people, in pursuit of new economic activities near the Mexico/USA border, or opportunities for legal or illegal transit into the USA. In Carpena-Méndez's case study village, one-fifth of the population had migrated in this way. Carpena-Méndez shows how this has fundamentally altered everyday, cultural life in the village. Some examples include:

- an influx of hi-tech commodities (TVs, cameras, CDs, DVDs) sent or brought back by migrants;
- new fashions, hairstyles and musical genres introduced by returning migrants;
- new home spaces - e.g. many families use money sent by migrant relatives to reconstruct traditional single-room homes into 'USA-style' multi-room houses;
- a breakdown of traditional family life as a result of new home spaces;
- new forms of inequality - e.g. between families who receive money and commodities from migrant relatives, versus those who do not;
- new youth identities and lifestyles - e.g. many returning migrants form gangs, with new cultures of 'hanging out', drug-taking and graffiti-tagging;
- new forms of social conflict - e.g. between generations, and between gangs of former migrants.

Box 9.6

Young people and HIV/AIDS in southern Africa

Since 1980, HIV/AIDS has caused the deaths of at least 25 million people. Approximately 33 million people currently live with HIV/AIDS. Around one-third of these people live in southern Africa, where the disease continues to have unprecedented social and demographic consequences. Research by Lorraine van Blerk and Nicola Ansell (2006) with children in Malawi and Lesotho has revealed some deeply affecting everyday consequences of this global pandemic. Some consequences of HIV/AIDS for children include:

- new everyday tasks, responsibilities and routines - e.g. feeding and administering medicines to dying relatives;
- displacement to new everyday contexts as a result of parents' deaths - e.g. relatives' homes, institutions or homelessness;
- experiences of isolation, alienation or homesickness as a result of displacement;
- psychosocial and embodied responses to grief and trauma - e.g. depression, anxiety, withdrawal, or behavioural difficulties;
- everyday experiences of stigmatisation - e.g. 'AIDS orphans' are often bullied, insulted and socially excluded;
- new everyday practices and social groupings - e.g. children orphaned by HIV/AIDS often take up work, begging or stealing to earn money.

event that have long preoccupied human geographers. However, in contrast with a great deal of research on these topics, the researchers highlighted in Boxes 9.5 and 9.6 have explored some everyday consequences of these global issues. In particular, they highlight, respectively, how neoliberalisation and the HIV/AIDs pandemic impact upon the everyday lives of children and young people (see also Box 12.4). Note, when reading about these researchers' findings, how an attention to everyday consequences reveals some deeply moving, but perhaps very easily overlooked, consequences. These poignant details remind us that all social/cultural issues should be understood as having consequences "amidst the routine practices and occurrences of everyday and everynight life" (Pred 2000: 118).

Reason 5: Acknowledging 'processual complexity'

As will be made clear in Chapters 10–13, focusing on everyday geographies alerts us to the immense, but often hidden or taken-for-granted, *complexity* that characterises all social-cultural practices, spaces and events. Cultural geographers increasingly acknowledge that this is true of even the most (apparently) simple, routine, passing everyday practice. Consider an example of this kind of practice: making a cup of tea or coffee (see Photograph 9.5). For millions of people across the world, making a hot, caffeinated beverage is a thoroughly routine part of everyday life. It is often done with little thought or ceremony: the kettle is boiling as I type this sentence; perhaps you are drinking a cup of tea or coffee as you read. As Angus *et al.* (2001) suggest, it is easy to dismiss our hot beverage as 'just a cuppa', with little inherent meaning, complexity or importance. And yet, as they demonstrate, 'the cuppa' is an example of a cultural object/practice whose apparent simplicity belies a formidably complex range of processes and geographies.

For example, Angus *et al.* (2001: 196) describe the complex processes and connections "between heres and theres, between humans, between humans and non-humans, between non-humans and non-humans" that are necessary for any cup of coffee to be made:

> Coffee plants had to have been planted in soil, fertilised, kept free of pests, tended, picked, dried, processed, bought, traded, packed in jars, labelled, advertised, transported, shelved, purchased, canned, paid for, brought home, shelved, scooped with a spoon, placed in the right receptacle . . . with the correct co-ingredients in the right proportion at the correct temperature. That hot water had come through the tap cold, through those pipes, via that purification plant . . . from that reservoir, collecting water from the catchment of that river, which had been rained on by those clouds, which had arrived as part of that weather system, which picked up water on its way from who knows where.

They continue at length in this vein, drawing attention to the complex processes behind boiling water, making ceramic cups, opening a fridge, producing milk, and so forth. In so doing, it becomes apparent that:

> It could all so easily have gone wrong if one connection had not been properly made: . . . if there had been a power cut; if the milk had gone off; if a plumber hadn't soldered [a] joint properly . . . ; if milk tanker drivers had been part of the recent fuel protest . . . ; if some crop disease had ruined

Photograph 9.5 Making a 'cuppa' involves considerable processual complexity.
Source: Dr Faith Tucker.

> a coffee harvest in Latin America; if the productivity of dairy cows in the UK declined; if [the] local shopkeeper had to close early that day due to ill-health; if the kettle element overheated and melted the plastic.
>
> (Angus *et al.* 2001: 196)

All of which might prompt us to reflect on the considerable invention, labour and skill involved in laying the pipes and cables that are infrastructural to any cup of tea or coffee being made (McCormack 2006). Or it might prompt us to reflect upon the global networks of trade, **production** and labour involved in producing a commodity such as coffee (see Chapters 2 and 10). Or it might cause us to reflect on the **bodily practices** and skills that are involved in making a drink (How do we learn how to make coffee just the way we like it? How do we avoid scalding ourselves in the process?) (see Chapter 12, also Roe 2006).

These perspectives make it plain that a seemingly simple act of **consumption** (see Chapter 3), such as making a cup of coffee, actually involves a formidable "processual complexity" (Paterson 2006: 7): that is, it involves an enormous range of processes, which are themselves reliant upon complex assemblages of **materials** and practices (see Chapter 10). Indeed, all of the social/cultural geographies described in this book (even waiting for a bus that never comes) could be understood as characterised by this kind of processual complexity. As Reasons 2, 3 and 4 suggested, cultural geographers now insist that it is in the complexities and connections of everyday life that social/cultural geographies *matter,* in practice and for real.

Reason 6: Critiquing politics of academic knowledge

Many social scientists have been drawn to everyday life precisely because it seems so out of place, 'other' and therefore challenging, in the context of contemporary academic discourses.

There are two aspects to this challenge. First, as this chapter has outlined, it is argued that everyday life has, on the whole, been overlooked in the work of many social scientists, including human geographers. In this context, explicitly attending to everyday life can be a kind of critical "slap in the face" to 'proper', 'important' academic concerns (McRobbie 1991: ix). A turn towards everyday geographies can therefore function as a critical challenge to prevailing notions of what 'counts' as 'important' and 'proper' within the prevailing disciplinary norms of human geography.

Second, developing this challenge, it is argued that certain forms and aspects of everyday life have been *more systematically overlooked than others* by social scientists. For example, as we noted in Chapter 1, work by earlier human geographers has been widely critiqued for marginalising popular culture, and the everyday lives and perspectives of many social groups. Given their frequent exclusion from academic discourses, it is argued that paying attention to these 'neglected other' (Philo 1992) kinds of everyday life should amount to a fundamental challenge to the prevailing norms, assumptions and hierarchies found within a discipline such as human geography.

To offer just one example, and to extend the point made in Reason 3, many feminist geographers have argued that women's everyday labour in diverse private, domestic spaces was overwhelmingly unheralded by earlier cultural geographers, which meant that cultural geographers had a poor grasp of contemporary forms of gender-based inequality, and a limited appreciation of the complex cultural lives ongoing within such spaces (on which, see Chapter 10).

Reason 7: Exposing limitations of key methods, concepts and representations

Perhaps the most difficult challenges posed by all of the above are the questions raised for *doing* cultural geography. In short: how might it be possible to effectively research, conceptualise and represent everyday life? Three related issues underlie this question.

First, everyday life poses challenging questions for research methods. For, if we accept that everyday geographies (waiting for the bus, making coffee) are worthy of investigation, what kinds of research method might be useful in researching them, in all their processual complexity? To a large extent, the methodological critiques and innovations that have characterised cultural geography in the wake of the **cultural turn** (see Chapter 1) have been driven by this problem.

Second, as we discuss in Section 9.4, attention to everyday life calls into question many previously accepted concepts that have been central to writing and understanding human geography. This kind of challenge – and the resulting **crisis of representation,** as discussed in the following section – has led many cultural geographers to explore and experiment with new concepts. Some of these explorations are charted in Chapters 10–13.

Third, underlying all this, it is argued that the nature of everyday life – its *everydayness* – calls into question many assumptions about our ability to represent the world, for example via research, concepts or language. This challenge is explored in the following section.

9.4 The everyday 'escapes'

Section 9.2 introduced the term 'everydayness' to describe the 'processual excess' that characterises everyday life. To reiterate, everydayness can refer to aspects of everyday geographies that are (i) overlooked as we go about our everyday lives, and/or (ii) get 'lost' in representations of everyday life. In an oft-quoted phrase, the literary theorist Maurice Blanchot (1993[1969]: 241) summarises this kind of 'overlooking' and 'loss' as follows: "The everyday escapes. That is its definition. We cannot help but miss it if we seek it through knowledge." The argument is therefore that, no matter how hard one might try to know or understand everyday geographies, there will always be some characteristic of them – their 'everydayness' – that 'escapes'. Or: "Everyday life escapes; it *exceeds*" (Seigworth 2000: 231). Or, again: "The world is more excessive than we can theorise" (Dewsbury *et al.* 2002: 474).

So, 'everydayness' is a name that some cultural geographers have given to those aspects of everyday life that 'escape' and 'exceed' our ability to know and theorise everyday geographies, both as human beings and as human geographers. As a tangible example, consider one of the questions that began this chapter: think about something that matters a great deal to you, in your everyday life. Try to write down how and why it matters so much.

Or attempt the following tasks.

- Think about the last time you danced. Then try to write down: (i) how it felt to dance; (ii) how you danced.
- Think about the last time you felt very sad. Then try to write down what this sadness felt like.

In effect, the idea that the everyday 'escapes' predicts that responses to exercises such as these will inevitably reveal a fundamental difference between the thing being described (something that matters; dancing; sadness) and the corresponding written description. In effect, it is argued that there is no way that writing can comprehensively, or even adequately, represent these kinds of features of everyday life. For there will always be a difference and a gap between 'word' and 'world' (see Dewsbury *et al.* 2002): that is, between aspects of everyday existence and the words used to represent them. This poses a significant challenge for conventional ways of writing and representing the world. For, it is argued, no matter how much we write, or how much detail is used, our responses to the above exercises will, somehow, never quite get to the heart of the things being described. Somehow, some things in life (things that matter; the vitality and energy of dancing; the feeling of sadness) exceed our ability to represent them. For the word 'sadness', or even a page of writing about 'sadness', can somehow never do justice to the actual feeling of *sadness* itself. It is precisely this kind of 'excess' to which the term 'everydayness' refers. It follows that written representations can only ever be a hollow, inadequate, somehow *lifeless* version of everyday life (see Thrift and Dewsbury 2000). Therefore the 'escaping', 'excessive' nature of everydayness should permanently challenge any assumptions that written representations can comprehensively represent everyday life.

We might ask: What does it matter if it is difficult to write about things that 'matter', dancing or sadness? This difficulty may seem obscure, and of marginal relevance. In fact, many cultural geographers now contend that this kind of difficulty should be understood as part and parcel of a much broader problem: the alleged inability of all systems of representation to adequately represent any aspect of everyday life. This problem – often referred to as the 'crisis of representation' – has also been a key concern in much art, philosophy and literature from the last century (see Box 9.7). This 'crisis' constitutes a profound critique of all social scientific theory, practice and knowledge that relies on an assumption that everyday life can be represented using some combination of texts, images, discourses, talk and so on.

Box 9.7
Attempting to represent everydayness: OuLiPo

How, then, might everydayness be grasped, understood and represented? This problem has been central to many methodological and theoretical debates in cultural geography (for example, see Chapters 11 and 12). It has also been a key concern in much art, philosophy and literature from the twentieth century. For example, consider the work of 'OuLiPo'. OuLiPo - short for 'Ouvroir de Littérature Potentielle' ('Workshop for Potential Literature') - was a Parisian collective of writers and mathematicians, founded in 1960. The group explored forms and limits of written language, through literary games and experiments. Much of their work concerned the inadequacy of written language when confronted with everyday places and events. The novelists Raymond Queneau (1903–76) and Georges Perec (1936–82) were key 'OuLiPans'. Consider the following examples of their work.

Queneau's bus ride

Picture the scene: on a crowded rush-hour bus, a man is jostled and then sits down. In *Exercises in Style* (1981 [1947]), Raymond Queneau demonstrates that is possible to represent this apparently simple, everyday event in 99 entirely different ways. For example, Queneau represents the bus ride as a sonnet, a musical score, a letter of complaint, a logical argument and a psychoanalysis. He also presents numerous narratives emphasising different aspects of the bus ride, including colours, textures, dimensions and personalities. Reading the book from start to finish is disorientating. It prompts challenging questions: which writing style is the 'best' representation of this bus-ride, and why? More than this, however, readers are left with a sense that *none* of the 99 accounts has really got to the heart of this everyday event.

The view from Perec's café

In *Species of Spaces* (1999[1974]: 50–53), Georges Perec challenges readers to sit in a local café and 'observe the street'. You could try this yourself, following Perec's instructions: "Take your time . . . Note down what you can see. Anything worthy of note going on . . . Force yourself to write down what is of no interest, what is most obvious, most common, most colourless." Perec attempts the exercise. He notes details of buildings, cars, paving slabs, lights, cigarettes, signs, posters, litter, women, shoes, dogs, birds, sewers, cables, and a complex sequence of events involving a man parking a car and buying a packet of jelly. The detail becomes excessive and exhausting. He tries to write down *everything*, but of course this is impossible, because more and more things happen, and he notices more and more complexity in his apparently uneventful surroundings. Perec becomes frustrated at the task's hopelessness, as his written notes fail to keep up with the scenes unfolding around him. Readers are left with a sense that the written word can never adequately 'capture' everything that is going on, even in a familiar, everyday scene.

Let us outline some key points of the 'crisis of representation' critique. As a point of departure, it is important to understand that many conventional ways of knowing and writing about the world rely on a range of assumptions about the extent to which the world can be represented. For example, the vast majority of social scientific research and theory has, to date, taken for granted the following assumptions:

- the existence of a world 'out there' (i.e. which exists straightforwardly and objectively) that can be researched and represented;
- the possibility of using particularly formulated methods to conduct research and gather data about this world 'out there';
- the possibility of using particularly formulated research tools, questions and empirical projects to develop theories, understandings and explanations about the world;
- the possibility of using written or spoken language, or other accepted systems of representation, to adequately write, theorise and represent the world;
- the possibility, ultimately, of exhaustively knowing aspects of the world through such methods, projects and representations.

Such assumptions have often been treated as broadly unproblematic throughout the histories of many social scientific disciplines, such as human geography. These assumptions have, in turn, habitually been founded upon a basic faith in language, explanation, understanding, interpretation and – especially – representation. That is, it is assumed that systems of representation, such as language, can be used unproblematically to explain, understand and interpret the world.

However, the 'crisis of representation' critique argues that if we really think about particular, tangible examples of representation in practice, each of these assumptions, and their foundations, is revealed as somewhat problematic. Box 9.8 attempts to outline

Box 9.8

Some elements of a crisis of representation: a cultural geographer (with hat) interviews a man at a bus stop

Picture the scene: a cultural geographer (with hat) and another man are waiting for a bus. The cultural geographer asks the man to tell him about his everyday life. The man looks around (A) and tells the cultural geographer what he sees (B).The cultural geographer makes notes based on what he hears (C). He eventually writes up this material as a book (D), which is read by a stranger in a university library (E).

Stages A–E caricature some of the processes and projects through which social scientists have often understood and represented the world: observation (A), language (B), interpretation and analysis (C), writing and reporting (D) and reading (E). The crisis of representation critique argues that each of these processes of representation is potentially problematic. Consider the following.

A Do we always accurately observe *everything* in our everyday contexts? Do we always understand everything that we observe? Is it not the case that much that happens in our everyday lives passes us by without us noticing? Do we not sometimes find ourselves in situations where "nothing (clearly) happens but something (obscurely) is and has been afoot" (Seigworth and Gardiner 2004: 140)? How much of our everyday lives is 'invisible' because it is unknown, or known too well (Law and Costa Marques 2000)?

B Can words (or any representations) ever *fully* describe our everyday lives? Can words ever comprehensively capture "how, moment by moment, we in fact conduct our practical everyday affairs" (Shotter 1997: 2)? Can we always articulate exactly what we are doing, and why? Is it not the case that "a lot of one's actions one can't give a coherent account of. Not afterwards and not at the time either, because one is too busy doing the particular thing one is doing" (Dewsbury 2000: 474).

C Are we always 100% accurate when we try to make sense of things we observe? When interpreting the world, how much do we overlook because it seems uninteresting, irrelevant, obvious or unhelpful (Law 2003)?

D How much gets lost when we attempt to translate the world, and thoughts and experiences, into coherent, conventional narrative forms? How much is 'cut away' and 'disregarded' in the process (Harrison 2002: 489)?

E Do other people always get the message in the way we intended? For example, have you remembered and understood everything you have just read?

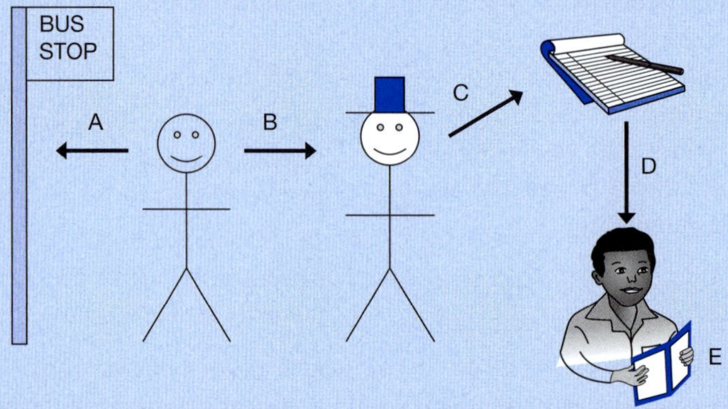

Figure 9.1 Some elements of a crisis of representation.

why this is the case. Since these foundations are problematic, it is argued that all bodies of work that rest on them must also be called into question. Hence: a crisis of representation.

Ultimately, then, for some cultural geographers, acknowledging everydayness – and recognising that a crisis of representation often follows such an acknowledgement – poses some major philosophical questions.

- Can we ever know *anything* for certain?
- What is the purpose and future of academic practice in the light of the crisis of representation?
- How might human geographers proceed in the absence of a basic faith in representation?
- How could/should social scientists deal with the processual excess of everydayness?
- More specifically, how might it be possible to do research and writing about everyday life, in the light of all the above?

Within cultural geography, these questions have been most explicitly addressed by a body of work known as 'non-representational theories' (or NRT). This term was coined by the geographer Nigel Thrift (2000c: 556) to describe "a radical attempt to wrench the social sciences out of their current emphasis on representation and interpretation". In practice, NRT is used as an umbrella term for plural, disparate lines of philosophical work. For example, Figure 9.2 shows Thrift's attempt to outline influential thinkers whose work is important for geographers interested in thinking on from the crisis of representation. Each of the key thinkers listed in the figure has produced a brilliant, challenging, fascinating body of work relevant to this context: any one of these key thinkers would be well deserving of your attention if you are interested in looking at some primary sources and exploring some higher-level material on this topic.

For Thrift (1997: 128), NRT should be understood as "provid[ing] a guide to a good part of the world which is currently all but invisible to workers in the Social Sciences and Humanities" through four related foci:

- "practices, mundane everyday practices" that are fundamental to everyday geographies, and which thereby "shape the conduct of human beings towards others and themselves in particular sites" (Thrift 1997: 128);

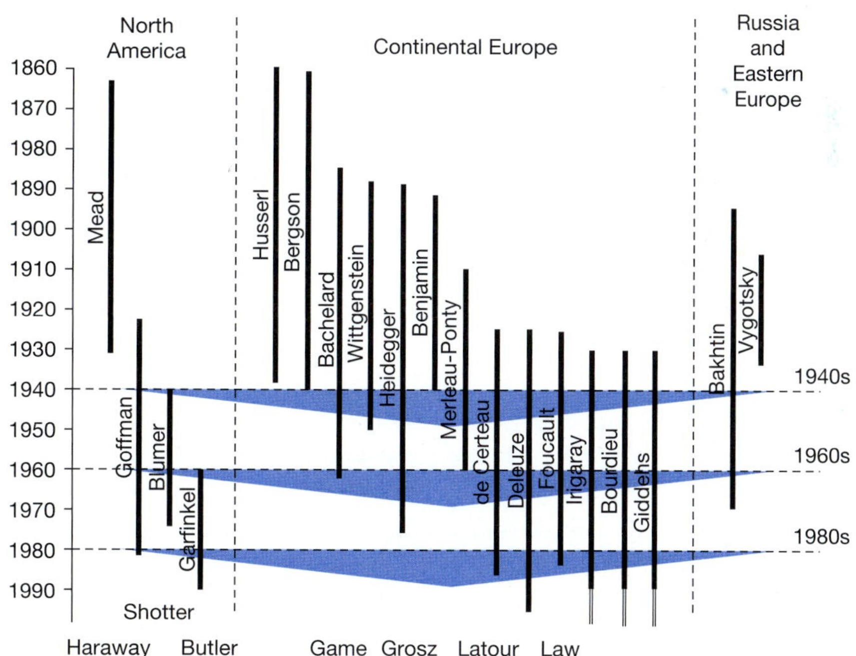

Figure 9.2 The life-time-lines of non-representational theory.

Source: from Steps to an Ecology of Place, *Human Geography Today*, p. 303 (Thrift, N. 1999), Blackwell, Oxford, Permissions granted by Nigel Thrift.

- **bodily**, **emotional** and **affective** aspects of human experience (see Chapters 11 and 12);
- the complexly **spatial** and **temporal** nature of everyday geographies (see Chapter 13);
- the complex 'technologies of being', such as **materials**, **technologies** and **architectures,** that are part and parcel of everyday lives (see Chapters 4 and 10).

The concerns and critiques of NRT have been widely influential – and controversial – within human geography. In Chapters 10–13 we provide a more detailed introduction to some of the key ideas contained within the bullet points above. Along with many other cultural geographers, we would argue that the ideas discussed in the following chapters should be understood as profoundly important – and in some senses revelatory – for cultural geographers. Certainly, at the very least, these concepts should prompt us to think carefully about the everyday processual complexities of cultural processes of production and consumption, and the diverse cultural phenomena discussed earlier in this book. However, in the chapters that follow we are mindful to show how concepts relating to bodies, emotions, material objects and spatiality are not solely the preserve of 'non-representational geographers', but actually have a longer heritage, and have been developed through the interplay of quite diverse sets of thinkers.

Summary

- Everyday life is fundamental to all human geographies and geographical issues. However, chief texts and approaches within human geography have overwhelmingly *overlooked* everyday life.
- One facet of the cultural turn has been a turn towards everyday spaces, practices and encounters as topics for academic research and enquiry. It is increasingly acknowledged that focusing on such everyday geographies can provide in-depth understandings of the tangible workings and consequences of social and cultural issues.
- Moreover, it is argued that attempting to write and theorise everyday life exposes the elitist norms of academic knowledge and some key limitations of many classic methods, concepts and representations.
- Everyday life is characterised by 'processual complexity' and 'excess'. That is: (i) everyday geographies are formidably complex and detailed (although this complexity is usually taken for granted); (ii) much of everyday life 'exceeds' or 'escapes' our attempts to understand it. The term 'everydayness' is used to refer to these two characteristics of everyday life.
- Everydayness can prompt some substantial philosophical questions, which have amounted to a 'crisis of representation'. In the wake of this 'crisis', non-representational theories (NRT) have attempted to address some of these questions.

Some key readings

Anderson, B. and Harrison, P. (eds) (2010) *Taking-Place: Nonrepresentational theories and geography,* Ashgate, Farnham.

An excellent collection of essays by geographers reflecting on the importance of NRT for their work. Contains an accessible and entertaining interview with Nigel Thrift about the inception and key elements of NRT.

Clay, G. (1994) *Real Places: An unconventional guide to America's generic landscape,* University of Chicago Press, Chicago.

In this fascinating, encyclopaedic book, Clay (an urban planning specialist) describes the easily overlooked spaces and features of 'generic' urban landscapes in North America.

Highmore, B. (2002) *Everyday Life and Cultural Theory: An introduction,* Routledge, London.

Highmore, B. (ed.) (2002) *The Everyday Life Reader,* Routledge, London.

An accessible introduction to key theorists of everyday life, and their concepts. The accompanying reader provides access to key primary sources by these theorists.

Rigg, J. (2007) *An Everyday Geography of the Global South,* Routledge, London.

Most scholarship about everyday life has been firmly rooted in the 'developed' Global North, in terms of its authors and subject matters. This book provides an alternative perspective, showing how attention to everyday life is crucial to understanding contemporary geographies of global development, poverty and inequality.

Rose, G. (1993) *Feminism and Geography: The limits of geographical knowledge,* University of Minnesota Press, Minneapolis.

This book summarises a range of geographical work regarding everyday lives and routines. In particular, it is a useful source of information about groundbreaking, but too often overlooked, early work by feminist geographers on this topic.

Sheringham, M. (2006) *Everyday Life: Theories and practices from Surrealism to the present,* Oxford University Press, Oxford.

A detailed synthesis of key lines of thought about everydayness. This book could be used to develop more in-depth understandings of the concepts introduced in the Highmore books listed above.

Valentine, G. (2001) *Social Geographies: Space and society,* Pearson, Harlow.

An excellent textbook, which is useful in this context for considering how everyday spaces are encountered differently by different social groups. Drawing on her work on gendered and heteronormative geographies, Valentine illustrates how social and spatial exclusion are interrelated.

10 Material things

Before you read this chapter . . .

Try to list all the objects and commodities around you at this moment.

Choose a thing: *any* thing. Try to answer the following questions.

- Where did it come from?
- Where/when/how were its parts made?
- Who made it?
- How did it get here?

Try to trace these stories as far back as you can, in as much detail as you can.

Chapter Map

- *Stuff* is everywhere
- 'Following the thing' and Marxian materialism
- Meaningful things and material culture studies
- Non-human geographies and heterogeneous materialities

 10.1 *Stuff* is everywhere

In Chapter 9 we explored how many human geographers have become increasingly attuned to **everydayness** within their work. One element of this turn towards everydayness has been an increasingly detailed recognition of the presence and importance of material objects – things, commodities, tools, 'bits and pieces', *stuff* – within everyday lives and geographies. If we pause to consider our surroundings (as in the questions at the start of the chapter), it is likely that – wherever and whoever we are – we shall find ourselves surrounded by all manner of material stuff (some of which may really matter to us, some of which we may hardly notice).

In this chapter, we outline three reasons why so many cultural geographers have focused on material objects, and we introduce three families of concepts that have often framed their work. First, we highlight geographical work that has 'followed the thing' to produce revelatory understandings of the processes and spaces through which commodities are produced in practice. We also introduce several key concepts from Karl Marx's theorisation of commodity capitalism that are widely used in this context to critique the murky geographical processes through which many commodities are produced. Second, we turn to geographical work that has analysed the importance of material stuff in the constitution of **spaces**, social practices, meanings and **identities**. Here, we introduce the concept of, and some key terms from, the field of 'material culture' studies. Third, we highlight how a consideration of 'materiality' has prompted many geographers to reconsider the very notion of 'human geography', via engagements with concepts such as 'cyborgs' and 'actor-networks'.

Dotted throughout this chapter you will also find short stories about objects that were on one author's desk while writing this book. We include these stories to demonstrate how material objects – even seemingly banal, everyday bits and pieces – can reveal complex, important, though often overlooked, cultural geographical issues and questions. In fact, before reading through the chapter, you may find it useful to look over these stories (see Boxes 10.1, 10.3 and 10.8) to get a flavour of some of the issues and questions we raise in this chapter.

10.2 'Following the thing' and Marxian materialism

When teaching and writing about the importance of material objects for the study of human geographies, we often return to a snappy quotation by the French social theorist Bruno Latour (see Section 10.4 and Chapter 9): "Consider things, and you will have humans. Consider humans, and you are by that very act interested in things" (Latour 2000: 20). Latour's comment neatly encapsulates the taken-for-granted centrality of material objects for everyday lives; he reminds us that material things are always present and important in human geographies (although they often go overlooked or unmentioned). As human beings, we are generally surrounded by material things; we continually do things with or through material things; material objects are central to all kinds of activities and elements of our lived experience, whether or not we realise or notice it. These may seem like pretty banal and commonsense observations. However, it is now widely argued that the importance of material objects has too often been underestimated and under-reported in geographical research, and that focusing on everyday material stuff can often prove to be revelatory.

One way in which a focus upon material objects has afforded new understandings of geographical issues has been through research exploring how particular commodities are produced. For example, the work of the geographer Ian Cook has been very influential here; seriously, we cannot recommend his work highly enough. Cook's modus operandi is to investigate the production of commodities and foodstuffs that are fairly commonplace – and considered pretty unspectacular – in contemporary Western societies: things like socks, bananas, mobile phones, papayas, cups of tea. Using multi-site ethnographic research, he works backwards from these items, to track the processes and spaces through which they are produced. In so doing, his work should prompt us to consider the stories, people and production processes that lie behind *all* commodities; indeed, he argues that any 'made in' label, or any list of ingredients, can serve as "starting points for geographical detective work", which can frequently begin to illuminate all manner of geographical, political, environmental and economic issues (Cook *et al.* 2007: 80). For one example, see Box 10.1, which attempts to trace the production of one mobile phone; for a much larger, and often updated, archive of examples visit Cook's website (www.followthethings.com).

This kind of research – what Cook calls 'following the thing' (see also the discussion of material culture in Section 10.3) – seems to us to be an especially engaging and affecting way of approaching the broad questions of cultural **production** and **consumption** outlined in Chapters 2 and 3. There is something particularly poignant and thought-provoking in the act of learning that *this* thing was made by *those* people under *those* conditions. (For example, in writing Box 10.1, it was unsettling to learn that *my* mobile phone was assembled in *that* factory with *those* degrading working conditions.) 'Following the thing' encourages a new and critical way of beholding the world: a new desire to learn about the human stories and real spaces that are involved in commodity production; a new set of questions to pose about hitherto taken-for-granted objects. A quotation by the novelist

Box 10.1

Things on our desk no. 1: a mobile phone

Photograph 10.1 Mobile phone on author's desk.
Source: John Horton.

Photograph 10.1 shows one author's mobile phone, which has been on his desk, within easy reach, while writing this book.

As you look at the phone, do you find yourself making some kind of judgement about the phone and its owner? Don't hold back. He has heard it all before: one of his students told him, 'Oh yeah, I *remember* that model'; another said, 'My *dad* used to have one of those.' So he is aware that it is considered an outmoded and bottom-of-the-range model, but he is not too bothered about it. He finds it slightly bemusing how much people seem to care about having the latest mobile device, and how it seems so important to their identities. (But, then again, proudly *not* having the latest mobile phone is part of the identity he constructs for himself, so perhaps he must care about it on some level. Anyway, we are getting side-tracked . . . As we said in Chapters 3 and 8, **identities** are complex practices.) Essentially, he likes to have the phone on his desk to send and receive text messages from family members, just in case there is any news.

Griffiths *et al.* (2009) provide an excellent set of resources for tracing the origins of mobile phone components. As an experiment in 'following the thing', the owner of the phone in Photograph 9.1 used these resources as a starting point to trace where, how and under what conditions that particular mobile phone was made. It proved difficult, particularly because the phone is made up of so many components, each with its own story and 'biography' (see Section 9.3). However, it was possible to map *some* of the production processes involved in making this one commodity.

- *Plastic casing* – Raw materials extracted from oil-based industries in Ecuador, Russia and Saudi Arabia. Transported to refineries in China and Mexico. Materials shipped to China for moulding. Plastic cases shipped to ports in Western Europe, then carried by freight lorry to Hungary or Poland for final assembly of phones.
- *Battery* – Cadmium, nickel and lithium mined in Chile, Japan and the USA. Batteries assembled in Mexican factories, then to ports in Western Europe, then carried by freight lorry to Hungary, Poland or Finland for final assembly of phones.
- *Speaker and microphone* – Raw materials sourced in Thailand. Electrical components sourced from ten suppliers in Malaysia. Materials and electrical components transported to Chinese factories for assembly.

- *Circuit board* - Electrical components created in five factories in Malaysia, Taiwan and Japan, then transported to Chinese factories for assembly.
- *Metals for electrical components* - Mined from South Africa, Chile and Democratic Republic of Congo, then shipped to SE Asia.
- *Screen, buttons, camera lens* - Raw materials extracted from oil-based industries in Ecuador. Components created in Mexican factories, then shipped to ports in Western Europe, then carried by freight lorry to Hungary, Poland or Finland for final assembly of phones.

As Griffiths *et al.* note, if you were to sketch these commodity chains on a map of the world (try it!), you would find a few intense clusters of activities: around China, south-east Asia and Mexico. This leads us to read reports by charities and NGOs investigating working conditions in mobile phone and electronics factories in these regions (e.g. CAFOD 2009). These reports present a catalogue of haunting and upsetting findings about the conditions under which our everyday commodities are made, often encompassing low wages, long hours, cramped and dirty factory spaces, dangerous and unhealthy working environments, and working conditions that are deeply, horribly, psychologically and physically exploitative. Reading this kind of research - beginning to understand those connections between the phone on my desk and the lives of those workers - is deeply challenging.

You might use Griffiths *et al.*'s resources, or the www.followthethings.com website, to try to map the production of your phone, or mobile device, in the same way.

Leah Hager Cohen (1997: 13–14) provides a lovely description of this kind of 'thing-following' mindset:

> Today, we want, and get, most things boxed and beribboned - sealed off from their origins . . . which are the stories of people, of work, of lives . . . Glass, paper, beans. Snapdragons, candles, incense. Caramel syrup, jazz, mittens . . . a shoelace, a nail clipper, jam. A postage stamp, garlic, an ice cube tray . . . the electricity that flows to your wall sockets . . . a square of sidewalk cement. The water that runs from our tap. All of these things are commodities, and they all contain stories encompassing geography and time, supply and demand, raw materials, market forces, and people. People with names and toes and sores and wages and fancies and memories. I look at the thing I touch: a doorknob, a can of soup, a ticket stub, a raisin . . . and I think 'did someone's hands touch this thing, did a worker's hands hold this?'

There are two reasons why 'following the thing' is often unsettling.

- In general terms, this kind of research highlights how detached we often are from producers of commodities and the circumstances in which they work.
- Specifically, this mode of enquiry can reveal our ignorance of deeply problematic and exploitative processes and geographies that characterise the production of many things we use every day.

As your answers to the questions at the start of this chapter may reveal (and as is evident in Box 10.1, and in our attempt to describe the production of this book in Section 2.1), there are often significant gaps in our knowledge when we attempt to describe how everyday objects were produced. As we go about our everyday business, we may habitually overlook how, where, and under what circumstances everyday objects and commodities came to be as/where they are. For example, the following oft-quoted discussion of the 'muteness' of many commodities asks us to consider our last meal, but Harvey's points could equally apply to all kinds of material object in our everyday lives:

> I often ask beginning geography students to consider where their last meal came from. Tracing back all the items used in the production of that meal reveals a relation of dependence upon a whole world of social labour conducted in many different places under very different social relations and conditions of production . . . Yet we can in practice consume our meal without the slightest knowledge of the intricate geography of production and the myriad social relationships embedded in the system that puts it upon our table . . . We cannot tell from looking at the commodity whether it has been produced by happy labourers working in a co-operative in Italy, grossly exploited labourers working under conditions of apartheid in South Africa, or wage labourers protected by adequate labour and wage agreements in Sweden. The grapes that sit upon the supermarket shelves are mute; we cannot see the fingerprints of exploitation upon them or tell immediately what part of the world they are from.
>
> (Harvey 1990: 422–423)

The advent of online resources (such as those archived on www.followthethings.com) has made it increasingly easy for many consumers to access reliable information about the intricate geographies and relationships through which everyday commodities and objects are produced. Reading research that has 'followed the thing' is often a troubling experience: as in Box 10.1, we are often confronted with wide-ranging evidence of exploitative, harmful and degrading working conditions, deeply dubious corporate practices, and environmentally damaging production techniques involved in the creation of commodities we may use and enjoy every day.

Several concepts from Marxian economics have been important for geographers exploring the production of material objects, and consumers' ignorance of systems and politics of production. Although Marx's *Capital, Volume I*, was published in 1867 – in a very different social-historical context – some key ideas from this book remain in common circulation among many cultural geographers. We have defined these terms in Box 10.2, and you will find that they are often

Box 10.2
Some key concepts from Marxian materialism

You will find that the following terms (from Marx 1976[1867]) are quite often used as a taken-for-granted vocabulary by geographers, and other social scientists, who write about commodification and material objects.

- *Materialism* (or *historical materialism*) – A broad term that describes Marx's overarching belief in the 'causal primacy' (i.e. the absolutely central, primary and impactful importance) of means of production, labour processes, class relations and economic inequalities in the development of human history (and geographies). Marx identified the analysis of commodity production/exchange as the key 'way in' to understand, and write materialist understandings of, wider systems of capitalism.
- *Value* – When writing about commodities, Marxian theorists differentiate between:
  - their value (or 'labour value'), which is the monetary worth of labour time, raw materials and tools involved in their manufacture;
  - their 'use value', which is the more qualitative sense in which a commodity is useful and meets consumers' needs;
  - their 'exchange value', which is the value that can potentially be attained for a commodity within the market;
  - their price, which is the actual amount demanded and paid for the commodity, in practice, at a particular moment.

  As many critics note, capitalist commodity producers endeavour to raise the exchange value of commodities far beyond their labour and use values.
- *Commodification* – The process of creating commodities (and/or transforming objects into marketable products), but also the creation of social and market contexts in which commodified objects are the norm. For Marxian critics, this process typically aims to maximise exchange value by:
  - exploiting labour (e.g. keeping labour costs down, rationalising production processes, not sharing profits with the workforce);
  - separating producers and consumers of products (e.g. by removing all traces of producers' processes from finished commodities), and reducing visibility and consciousness of this exploitation;
  - fostering 'false needs' (e.g. making new commodities appear to have an urgent use value, or creating desires for commodities that are not strictly useful).
- *Fetishism* – Traditionally, this was an anthropological term to describe how certain material objects were understood to have divine or magical power, or were invested with such power through rituals and belief systems. Marx applied this concept to the production and use of commodities in capitalist societies: commodities are fetishised in that they come to be invested with remarkable significance (e.g. a 'must have' product, or the particular cachet, and price tag, attached to certain brands) that far exceeds their use value. Commodity fetishism means that consumers attach value to commodities themselves, rather than to the labour involved in making them.
- *Reification* – From the Latin *res facere*: to 'make a thing'. The term is used by Marxists to refer to the way in which particular people can come to be seen as objects, and particular processes can come to be seen as inevitable and – like objects – 'just so'. The term is particularly

applied to the way in which humans and their labour can come to be objectified (perhaps represented as figures on the balance sheet of global commodity producers). For example, as discussed elsewhere in this chapter, we may have only a detached and limited sense of who produces commodities that we use every day; consequently, research that 'follows the thing' can be deeply unsettling when it reveals some of the particular individuals, stories and spaces involved in commodity production. Reification is often a flipside, and consequence, of commodity fetishism; it is often the case that the more that commodities themselves are invested with special value, the more the consumers of those commodities are distracted from considering the labour processes behind them.

- *Alienation* – A central concept of Marxian theory, and a consequence of commodification and commodity fetishism. Marx argued that alienation – feeling somehow 'foreign' in the world one is living in – was a characteristic malaise in capitalist societies. Marx developed a particularly poignant understanding of how the makers of commodities experience multiple forms of alienation, often becoming:
  - distanced from the products of their labour (e.g. workers on an assembly line may not see the finished item they have worked on);
  - distanced from the profits of their work (as already noted, profits of commodification are seldom shared with the individuals who actually make the commodities);
  - bored, unhappy and exploited during the process of production itself;
  - painfully but futilely self-aware of how their labour is exploited within the system of commodification;
  - distanced from the end-users of the products they are making.

used in many geographers' discussions of cultural production and politics. These terms have been important in providing a vocabulary with which to critique, analyse and politicise processes of cultural production, by directing attention towards the commodification of objects, the fetishisation of commodities, and the reification and alienation of human labour (all of which have been brought to light, in diverse contexts, by geographers who have 'followed the thing').

Research that 'follows the thing' is typically engrossing and unsettling, and should make us pause for thought as it reveals:

- the often 'invisible', hidden, unnoticed, unappreciated labour that goes into making taken-for-granted objects;
- how any purchase or use of a commodity connects us to "intricate and wide reaching networks of people and machines" (Cook *et al.* 2007: 80);
- the forms of exploitation, unethical practices and environmental degradation that frequently feature in these networks;
- how "in the contemporary world, we are all unavoidably and increasingly interconnected with the lives and landscapes of people and places around the globe . . . wherever we are, our lives impact on far distant others" (Desforges 2004: 1);
- how (we as) consumers ought to feel a sense of responsibility to distant strangers – "these stories . . . should provoke moral and ethical questions for those unknowingly participating in this exploitation" (Cook 2005: 3).

10.3 Meaningful things and material culture studies

In the last section we noted that human geographies are characteristically full of material objects (although the presence and provenance of these objects might sometimes be overlooked). It is also the case that material objects can really *matter* to people. Material objects make things possible (even though this making-possible can be taken for granted); they can be important to us in all kinds of ways; indeed, we might care about particular material objects, and feel lost without them (see the sad story of 'Monkey' in Box 10.3).

Material objects can have value and meaning in different ways: their importance can be practical (like the phone in Box 10.1, which facilitates communication with far-away family members); sentimental or emotional (as in Box 10.3 – a photograph that brings back a particular memory); symbolic (a particular possession

Box 10.3

Things on our desk no. 2: photograph of 'Monkey'

Photograph 10.2 Photograph of author with mother and Monkey.

Source: John Horton.

One of the authors has a photocopy of a photograph pinned up next to his desk: it shows a woman, holding a boy, holding a toy monkey, on the Thames embankment in London, sometime in the late 1970s (Photograph 10.2). He has used this image quite often when lecturing about the importance of material objects. It both illustrates the use of family snaps to make spaces (in this case a work space) feel homely (see Box 10.4) and provides an opportunity to talk about the emotions and memories that can be attached to particular material objects and commodities, because he feels a vivid pang of nostalgia and sadness when he thinks of (the imaginatively named) 'Monkey'. This cuddly toy was about six inches tall; he was light orangey-brown in colour, with a bobbly fabric body and soft velvety face; he wore a little blue bow tie. Monkey was bought by a parent just after the author was born. For the first few years of his life, Monkey and the author went everywhere together: to the park, to bed, to the hospital (Monkey was given a bandage to match the one on the author's arm), on daytrips and on holiday.

Here and now, the author and his family look at this photograph with a sad smile. Many aspects of this trip have been forgotten: the family can remember very little about what they did, or places they saw, or what they felt on the cold, wintery, rainy day shown in the photograph. What they do remember is that this was *the holiday when Monkey was lost*, never to be seen again. They remember the details of that incident with a sharp clarity: Monkey was left behind in the hotel bed, but when the family returned from a day of sightseeing, he was gone (maybe stolen, maybe accidentally taken to the laundry room with the bedsheets). The author remembers feeling panicked, trying not to cry, going into the bathroom and *praying* for Monkey's safe return, while the very kind hotel manager helped with the search. But he never did turn up. For many years, before he knew grief or heartache, that was probably the saddest day of his life.

The story of Monkey is part of the author's family folklore. It illustrates how mass-produced material objects can be invested with special significance (it was not just a toy monkey – *he* was *Monkey*), and can prompt vivid memories, nostalgia and emotions, even decades later. It also reminds us of the uses of photographs within many Western families (see Box 10.4). To produce the image shown in Photograph 10.2 the original photograph had to be very carefully removed from a particular page in a particular photograph album (stored in boxes, chronologically ordered, in a cupboard in the family home), to which it was returned as soon as it had been scanned. The family love looking back at these photographs, reminiscing about particular events and places, and retelling the old stories: whenever they turn to the photograph of Monkey there are sad sighs, rueful smiles, as someone says "Poor old Monkey – I wonder where he is now."

might 'say something' about us); economic (in terms of how much an item is 'worth', or how it affords economic success); or all of the above. Thinking about how particular objects matter to us can reveal the important role of material stuff in the constitution of **spaces** (Chapter 13) and **identities** (Chapters 3 and 8). As Cresswell (2004: 2) describes beautifully at the beginning of his book *Place: A short introduction*, the manipulation of material objects is often a key way in which people modify spaces, to make themselves feel comfortable and 'at home' (and in so doing leave material traces of themselves) therein.

> Cast your mind back to the first time you moved into a particular space – a room in college accommodation is a good example. You are confronted with a particular area of floor space and a certain volume of air. In that room there may be a few rudimentary pieces of furniture . . . These are common to all the rooms in the complex . . . Now what do you do? A common strategy is to make the space say something about you. You add your own possessions, rearrange the furniture within the limits of the space, put your own posters on the wall, arrange a few books purposefully on the desk. Thus space is turned into place. Your place.

This importance of material objects in making spaces homely, comfortable, functional or connected to other people and locales is widely recognised by geographers interested in home and **architectural geographies** (Chapter 4). There is now a wonderfully rich body of geographical research exploring how we human beings habitually surround ourselves with material objects that mean something to us (perhaps mementoes, ornaments, decorations, photographs), or which make possible certain practices and lifestyles. For example, Rose's research on practices involving family photographs (Box 10.4), Tolia-Kelly's research on the importance of material objects in the homes of British-Asian women (Box 10.5), and Hallam and

Box 10.4
Rose on family photographs in domestic spaces

Here are two quotations from Rose's (2003) research with mothers in South-East England:

> Yeah, I moved in, just sort of put pictures [i.e. family photographs] up. Just to make it feel like home, you know.
>
> (p.6)

> I suppose because they [i.e. family photographs] are in that bay window, I have to look at those. You know you can't help, but you know, you draw the curtains at night. I look at them maybe only for one second, but you know my eye catches them and I think, well I'll look at them.
>
> (p.10)

Rose's research is full of stories about how material objects are important in the making of homely spaces, and family photographs are invariably described as central to the process of transforming 'a house into a home'. As in the quotations above, she notes that family snaps are often among the first things to be displayed when moving into a new house, and that they are frequently located so as to almost subliminally 'catch the eye' during daily chores and routines.

Rose argues that family snaps are important *because of what is done with them*, as much as what they show (see also Rose 2010). Thus she describes the considerable (and highly gendered – i.e. overwhelmingly done by mothers) work involved in sorting and storing photographs (in envelopes, albums, boxes, cupboards, digital files), framing, displaying, collaging or pinning them to noticeboards, and exchanging them with family members and close friends. Rose observes that doing family photography in these ways is important for families in a number of ways:

- creating and 'curating' a sense of the family, its history, and its landmarks and key moments;
- 'stretching' domestic space, and the feeling of being 'at home' (e.g. by carrying photographs in purses and wallets, or displaying them at work);
- maintaining family and friendship networks (especially via exchanging photographs);
- enabling spaces and moments of 'togetherness' (e.g. through the act of sharing, showing, giving, receiving snaps), even between geographically distant relatives;
- facilitating 'ritual' family practices (e.g. 'putting faces to names' of relatives and ancestors, keeping a record of rapidly changing faces and features, reliving and recounting special moments).

Box 10.5
Tolia-Kelly on artefacts in British-Asian homes

Divya Tolia-Kelly (2004a, 2004b) has conducted rich qualitative research with groups of British-Asian women living in and around north-west London, many of whom migrated to the UK in the 1970s. Through individual and group interviews and home visits her work reveals the importance of material objects (religious icons, ornaments, photographs, souvenirs, representations of landscapes, items from pilgrimage sites) in 'making home'. This notion of 'making home' simultaneously involves:

- making a house feel homely;
- making oneself feel 'at home' in the years and decades following migration;
- constituting identities and making/extending connections with diasporic communities.

For example, Tolia-Kelly highlights the importance of the mandir in the homes of Hindu women. Within Hinduism, mandir - from the Sanskrit *mandira*, meaning 'home' or 'dwelling place' - is the name given to a shrine or temple to a particular deity. Mandirs can be monumental in scale, but it is also traditional for Hindu families to maintain a small mandir in their home (often in a cupboard or cabinet, or else in a room devoted to the shrine).

Tolia-Kelly describes how religious icons and relics in mandirs - commonly including murtis (devotional statues, sometimes from pilgrimage sites), gangajaal (phials of water from the River Ganges), vibhuti (incense dust from temples or pilgrim sites), images of gurus - are typically surrounded by a changing 'collage' of personal and familial objects. Thus:

> [mandirs] allow growth of a collection of pieces that are sacred and blessed. These are not limited to religious sacred objects and icons . . . The shrines become places where . . . important family objects are placed. The shrines incorporate family photos of those who have passed away - grandfathers, great aunts, grandmothers and great uncles. The shrine becomes collaged, continually superimposed with objects reflecting intimate moments and sacred life moments . . . : a rosary given by a father, a piece of gold received on a wedding day, . . . sea shells . . . that were in the family home prior to marriage are all composed and united in the laying out of the shrine . . . Sometimes families place letters at the shrine - job applications, offers, letters of achievement and even travel tickets, so that all these possible journeys can be blessed. Symbolic representations of the first breath of a baby, the first job, the wedding, and all the rites of passage and major life stages can sometimes be traced . . . Intensely personal prayers are recited here - celebrating, requesting, and proffering. These are a set of intense moments, and small things; miniature icons representing larger things, moments and connections.
>
> (Tolia-Kelly 2004a: 319).

In the process, mandirs come to signify, simultaneously:

- religious iconography (images of figures, stories and messages from sacred Hindu texts);
- religious observance (a space for prayer and reflection);
- family genealogy and relationships;
- particular events (e.g. the event of receiving the rosary from one's father);
- community membership and identity (containing symbols that connect to, and are meaningful for, the family, local community and wider diasporic community);
- past, present and future journeys (as well as the example of 'possible journeys' above, objects associated with mandirs can also memorialise family members' homes prior to migration).

Hockey's work on memorialisation practices in the aftermath of a death (Box 10.6), each provides deeply affecting qualitative data on this topic. Spend some time reading the boxed extracts from these pieces of research (we hope you will be inspired to explore these authors' work in more detail). Although these studies deal with diverse topics, they share a sense that:

- material objects (often very small, seemingly inconsequential material objects) can have profound importance for individuals and communities;
- spatial practices involving objects can be very important to individuals' identity and sense of self (e.g. in terms of family belonging, diasporic community, or grief and loss in the examples in Boxes 10.4–10.6).

There are important connections here to the discussions of cultural **consumption** and **identities** elsewhere in this book: as we outlined in Chapters 3 and 8, the possession, display and use of particular objects (e.g. garments, commodities, texts) is

Box 10.6
Hallam and Hockey on the decoration of cemetery headstones

Hallam and Hockey (2001) explore the importance of material culture in processes of grief and commemoration, past and present. They note the huge range of material objects that have commonly been central to coping with death: from monumental material commemorations (e.g. a war memorial), to smaller memorial sites (e.g. memorial benches, or roadside memorials to victims of traffic accidents), to the material objects involved in funeral rituals (e.g. coffins, headstones, wreaths), to the role of smaller objects (e.g. finding a deceased grandparent's slippers, or clearing out cupboards in their home) in prompting raw grief or comforting memories (see also Maddrell 2009b; Wylie 2009; Maddrell and Sidaway 2010; Horton and Kraftl 2012a).

Material practices associated with death vary significantly over time and space. As one example of the (changing) role of material culture in the aftermath of bereavement, Hallam and Hockey observe the material objects left by visitors to a cemetery in Nottinghamshire, UK. They note that, since the 1990s, visitors to the cemetery:

> have increasingly brought gifts and personal possessions to the graves of their . . . relatives and friends. Clustered around and expanding out from the headstones are potted plants and vases of flowers, letters and cards, toys and small ornaments, decorative windmills and wind chimes, poetry in small frames, lanterns and candle holders, a can of beer.
>
> (Hallam and Hockey 2001: 147)

Hallam and Hockey argue that what is particularly important, and touching, about these decorative material is what visitors *do* with them. They note that visitors:

- frequently tend, repair, wash and weed the decorations around the headstone they are visiting;
- rearrange and add to the decorations, tailoring them to the interests, hobbies and personality of the deceased person;
- use the material objects for prompts for talking with other visitors, and swapping little family stories or experiences of bereavement;
- mark anniversaries, holidays, birthdays and significant dates by adding to, or rearranging, the displays around the headstone.

Through these practices with material objects (perhaps a toy, flower, candle, stone or beer can) some sense of contact, care, reflection - and what Hallam and Hockey call 'liveliness' - is maintained around the headstone. They note that these practices are especially intense - and produce particularly large displays - around the graves of children.

frequently central to the development and maintenance of cultural identities, whether via spectacularly subcultural styles (see Section 3.4) or more subtle practices (as in the wearing of contact lenses in Box 10.8).

Many geographers exploring the importance of material objects have drawn on a set of concepts and approaches known as 'material culture' studies. Formally and traditionally, this term describes a specific tradition of anthropological, archaeological and historical work focusing on material objects (indeed, you may remember from Chapter 1 that many **traditional cultural geographers** were interested in material culture, in this sense). More recently and broadly, the term has been extended to label a renewed 'materialised' sensibility in many areas of social scientific research (the geographical research summarised in this chapter is very much part and parcel of this sensibility). As Celia Lury (1996) explains, the term 'material culture' conveys how:

- cultures are always material – i.e. the implicit or explicit importance of physical, material objects in the everyday lives and cultural practices of individuals, communities and societies;
- material objects are always cultural – i.e. norms, meanings, and values invariably are attached to material objects in any particular time and location;
- these two facts (the material nature of culture, and the cultural nature of material objects) are closely and complexly interrelated and manifest in, for example, customs, social structures, geographical and historical patterns and artistic, religious and philosophical movements.

Classic conceptualisations of material culture by anthropologists such as Arjun Appadurai and Igor Kopytoff have shaped a concern with 'objects in use' and 'person–thing relations' in diverse research contexts. As Lury (1996) neatly describes, these theorists have been influential in making clear that: (i) 'social lives have things'; and, simultaneously, (ii) 'things have social lives'.

Let us examine each of these claims in turn. The first observation – social lives have things – describes how material objects are so often directly or implicitly involved in social relations and interactions. As we shall discuss in relation to status objects and gift objects below (and as is very evident in Boxes 10.4–10.6), material objects are frequently centrally involved in social relations and cultural practices. Theorists of material culture have essentially provided a vocabulary with which to describe this importance of objects in social structures, past and present. For example, archaeologists like Schiffer (1992) and Graves-Brown (2002) have attempted to describe the multiple functions of objects in societies, identifying how an object may have:

- *a technofunction* – a pragmatic purpose (e.g. a chair supports the weight of a human body);
- *a sociofunction* – a particular kind of object may have a broader meaning or purpose (e.g. in addition to its technofunction, a particularly luxurious designer chair may also serve to symbolise and show off its owner's wealth and status);
- *an ideofunction* – certain objects can represent broader ideas (e.g. a throne may be a comfortable seat, and may advertise its owner's wealth, but it also symbolises the very notion of monarchy and power itself, as manifest in notions such as 'the power of the throne').

Similarly, the anthropologist McCracken (1988) lists some of the practices through which material objects can be used or manipulated in rituals that play important roles in identity formation and social interactions. These rituals include:

- *possession rituals* – activities involving the accumulation, storage, display, curation or maintenance/repair of objects; acquiring certain objects can be important in **identity** formation (see Chapters 3 and 8), and bringing together and sorting out particular objects is an important task in the production of home and personal space (as discussed elsewhere in this chapter);
- *gift rituals* – as described later in this chapter, activities involving the choice, presentation, exchange and receipt of gifts can be an important mode of interpersonal communication and influence;
- *divestment rituals* – activities that involve either 'emptying' an object or space of meaning (e.g. clearing out a room when moving out, or redecorating when moving in) or erasing meanings that were previously attached to an object (e.g. persuading oneself that an object is no longer important, and can be disposed of).

Not all geographers employ exactly these terms, but what is important for geographers is the process of recognising and analysing the substantial and complex importance of material objects for social practices and cultural geographies (very evident in the examples in Boxes 10.4–10.6).

Lury's (1996) second point – that material things have social lives – may seem counter-intuitive. It can feel foolish to think that an object (like the mobile phone, monkey toy and contact lens solution bottle featured in boxes throughout this chapter) has a 'social life'. However, theorists of material culture urge us to pay close attention to the 'life stories' or 'biographies' of material objects. Kopytoff (1986) argues that objects should be understood as having biographies in two senses.

- Individual objects have a particular life story, each being made, owned and used in particular ways, by particular people in particular spaces (e.g. the monkey toy photographed in Box 10.3 was purchased by a particular parent and loved by a particular child, before getting lost in a particular London hotel one day in the 1970s).
- More broadly, types of object have a social history, and their form and characteristics reflect wider cultural-historical changes (e.g. in Box 10.3 Monkey can also be interpreted as exemplifying the phenomenon of the 'cuddly toy', popular in many

parts of the world since the mid-nineteenth century, and closely related to a particular construction of early childhood, and his particular representational form perhaps relates to the popularity of the movie *Jungle Book*, there and then).

In a hugely influential essay on 'the social life of things', Appadurai (1986: 3–5) argues that tracing these kinds of individual or social history reveals how:

> commodities, like persons, have social lives . . . [W]e have to follow the things themselves, for their meanings are inscribed in their forms, their uses, their trajectories. It is only through the analysis of these trajectories that we can interpret the human transactions and calculations that enliven things.

Appadurai's discussion of 'following the things themselves' – and the social 'liveliness' of things that is thus revealed – has prompted many geographers to notice and discuss 'things in use' and 'things in motion' in the diverse contexts in which they work (see Bridge and Smith 2003). Again, we refer you to the case studies in Boxes 10.4–10.6, as well as Ian Cook's work on 'following the thing' (see Section 10.2) for some fascinating illustrative examples. Moreover, given the importance of material objects to all human geographies, we invite you to consider the importance of material objects in one space, or for one geographical issue, that interests you. The following questions draw upon some key ideas that have emerged from material culture studies: use them as prompts to reflect upon the significance of material objects for your chosen space or issue.

Which material objects are valued, and why?

As already noted in this chapter, material objects can have financial, symbolic or sentimental value, and they can be valuable to us as consumers, or in making us feel at home, or in terms of identity formation, or in many other ways besides. The particular ways in which objects are valued varies substantially in different places/times.

Which material objects have high status?

The symbolic status of particular objects (e.g. a new 'must-have' commodity) can be profoundly important in shaping social interactions. As a small example, in Box 10.1 having an 'out of date' mobile phone was noted as a potential cause of social embarrassment in contemporary UK consumer culture. Countless value judgements of this kind are made, to the extent that the material objects we own (the clothes we wear, the car we drive, the stuff we buy, the stuff we cannot buy) can be an important basis for identity formation and social exclusion.

How are material objects involved in social practices and relations?

As anthropologists have long recognised, practices such as giving gifts, writing letters or lending objects are often crucial to the maintenance of friendships, family ties and social relations. These practices typically have culturally specific ritual forms (e.g. the annual ritual of Christmas morning in the UK and North America). More generally, we can say that material objects can be crucial *intermediaries* in social relationships (consider the importance of phones, pens, laptop computers and so on in facilitating communication).

How are material objects important in the production of space?

As already illustrated, material objects are often important in the constitution of spaces such as home or the desktop workspace. More generally, as **architectural geographers** have made very clear (see Chapter 4), material features and components are central to the construction and experience of spaces. The material characteristics of a space can be manipulated to create particular atmospheres (Chapters 11 and 13), to affect behaviour, to present a certain image, or to include some people and exclude others (as a simple example, consider the building of material features such as walls and barriers).

Which kinds of objects have emotions or memories attached to them?

As the story of Monkey (Box 10.3) illustrates, mass-produced material objects can be invested with significant emotion, and objects like photographs can stir up all kinds of memories (see also Box 10.4). Many common practices involving material objects engage with these kinds of memories and emotions, as in the display of ornaments (Box 10.5), the maintenance of memorials (Box 10.6), the purchasing of souvenirs, the exchange of romantic gifts, or the production of 'home' (see the earlier Cresswell quotation).

How are material objects classified and categorised?

As the following quotation illustrates, human beings frequently undertake different kinds of classification work: the ways in which material objects are sorted, categorised and valued typically reflects personal and societal norms and ideals (e.g. about which tasks should be completed first, or which objects should (not) be on display).

> To classify is human . . . We all spend large parts of our days doing classification work, often tacitly, and make up and use a range of ad hoc classification in order to do so. We sort dirty dishes from clean, white laundry from [coloureds], important email . . . from e-junk . . . Our desktops are mute testimony to a kind of muddled folk classification: papers which must be read by yesterday, *but which have been there since last year*; old professional journals which really should be read and even in fact may someday be, *and which have been there since last year*; assorted grant applications, tax forms, various work-related surveys and forms . . . These surfaces may be piled with sentimental cards which are already read *but which can't yet be thrown out* . . . Any part of the home, school or workplace reveals some . . . system of classification: medications classed as not for children occupy a higher shelf than safer ones; books for reference are shelved close to where we do the Sunday crossword puzzle; door keys are colour-coded.
>
> (Bowker and Star 1999: 1)

Which material objects generally go unnoticed?

Attfield (2000) notes that, even in the field of material culture studies, there is a tendency for certain kinds of object to go unnoticed or be taken for granted in everyday spaces. It is relatively easy to notice and explore dramatic, spectacularly meaningful, visible, 'celebrity' objects within a cultural context, but Attfield reminds us to look for other kinds of object (and also to note what happens to visibly high-status objects once they have slipped 'off the radar' or become obsolete).

What material objects are treated as undesirable, waste or rubbish?

Relatedly, Attfield (2000: xv) argues that "there is so much to learn not only from the things we value, but from the rubbish, detritus and discarded things": she calls our attention to geographies of waste, rubbish, clutter, mess, and the world of objects that are considered taboo or undesirable in a particular context.

What can material objects tell us about wider social trends and contexts?

As much research on material culture has argued, the social history of particular kinds of object can reveal broader cultural-historical trends. For example, social scientists interested in childhood have used the changing shapes and functions of beds, clothes, toys and rooms designed for children in different historical eras as an indicator of shifting societal attitudes towards children.

What activities do material objects make possible?

A central contribution of material culture studies is to demonstrate how material objects make possible – or 'afford' – all manner of activities and interactions. As already discussed, material objects frequently serve as intermediaries between individuals and communities. It is also literally the case that material objects such as tools, devices, commodities and construction materials make possible all kinds of activity. Finally, as the poem *Football Stone* nicely illustrates (Box 10.7), material objects can afford play and imagination: in this case, imaginatively transforming a suburban street into a football stadium with big-match atmosphere.

10.4 Non-human geographies and heterogeneous materialities

Let us reiterate two key points that have emerged through Sections 10.2 and 10.3. First, we have established that material objects are profoundly important, although often taken for granted, in everyday geographies:

> We live so closely with things; we take them for granted, and do not regard our relationship with the washing machine or the chair as in any way important or problematic. Once acquired, we often do not notice these objects until they break or are commented on by someone else. Until they are put in a museum or turn up in a strange context, we do not notice that they are culturally distinctive, that they are part of *our* lives alongside the people we live with. Much more of our daily lives is spent interacting with material objects than interacting with other people. Even when not actually handling them, our contact with objects is often continuous and intimate in comparison with our contact with people (think of the chair you are sitting on and the range of objects in sight, within a short reach that are there, ready for when you need them).
>
> (Dant 1999: 15)

Box 10.7

Football stone: imaginative play afforded by a material object

Photograph 10.3 A stone on a pavement can afford play and imagination.
Source: Pat Bond.

On my way home, on the ground
I see a stone, small and round
I give the stone a mighty whack
It hits the wall and comes straight back
Another touch, nice control
A gentle tap, a steady stroll
Up the street, heading home
Keeping it moving, football stone
...

I start to run, I'm a star
Round the lamppost, watch the car
Past the pillar-box, shoot to win
Blistering strike, it must be in
The ref blows loud, end of play
I'll leave it there for another day
Up the street, heading home
Keep it moving football stone

Source: Unknown

Indeed, the ubiquitous presence of objects can make us overlook their importance in shaping our world:

> Somehow there is something quite natural about 'things'. Things seem always to have been there, defining the world physically - through such objects as walls . . . and doors . . . Things are ubiquitous like furniture that gives support to our bodies when we sit down, and paths that direct our feet where to walk, vehicles that transport us on the daily round. Things lend orientation and give a sense of direction to how people relate . . . to the world around them.
>
> (Attfield 2000: 14)

Second, we have noted that 'following the thing' draws our attention to the complex biographies and production processes of material objects. Box 10.8 provides another example of this kind of complexity; as we found when trying to write it, a short ingredients list on a bottle can lead to countless geographical stories (see also Box 10.1). This is an example of what Miller calls the 'explosive' and spiralling complexity of material culture:

> Material culture virtually explodes the moment one gives any consideration to the vast corpus of different object worlds that we constantly experience. Within an hour of waking we move from the paraphernalia of interior furnishing . . . through choices of apparel, through the moral anxieties over the ingestion of foodstuffs, out into the variety of modern transport systems held within vast urban architectural and infrastructural forms.
>
> (Miller 1998: 6)

Box 10.8

Things on our desk no. 3: half-empty bottle of contact lens fluid

Photograph 10.4 Bottle of contact lens fluid on author's desk.

Source: John Horton.

One of the authors keeps a 120ml bottle of branded 'advanced conditioning solution for gas permeable contact lenses' on his desk (Photograph 10.4). Essentially, he squirts this colourless fluid onto his contact lenses when a piece of dust, or similar, gets stuck on the lens surface (incidentally, as lens wearers will know, it is astonishing how *tiny* material objects, like a microscopic dust speck, can cause substantial ocular discomfort). He has worn contact lenses since his late teens. It made a big difference to his self-confidence and **identity** (Chapter 8) when he changed from wearing glasses to wearing contact lenses; indeed, he used to feel shy and embarrassed about wearing his glasses in public (see Horton and Kraftl 2006b), although he feels far less hung up about this now.

Wearing contact lenses involves an intricate (and rather expensive) daily routine involving various equipment and products (mainly bottles of fluid). Lens wearers tend to have detailed and decided opinions about their favoured lens type, cleaning regimes and eyecare techniques (incidentally, again, lens wearers also tend to become very skilled and used to messing with their eyes, developing a set of bodily practices that non-wearers tend to find inconceivable and repulsive). However, the author realises that, although he has regularly put this 'advanced conditioning solution' in his eyes for nearly 20 years, he has never really considered what it actually is. So here is the list of ingredients:

> polyaminopropylbiguanide, chlorhexidinegluconate, disodium edetate, polyquaternium 10, cellulosic viscosifier, polyvinyl alcohol, triquaternary phospholipid, derivatised polyethylene glycol.

Following Cook (2004, 2005) and Cook *et al.* (2007) (see Section 10.2), we have tried to use this list as a set of 'starting points for geographical detective work'. To be honest, this is challenging for cultural geographers with limited chemical understanding and vocabulary. However, a day of online searches for information about these ingredients does reveal some thought-provoking geographical issues and phenomena.

First, it is clear that these ingredients are the products of a vast global industry devoted to the manufacture or derivation of chemical compounds: most of these chemicals are produced and exported at a rate of millions of metric tonnes per annum, with particular clusters of production in China and south-east Asia.

Second, it is particularly difficult to find information about how, where and under what conditions these substances are produced: the way they are named and listed suggests that they are essential, naturally occurring elements, giving no clue about what they actually are, or their means of production.

Third, there exists a well-developed array of online consumer forums devoted to sharing and archiving information about ingredients such as these: these online spaces allow consumers to query, understand and critique the use of particular ingredients in everyday products such as cosmetics, medicines, cleaning fluids, shampoos and foods.

Fourth, the origin stories of these substances, which are hidden by the rather lifeless and blank chemical nomenclature, are often remarkable. Indeed, some of these ingredients originated as/in living organisms: for example, molecules used in polyaminopropylbiguanide production originated in a rather beautiful herbaceous plant (French Lily), viscosifiers are manufactured from the cell walls of algae or bacteria, and triquaternary phospholipids are derived from a substance in cocoa plants. Notably, compounds of these substances have been copyrighted, and marketed under trade names, particularly by pharmaceutical companies.

Clearly, there is a lot going on here. And this is without considering the mixing and bottling of these ingredients, the crafting and fitting of contact lenses themselves, the manufacture of the lens solution bottle . . . The more one attempts to follow the thing, the more questions are raised.

So, we can safely say that material objects: (i) are important for everyday geographies; and (ii) are complex. Moreover, many experts on material culture argue that material objects are becoming *increasingly* important and *increasingly* complex in many current contexts. Social scientists such as Scott Lash have argued that there are increasingly 'technological forms of life' in many parts of the world: where commodities and technologies have become increasingly important to many people's identities and lifestyles (Lash 2001), and where these commodities and technologies are being created via increasingly complex, globalised systems of production (Lash and Lury 2007).

Many geographers argue that the importance and complexity of material objects demand some new ways of thinking about human geography. Four key elements of this argument are summarised below. In each case, we introduce some of the broader concepts that have been important for many geographers. In this context you will often see the term 'materiality' used, to describe the complex materialness of the world.

Point 1: Things, and 'non-humans', can act back

It is now quite widely argued that human geographers, and other social scientists, have tended to over-privilege human agency. That is, they have overwhelmingly focused on *what people do,* and the reasons for, and effects of, their actions. However, it is now recognised that this is a rather limited view of the world. Human beings are *not* the only ones shaping and affecting geographical issues. There are other forms of (what is slightly clumsily termed) 'non-human agency' going on in the world too: that is, *things other than human beings* do stuff and impact upon spaces.

Some of the most vivid examples of non-human agency can be found in the work of human geographers who are interested in animals, plants and organisms; indeed, it may surprise you to learn that there is a subfield of social and cultural geography devoted to 'animal geographies'. Animal geographers often highlight the ways in which animals "often end up evading the places to which humans seek to allot them" (Philo and Wilbert 2000: 14). As anyone who has owned a pet or worked with livestock will know, animals can create considerable work for humans: they can require humans to do things; they can transform the spaces they inhabit in all kinds of ways; they can 'misbehave', 'act back' and transgress our rules and boundaries.

Philo and Wilbert cite as a direct and dramatic instance of this kind of transgression the example of Cholmondeley the chimpanzee, who escaped from his enclosure at London Zoo in January 1951 and briefly ran amok (Photograph 10.5). Through his action, Cholmondeley thus profoundly unsettled the neat enclosures, boundaries and expectations of the zoo space (as idealised in Photograph 10.6).

But this is not the only form of non-human agency. As the example of the scallops of St Brieuc Bay suggests (Box 10.9), even less obviously active beasts can be important in the subtle shaping of geographical contexts and issues: in St Brieuc Bay, the gradually changing scallop ecosystem prompted a considerable amount of human activity and behaviour change.

We may question whether this really counts as agency: Cholmondeley the Chimp probably meant to escape (although we cannot be sure – perhaps he was just momentarily freaked out), but the scallops were almost certainly unaware of their role in the development of marine conservation and biological science. Nevertheless, these non-humans were important in shaping their environment and prompting humans

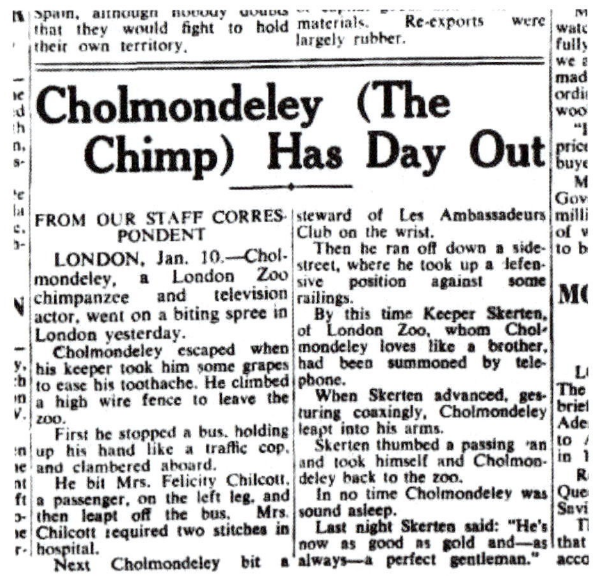

that they would fight to hold their own territory.

materials. Re-exports were largely rubber.

Cholmondeley (The Chimp) Has Day Out

FROM OUR STAFF CORRESPONDENT

LONDON, Jan. 10.—Cholmondeley, a London Zoo chimpanzee and television actor, went on a biting spree in London yesterday.

Cholmondeley escaped when his keeper took him some grapes to ease his toothache. He climbed a high wire fence to leave the zoo.

First he stopped a bus, holding up his hand like a traffic cop, and clambered aboard.

He bit Mrs. Felicity Chilcott, a passenger, on the left leg, and then leapt off the bus. Mrs. Chilcott required two stitches in hospital.

Next Cholmondeley bit a steward of Les Ambassadeurs Club on the wrist.

Then he ran off down a side-street, where he took up a defensive position against some railings.

By this time Keeper Skerten, of London Zoo, whom Cholmondeley loves like a brother, had been summoned by telephone.

When Skerten advanced, gesturing coaxingly, Cholmondeley leapt into his arms.

Skerten thumbed a passing van and took himself and Cholmondeley back to the zoo.

In no time Cholmondeley was sound asleep.

Last night Skerten said: "He's now as good as gold and—as always—a perfect gentleman."

Photograph 10.5 Newspaper report about Cholmondeley the Chimpanzee.

Source: from Cholmondeley (The Chimp) Has Day Out, *Sydney Morning Herald* 11/01/1951, 3.

to do stuff. In this context, non-humans are often described as 'actants': they have an effect, prompt activity, and transform spaces, even though they may not intend to do so (or, indeed, they may lack any capacity for, or concepts of, intentional action).

Extending this argument, some geographers have explored the 'actancy' of trees and other plants (Cloke and Jones 2004). Extending the argument even further, it can also be claimed that entirely inanimate objects can be actants too: they can have an effect, prompt activity, and transform spaces, even though they are 'just' things. Consider the case of Ozzy the owl in Box 10.10: it is just a small earthenware jug, but 'he' brings liveliness and activity to the gallery space where he is exhibited.

Thrift (2000a) memorably talks about how material things can 'act back': for example, commodities may be made and used by human beings, but they can act back when they break or malfunction. A malfunctioning material thing (perhaps a broken phone, a lost toy, a gritty contact lens) can have an effect, prompt

Photograph 10.6 1950s visitor map of London Zoo.

Source: from 1950s visitor map of London Zoo, Zoo Guide © Published in 1958 by The Zoological Society of London.

Box 10.9
The scallops of St. Brieuc Bay

Photograph 10.7 The king scallop (*Pecten maximus*).
Source: FLPA Images of Nature/DP Wilson.

St Brieuc Bay, in Brittany, is one of Europe's foremost locations for the fishing of edible scallops (Photograph 10.7).

In a paper that is recognised as a classic example of actor-network theory, Michel Callon describes a meeting of marine biologists to discuss declining scallop stocks: "a handful of researchers discuss a few diagrams and a few tables with numbers in a closed room . . . in a conference room at the Oceanographic Centre of Brest [Brittany] in November 1974" (Callon 1986: 218). These diagrams and data had significant consequences: they led to new scallop fishing practices and marine conservation measures in the bay.

In a similar move to Latour and Woolgar's *Laboratory Life* (see Box 9.3), Callon shows how these scientific facts were produced through complex interactions between three marine biologists (and a global network of specialists), elected representatives of local unions of fishermen (and the workers and communities they represented), a sample of 100 specimens of the king scallop, *Pecten maximus* (used in the biologists' experiments), and numerous pieces of laboratory equipment (including, for reasons, we shall not go into, nets made from straw, broom, moss and palm fibre).

Producing those diagrams (out of all that complexity), and persuading local fishermen of their implications, involved considerable work, most obviously on the part of the marine biologists. More than this, though, Callon shows how the scallops, biologists and fishermen of St Brieuc Bay were just part of a much broader context of interconnected processes and geographies. For example, the lives of the biologists, fishermen and scallops were affected by: shifting ecological niches of different scallop species around the Normandy and Brittany coasts; invasions of predatory starfish in particular ecological niches; impacts of hard winters upon scallop reproduction and habitats; innovations in fishing practices; shifting labour relations, union politics and local political administrations in local fishing communities; the changing tastes of French scallop consumers (increasingly, scallops were popular at Christmas, and scallops with roe seen as a particularly delicacy) linked to the contemporary marketing and commercialisation of shellfish; increasingly intensive fishing practices to meet consumer demands in certain coastal areas; increasing anxiety about falling scallop stocks in some areas . . . We could go on: Callon's point is that the activities of the marine biologists were just part of a remarkably complex range of associated processes.

Rather poignantly, too, Callon also makes a point about the precarious nature of actor-networks. Despite the existence of scientific data, despite the work of the marine biologists, and despite the community's decision to implement a marine conservation scheme, he notes that the scheme was quickly abandoned. A complex combination of diseases to marine organisms, changing predation patterns, unseasonal tides, currents and water temperatures, changing fishing union representatives, and withdrawal of local funding for marine conservation meant that the scheme did not 'work' in its first 12 months.

Box 10.10
Ozzy the owl

Photograph 10.8 Ozzy the owl.
Source: Stoke Potteries Museum.

'Ozzy' the Owl (Photograph 10.8) is a seventeenth-century slipware jug, on display in the internationally renowned ceramics gallery in Hanley City Art Gallery, Stoke-on-Trent, UK.

The gallery's priceless collection of English ceramics was assembled in the 1930s, and the gallery was laid out, in its current form, in 1979. However, Ozzy is a more recent acquisition. In a charming paper, Hetherington (1997: 206) describes how Ozzy came to reside in this space.

> In 1990 . . . he made national news. Ozzy turned up in Northampton on the BBC TV's *Antiques Roadshow*, a long-running television programme in which members of the public bring in their antiques to be assessed and valued by experts . . . Ozzy was brought in to the show in a plastic carrier bag by a woman whose family had owned him since the 1930s, when he had been bought cheaply in a Birmingham antiques shop. His owner, who had been using him as a flower pot, thought him unremarkable but loveable . . . In fact, the expert . . . valued him at over £20,000.

Ozzy was sold at auction and acquired by the ceramics gallery. Although he is surrounded by thousands of more prestigious, valuable, and academically and historically significant artefacts he has become the gallery's star exhibit: he features on all of its publicity materials and, as Hetherington records, many visitors come especially to see him. Despite being fairly small and crudely made, he has a kind of 'charisma'. He has a name, celebrity status, a 'cheeky' smile, television appearances, and a charming and humble back story (bought in a junk shop, transported in a carrier bag, used as a flowerpot, loved by his owner).

Hetherington describes how the presence of this engaging object has transformed the space of the gallery: many visitors are drawn to Ozzy; they visit him first, disregarding the intended, chronological sequence of exhibits; they 'connect' with him, in a way not seen in other parts of the gallery; there are smiles and laughter around his display case; many visitors disregard the gallery's technically finest, and highest-status, exhibits and prefer to look for items that resemble Ozzy. Hetherington thus suggests that inanimate material objects can transform spaces and afford new emotions and spatial practices.

activity, and transform spaces too. To acknowledge non-human agency is to understand that human geographies are always 'co-fabricated' by the interplay of humans and all manner of non-humans; indeed, it should make us recognise that 'human agency' is really just one form of actancy among many, in a hugely complex world (Whatmore 2006).

Point 2: 'Human' geographies do not only involve human beings

In this chapter we have already considered how material objects can be important in mediating social relations (e.g. via gift exchange and status objects). However, as noted in Point 1, the acknowledgement of non-human agency requires us to recognise how humans and non-humans (and their diverse forms of agency and actancy) are complexly interrelated in all human geographies.

A body of concepts known as actor-network theory (or ANT) has been important for geographers thinking through along these lines. Actor-network can be thought of as a way of looking at the world, acknowledging that "society, organisations, agents and machines are all effects generated in patterned networks of diverse (not simply human) materials" (Law 1992: 2). That is, geographical issues and contexts – and more or less everything in the world – must be understood as complex and (to use ANT terminology) *heterogeneous* associations of humans and non-humans. Heterogeneous means dissimilar, differentiated, with multiple constituent parts (i.e. the opposite of homogeneous).

In Chapter 9 we introduced Latour and Woolgar's work on *Laboratory Life* (see Box 9.4). Latour is a key thinker in ANT, and his work on the laboratory is considered an early classic of ANT. As we discussed, *Laboratory Life* explores the practices and work that produce scientific 'facts'. Along with a number of other sociologists of science, Latour and Woolgar (1987) reveal how scientific knowledge is invariably "the end product of a lot of hard work in which heterogeneous bits and pieces – test tubes, reagents, organisms, skilled hands, scanning electron microscopes, radiation monitors, other scientists, articles, computer terminals, and all the rest . . . are juxtaposed into a patterned network" (Law 1992: 2). Latour's work shows how scientific data emerge only through hard, material work and processes of 'translation' and ordering (practices, such as taking readings, doing equations, writing reports, which translate the complex heterogeneity of laboratory life into neat, legible, singular and powerful outputs).

Actor-network theorists argue that this way of looking at the world – acknowledging complex heterogeneous networks and practices – must be applied to all social institutions, organisations, hierarchies, issues and phenomena (basically, all human geographies). To make sense of this argument, it is useful to look at a case study. Box 10.9 outlines a famous example of ANT by the sociologist Michel Callon, which introduces us to the scallops, fishermen and biologists of St Brieuc Bay in Brittany. The case study opens with an example of apparently straightforward human agency: three marine biologists present some data, which results in changes to fishing practices in the bay. However, Callon's very detailed and careful analysis reveals the complex processes – and heterogeneous forms of human and non-human actancy – that intersected, there and then.

This realisation – that apparently straightforward issues and spaces are actually complex, heterogeneous associations of humans and non-humans – is perhaps the most important point of actor-network theory for geographers. To reiterate, this realisation is challenging for a range of reasons, because it requires us to:

- acknowledge that "the stuff of the social isn't simply human" but is co-constituted *with* all manner of non-humans, machines, texts, architectures, and everything else (Law 1992: 2);
- avoid oversimplifying the world's heterogeneity;
- refuse to be tempted by accounts of the world that put human beings at the 'centre of the universe' and prioritise human agency (but overlook multiple forms of actancy);
- think critically about how social scientific accounts and methods that prioritise human agency have come to seem normal and natural.

Point 3: Humans and non-humans are intimately associated

Humans and non-humans are often very intimately interrelated. Think about the example of contact lenses (see Box 10.8). Many people have these small, manufactured, commercialised objects (and their associated fluids and products) in their eyes for hours on end, every day. They can become part of our daily routines and spaces, part of our identities and, essentially, an extension of the visual apparatus of our eyes. With a little practice, it is possible to wear them without really noticing they are there: we just *see*, paying little notice to the fact that an additional synthetic lens resting on our cornea is modulating our eyes' focal length.

Contact lenses are just one case study of a very close association between humans and non-humans.

Further examples come easily to mind: for example, clothes, shoes, hearing aids, hip replacements, spectacles, sunglasses, prosthetic limbs, wigs, jewellery, hats. Each of these objects can be closely associated with/in human bodies. It is also the case that these objects can extend the capacities of human bodies: contact lenses and spectacles optimise vision; clothes afford bodily comfort and survival in all weathers; prosthetic joints and limbs enable people with injuries to walk or live more independently; and so on (see Pels *et al.* 2002).

We might also think about other kinds of intimate connection between humans and technologies: wearing headphones; staring at screens; typing, or writing a mobile phone text message (perhaps not really needing to look at the keypad); driving (perhaps not really having to think about every little action or manoeuvre); operating machinery (experienced digger drivers talk about how the levers and parts of the machine feel like an extension of their body); working out in a gym.

These close associations between humans and non-humans have prompted many geographers to think critically about the distinction between humans and non-humans. As already mentioned, human geographers and other social scientists have often prioritised human beings, and implicitly understood us as somehow separate from (and worthy of more attention than) non-humans. But this assumption has been increasingly unsettled since the cultural turn, and is now considered untenable by many cultural geographers.

The feminist theorist Donna Haraway's notion of 'the cyborg' has been very influential here. Haraway (1991: 150) argues that the world is increasingly full of increasingly complex "couplings between organism and machine". She uses the term 'cyborg' to describe this kind of 'hybrid' coupling, where bodies, organisms, objects, technologies, humans and non-humans are closely associated. Haraway argues that three 'crucial boundary breakdowns' in contemporary science, medicine, society, culture and politics have produced multiple hybrid associations of this kind, and created a kind of 'leakiness' between categories that were traditionally seen as mutually exclusive. Her three boundary breakdowns are as follows.

- *Breakdown of the boundary between humans and animals* – Haraway argues that: advances in biological and evolutionary theory show that *Homo sapiens* is not exceptional or distinctive, but just one animal species among many; animal rights movements demonstrate that there are no philosophical or practical reasons for believing that humans are separate from or superior to other animals; ecofeminist and environmental movements demonstrate that distinctions between nature and society are untenable, and human cultures must be understood as part of ecosystems and biocycles.
- *Breakdown of the boundary between humans and machines* – As well as drawing attention to the kinds of association between humans and non-humans already mentioned in this section, Haraway argues that the development of increasingly 'lively' machines (e.g. forms of artificial intelligence), new forms of cybernetic technology, and increasingly sophisticated interfaces built into commonplace technologies are increasingly reducing the sense of separateness between humans and machines.
- *Breakdown of the boundary between the physical and non-physical* – Haraway draws attention to the ways in which 'virtual' technologies and spaces are important for many human geographies. She argues that apparently non-physical connections and invisible structures (e.g. virtual connections between computers) increasingly have real meaning and importance, and can structure, empower or disempower our lives.

Haraway's work has prompted a rich vein of geographical work exploring, for example, the commodification of bodily substances, organs and genetic resources (Greenhough 2006; Greenhough and Roe 2006), machinic extensions of human bodies and imaginations, the trademarking of biological and molecular substances as intellectual property, and controversies and anxieties surrounding hybrid forms such as genetically modified organisms (Whatmore 2002, 2006).

Point 4: Human bodies are complexly material

Finally, we should make a link to a point made in Chapter 12: that **bodies**, themselves, are complex material objects. It is easy to fall back on a view of the world that sees human bodies as singular objects, with neat, coherent characteristics and identities. But, as we shall explore in Section 12.5, this is problematic, because it overlooks the significant and shifting complexity, materiality and heterogeneity of our bodies. As a point of connection with the next two chapters, we shall close this section with a quotation from the political scientist Jane Bennett that wonderfully evokes the heterogeneous assemblage of actants that make up any bodily practice (like sitting, typing out a book manuscript):

> The sentences in this book emerged from . . . : 'my' memories, intentions, contentions, intestinal bacteria, eyeglasses, and blood sugar, as well as from the plastic computer keyboard, the bird song from the open window, or the air or particulates in the room, to name only a few of the participants.
>
> (Bennett 2010: 23)

Summary

- Geographical work that 'follows the thing' can produce revelatory understandings of the processes and spaces through which commodities are produced in practice. Karl Marx's theorisation of commodity capitalism offers a vocabulary and framework for thinking about the politics of these spatial processes.
- Many cultural geographers have explored the importance of material objects in the constitution of spaces, meanings and identities. The interdisciplinary field of 'material culture' studies provides useful concepts and prompts for reflection in this context.
- Relationships between people, material objects and other 'non-humans' are so complex that they require us to think and write about the world in new ways. Many geographers find the language and concepts of actor-network theory and cyborg hybridity useful in this task.

Some key readings

Actor-Network Resource (www.lancs.ac.uk/fass/centres/css/ant/ant.htm)

For anyone wishing to explore actor-network theory, this is the place to begin. A definitive list of resources and readings, helpfully annotated to cater for all interests and levels of confidence.

Anderson, B. and Tolia-Kelly, D. (2004) Matter(s) in social and cultural geography. *Geoforum*, **35**, 669–674.

A very effective summary of the implications of ideas of material culture and materiality for cultural geographers.

Bennett, J. (2010) *Vibrant Matter: A political ecology of things*, Duke University Press, Durham, NC.

As signalled by the quotation at the end of this chapter, this book develops connections between themes of materiality, emotional and affective geographies (see Chapter 11), and bodily geographies (see Chapter 12).

Clark, N., Massey, D. and Sarre, P. (eds) (2008) *Material Geographies: A world in the making, Sage, London.*

A useful collection of essays that apply many of the ideas covered in his chapter to 'real life' case studies of geographical issues. www.followthethings.com

A fascinating, troubling, wide-ranging website run by Ian Cook *et al.* The site compiles research and resources that trace the production of everyday commodities. Resources are grouped into 'departments' – grocery, electrical, fashion, pharmacy, and so on – covering most aspects of contemporary Western domestic life and consumer culture.

Lash, S. and Lury, C. (2007) *Global Culture Industry,* Polity Press, Cambridge.

This accessible book neatly connects theories of material culture and materiality to the debates on cultural production and consumption we discussed in Chapters 2 and 3. The book includes a wide range of engaging case studies exploring the circulation of popular culture texts, objects and media.

Miller, D. (2008) *The Comfort of Things,* Polity Press, Cambridge.

The anthropologist Daniel Miller is a key thinker on material culture, and this book reflects upon the importance of material objects for the residents of one 'unremarkable' street in London.

Whatmore, S. (2002) *Hybrid Geographies: Natures, cultures, spaces,* Sage, London.

An outstanding, challenging and thought-provoking book that discusses the implications of actor-network theory and Haraway's hybridity for geographers.

Emotional and affective geographies

Before you read this chapter . . .

- Identify a space that affects how you feel. Why is this so?
- Consider any major geographical issue that you have studied. How does it *matter* for people affected by it? To what extent do *you* care?
- Think about a time when you were happy. What does 'happy' feel like? (Indeed, how do you *know* when you are happy? How easily can you put this into words?)

Chapter Map

11.1 Introduction: an 'emotional' moment

As an introduction to this chapter, consider the following quotation, in which the geographer Kye Askins describes a visit to a community centre in north-east England:

> I felt it the moment I walked through the door . . . I can't say what 'it' is exactly. Let's call it 'emotion' for now.
>
> (Askins 2009: 4)

For Askins – who had worked closely with the centre, and developed close relationships with its staff and users – this was a moving moment: '*I felt it the moment I walked through the door.*' In hindsight, she finds that some characteristics of this moment can be recalled and put into words:

> intangible warmth . . . metaphorical embrace. A certain kind of happiness, of pleasure . . . [S]miles on all our faces . . . hugs and greetings . . . renewed hope, renewed anxiety. What was I doing here? How could I make a difference?
>
> (Askins 2009: 1)

Askins' description of her feelings as she walked through that door illustrates how complex, deeply felt 'emotions' can be evoked, at particular moments, by certain spaces, encounters, journeys and relationships.

Indeed, the preceding quotations might prompt us to recall similar 'emotional' moments from our own lives, and to consider how our everyday geographies are always emotional in some way (wherever we are, and whatever we are doing, we are always feeling *something*).

In the following section we suggest that human geographers have often tended to overlook or underestimate the importance of 'emotions' for **everyday geographies** (Chapter 9), **landscapes** (Chapter 5) and **identities** (Chapter 8). We then describe a turn to acknowledge and study the importance of 'emotions' in diverse social-cultural issues, and consider some key geographical work in this context. You may have noticed that – as in Askins' opening quotation – we have placed the word 'emotion' in inverted commas in this introduction. This is to indicate that – while it may be widely used, both popularly and by geographers – this term and concept is considered problematic by some critics: for an explanation, see Section 11.4 entitled *'emotions' or 'affects'?*

11.2 Trouble expressing 'emotions'?

Many social and cultural geographers, like Bondi *et al.* (2005: 1), now argue that the discipline of human geography "often presents us with an emotionally barren terrain, a world devoid of passion", and that "geography, like many of its disciplinary siblings, has often had trouble expressing its emotions". Since the **cultural turn** (Chapter 1), this 'trouble expressing emotions' has been recognised and critiqued as "deeply buried in the history of Western thought" and, as such, evident in many domains of social scientific enquiry (Williams 2001: 2). Five related root causes have been suggested for this 'trouble expressing emotions'. Note that these factors are closely linked to the longstanding absence of **bodies** and **embodiment** from much work in human geography and other social sciences (see Chapter 12).

First, it is argued that current assumptions about what counts as 'good' academic work continue to be shaped by a specific set of ideals about scholarship and the nature of knowledge that – it is commonly argued – originated during the 'Enlightenment' in eighteenth-century Western societies. The term 'Enlightenment' is used to describe a social-historical period that marked the inception of many modern academic institutions and disciplines, and which was characterised by intellectual movements highlighting the importance of rational scientific enquiry. In this context, ideals like 'objectivity', 'truth', 'reason' and the 'detached scientific mind' were privileged, and, it is argued, emotions (being supposedly base, bodily, animalistic and prompting 'irrational' behaviour) were considered "the very antithesis" of these ideals (Williams and Bendelow 1998: 131). More broadly, this period witnessed diverse social reforms in relation to democratic processes, religious tolerance and human rights: these ideals of 'civilising progress' were often celebrated as a "move on" from earlier – more 'irrational', *'emotional'* – forms of society (Greco and Stenner 2008: 5).

It is argued that these understandings of emotion remain influential in many aspects of Western societies. For example, they perhaps underlie popular assumptions about research. For the notion of the 'the detached scientific mind' continues to be valued and privileged; meanwhile, in popular and academic **discourses** (see Chapters 2 and 6), emotion continues to be represented, or implicitly understood, as "a source of subjectivity which clouds vision and impairs judgement, while good scholarship depends on keeping one's emotions under control and others' under wraps" (Anderson and Smith 2001: 7).

Second, many feminist social scientists have argued that "discussing one's emotions appears to be perceived as illegitimate or irrelevant to academic debate (within the male-dominated academy)" (Widdowfield 2000: 200). Such critiques identify both a lack of research acknowledging the personal, private, emotional lives and work of many women, and a broader "gender politics of research in which detachment, objectivity and rationality have been valued, and implicitly masculinised, while engagement, subjectivity, passion and desire have been devalued and frequently feminised" (Anderson and Smith 2001: 7). This latter point reflects dualistic cultural norms and assumptions in contemporary Western societies, where "the mind/body, public/private, culture/nature, reason/emotion, concrete/abstract dichotomies are mapped onto gender differences so that the inferior of the two attributes is, in each case, assumed to be feminine and . . . excluded from theoretical investigation" (McDowell 1992: 409).

Third, culturally, the disclosure of emotion remains taboo in many contemporary societies. Indeed, social scientists concerned with health and well-being have long recognised a tendency for emotion to be 'pathologised', such that individuals demonstrating 'excessive' emotion are widely perceived and represented as irrational, defective or problematic. This is evident, for example, in contemporary taboos around the disclosure of mental health conditions (Parr 2008). Echoing an earlier point, it is argued that representations of 'excess' emotion as problematic are overwhelmingly gendered. That is, emotions "have tended to be dismissed as private, 'irrational' inner feelings or sensations, tied, historically, to women's 'hysterical' bodies and 'dangerous desires'" (Williams 2001: 2).

Fourth, the suppressions of emotion described in the preceding points are arguably manifest and reproduced in normalised everyday practices of contemporary academic research. Critics suggest that, for example, they can be detected in conventions of academic writing (where a neutral, 'objective', 'reasoned', 'factual' style remains the norm) and research (where the researcher is widely expected to function as a 'neutral', 'rational' collector and interpreter of data). In this context, emotions have often been represented as relatively unimportant compared with 'proper' academic concerns: as "a kind of frivolous or distracting background" to the "'real', really important business" (Thrift 2004: 57). This suppression of emotions could also be said to be present in norms of behaviour during learning or teaching within academic institutions: in the calm and quiet expected and enforced during most classes, for example, or in the distant and circumscribed personal relationships that conventionally exist between teachers and pupils.

Fifth, moreover, the establishment of subjects such as sociology and human geography as 'reputable' and self-sustaining academic disciplines involved emphasising their separation from other, longer-established ways of researching human beings. In particular, it is argued that early sociologists and geographers gained legitimacy by distancing their work from major extant domains of work regarding human physiology and psychology: that is, their work was able to take shape and gain prominence *precisely because* they differentiated themselves from work on human bodies and emotions (Greco and Stenner 2008). This disciplinary identity – and this kind of separation from the emotional, bodily aspects of human beings – has traditionally been reproduced through the relatively narrow framework of literatures and research that geographers are exposed to during their education and training. In practice, too, the complex, specialist languages, concepts and methods used in, for example, research on the neuroscience and physiology of emotions make these domains of work hard to access for other social scientists.

Increasingly, then, it is recognised that human geographers have, by and large, tended to "deny, avoid, suppress or downplay" 'emotion' in their research, writing and concepts (Bondi *et al.* 2005: 1). Moreover, it is argued that this 'trouble expressing emotions' has often constituted fundamental gaps and limitations in geographers' understandings of social and cultural spaces. For the 'silencing' of 'emotion' – resulting from the above points – has surely produced "an incomplete understanding of the world's workings . . . , exclud[ing] a key set of relations through which lives are lived and societies made" (Anderson and Smith 2001: 7).

11.3 An 'emotional' turn?

In reaction to the contexts outlined in the preceding section, a great deal of recent work by geographers has implicitly or explicitly considered the importance of 'emotions' in geographical issues, and for geographical research. Indeed, many geographers identify something of an 'emotional turn' (Bondi *et al.* 2005) within their discipline in the wake of the **cultural turn** (see Chapter 1). Geography's 'emotional turn' has consisted of diverse lines of theory and research, and should be understood as one component of a much broader turn to consider emotion within practically every domain of the social sciences (see Bendelow and Williams 1997; Williams 2001; Greco and Stenner 2008). Four key constituents of the 'emotional turn' in human geography have been as follows.

- During the 1970s, diverse work by humanistic and behaviouralist geographers critiqued the 'dehumanising' tendencies of Marxist analysis and quantitative spatial science. In essence, they argued that Marxian and positivist approaches – which were of prevailing significance within the work of many contemporary human geographers – were problematic because they said little about personal, taken-for-granted human experiences, "feelings and ideas as regards space and place" (Tuan 1976: 266). In order to understand these 'feelings' better, and their significance, it was argued that human geographers should engage with concepts and methods drawn from a variety of traditions, including phenomenology, existentialism, psychoanalysis and social/environmental psychology (see Cloke *et al.* 1991: 57–92; Ekinsmyth and Shurmer-Smith 2002). Although humanistic and behaviouralist approaches have latterly been critiqued – indeed, relatively few geographers would now adopt these methods uncritically – they were important in fostering an openness towards considering 'emotion' within geographical contexts.
- Since the mid-1970s, diverse feminist geographers have been strongly critical of the limited engagement with 'emotions' evident in countless aspects of the discipline of human geography. Four key examples are as follows.
    - Feminist critics of academic workplaces highlighted a shortage of empathy, care, support and interpersonal communication between (overwhelmingly and traditionally male) colleagues therein. They sought to transform, subvert or cope with these contexts by fostering working practices based on collaboration, peer support and shared acknowledgement of emotional experiences.
    - Reflections on the importance of positionality in research made it plain that – quite contrary to the aforementioned Enlightenment ideal of a 'detached scientific mind' – "not only does the researcher *affect* the research process, but they are themselves *affected* by the process" (Widdowfield 2000: 200).
    - Feminists' research on social and cultural geographies has frequently understood emotions as "alternative sources of knowledge" (WGSG 1997: 87). That is, their research has made it clear that using 'emotions' as a starting point or focus in any given research project is likely to "[open] up different understandings of everyday human lives" (Parr 2005: 473) in the given research context (see, for example, the following bulletpoint). To this end, feminist geographers have been important in developing research methods attuned to emotional, lived experiences in geographical contexts: for example, the pioneering work of feminist geographers has ensured that qualitative methods such as focus groups, in-depth interviews and lifecourse narratives have become staple tools for cultural geographers. Moreover, feminist geographers have been instrumental in creating new research agendas around emotive – and perhaps previously taboo – topics such as sexuality, activism, care, home, motherhood and family practices.
    - As noted in Chapter 12, feminist geographers' pioneering work on **bodies** and **embodied** experiences has opened up radical new ways of conceptualising how bodies *feel* in particular spaces, and the nature and importance of 'feelings' themselves. In each of these ways, feminist geographers have had a transformative impact upon their wider discipline, introducing new understandings, concepts and methods, and calling for new, emotionally attuned research in many social and cultural contexts.
- A lineage of research by feminist and social geographers focusing on 'disability', 'illness' and 'mental health' (see Butler and Parr 2004) is perhaps the most sustained attempt to investigate 'emotional' geographies empirically. This work has been important within geography's 'emotional turn' in three senses.
    - It has highlighted, and investigated in considerable detail, how emotions are socially constructed. Here, geographers have undertaken historical archival research (Philo 2004) and analyses of more recent policy and popular discourses (Parr 2008: 29–30) to investigate how particular emotional and physiological states have come to be categorised as problematic impairments (e.g. 'madness', 'depression') in need of treatment,

confinement or concealment. Their work reveals a long history of medicalised and popular discourses that have cast such impairments as 'other' in Western societies.

- In this context, geographers' research has revealed everyday forms of social and spatial exclusion experienced by people with diverse impairments, disabilities and illnesses. Thus they have explored how, in practice, particular individuals and groups are 'othered' as a consequence of their emotional and physiological condition, as well as the emotional experience of living with an 'othered' condition.
- Geographers have considered the capacity of particular kinds of space and social interaction to afford (or indeed restrict) emotional support, care, respite, and solidarity for people experiencing such conditions.

- As noted in Chapter 9, geographers' engagements with **non-representational theories** have often considered the implications of bodily, sensuous and physiological aspects of human being for classic geographical understandings of the world. An examination of the embodied and cognitive intensities and sensations that are commonly described as 'emotions' has been key to this work. Thus much non-representational geographical research might be read as a succession of case studies of relationships between 'emotions' and everydayness. More than this, however, non-representational theorists have called for greater critical reflection on the very notion of 'emotion' within geography's 'emotional turn'. Indeed, they note that calls for geographers to engage with 'emotion' have seldom defined what, exactly, is meant by 'emotion', or what particular 'emotions' consist of in practice (see Anderson and Harrison 2006). As suggested in Chapter 9 on **everydayness,** this latter problem is particularly important for non-representational theorists. The way in which 'emotions' threaten to 'overwhelm language' and 'resist words' (Harrison 2007) – the difficulty of finding adequate words to describe 'love', 'happiness' or 'grief', for example – neatly illustrates the capacity of the world to exceed conventional representations. Non-representational theorists have consequently questioned the notion of 'emotion' deployed within the lines of work outlined above, and have argued that the notion of 'affect' is more instructive (see Section 11.4). Feminist and social geographers' ripostes to these critiques have, in turn, opened up new debates about relationships between 'emotion'/'affect' and social and cultural issues (see Nash 2000; Thien 2005; Tolia-Kelly 2006; Colls 2012).

Note, then, that the 'emotional turn' in human geography has actually comprised multiple lines of theory and practice, encompassing the work of many geographers, whose work has coincided, interrelated or conflicted to varying degrees at different times. In practice, these lines of work have resulted in a remarkable variety of investigations of the importance of 'emotions', ranging across diverse social and cultural contexts. Pile (2010: 6) neatly illustrates the breadth of these investigations, noting that during the first decade of the 21st century:

> geographers have described a wide range of emotions in various contexts, including: ambivalence, anger, anxiety, awe, betrayal, caring, closeness, comfort and discomfort, demoralisation, depression, desire, despair, desperation, disgust, disillusionment, distance, dread, embarrassment, envy, exclusion, familiarity, fear (including phobias), fragility, grief, guilt, happiness and unhappiness, hardship, hatred, homeliness, horror, hostility, illness, injustice, joy, loneliness, longing, love, oppression, pain (emotional), panic, powerlessness, pride, relaxation, repression, reserve, romance, shame, stress and distress, suffering, violence, vulnerability, worry [etc.]

Summarising the contributions of non-representational geographers to this context, Pile's list continues:

> Even those [geographers] who declare themselves suspicious of the language of emotions . . . have nonetheless attended to anger, boredom, comfort and discomfort, despair, distress, enchantment, energy, enjoyment, euphoria, excitement, fear, frustration, grace, happiness, hope, joy, laughing, liveliness, pain, playing, rage, relaxation, rhythm, sadness, shame, smiling, sorrowfulness . . . surprise, tears . . . touching, violence, vitality.

Anderson (2009a: 188) identifies two broad thematic foci within all this work: first, a concern to acknowledge emotions as an "intractable aspect of life and thus potentially a constitutive part of all geographies"; second, a concern to investigate how "emotions have long been manipulated and modulated as a constitutive part of various forms of power". That is, on the one hand, work that seeks to consider the importance of 'emotions' to different geographical issues; on the other hand, a more specific set of work regarding the manipulation of emotions by those exercising forms of political and/or cultural power. Examples of each of these types of emotionally attuned geographical work are introduced in Sections 11.5 and 11.6 respectively. By way of introduction, the following section clarifies the different, sometimes opposing, terminologies used in these bodies of work.

11.4 'Emotions' or 'affects'?

The terms 'emotion' and 'affect' are now widely used by human geographers. Often, these terms are left undefined, or are used vaguely, inconsistently or interchangeably, much to the frustration of theorists, who strongly advocate distinguishing between them. The conceptual distinction between 'emotion' and 'affect' can perhaps be most readily grasped with reference to an example: see Box 11.1 before reading on.

Consider romantic love, then: what is it *like* to 'fall in love'? Pondering this question may prompt reflection upon a series of bodily and mental sensations, intensities and experiences: perhaps 'butterflies

Box 11.1
'Love' is all around

Photograph 11.1 Valentine's Day gifts: cultural representations of feeling 'in love'?
Source: Rex Features/Per Lindgren.

This chapter was written a few weeks before Valentine's Day. At the time of writing, it was possible to buy the following items from one UK-based online retailer:

- 402,804 different 'love songs' in mp3 format;
- 47,093 different CD compilations of 'love songs';
- 18,755 different anthologies of 'love poetry';
- 27,011 different works of 'romantic fiction' ("ideal for Valentine's Day");
- 3,593 'romantic comedies' on DVD ("for a night in with a loved one");
- 293,002 different Valentine's Day cards;
- 22,317 'romantic gifts' (including a fountain pen engraved with 'love always'; a red rose in a 'luxury velvet presentation case'; a teddy bear with embroidered 'I♥U' motif; a cup with hand-painted 'good morning handsome' message; a 'bronze sculpture of mermaid

Box 11.1 continued

and lover'; and much else besides) (from Amazon.co.uk, January 2012).

Reflecting upon the remarkable array of texts, commodities and objects devoted to romantic love in Western popular culture (of which the above list is just a small sample), the cultural theorist Doug Aoki (2004: 97) writes: "Although each and every person is authorised to speak of love, each and every speaking must prove to be insufficient." That is, despite the vast proliferation of representations of 'love' in Western cultures – and the countless words, phrases, metaphors, concepts and clichés therein – 'love' *still* "defies being said (or written)" (Aoki 2004: 97): arguably none of the representations quite does justice to the feeling itself. So he argues that something about 'love' remains elusive, no matter how many love songs we download, love poems we read, or romantic gifts we buy and give. This elusiveness – and this mismatch between representations of a feeling, and the feeling itself – is a useful example to have in mind when trying to understand the distinction between 'emotion' and 'affect'.

in the stomach', blushing cheeks, holding hands, kissing, a kind of contentment, warmth and light-headedness that makes one feel different about oneself and the world; something like that. Or perhaps this question prompts a sigh, or some deep feelings of regret, loss, absence, disappointment. Or it may prompt anger at the enduring presence of the normative, historically specific, frequently heterosexist social construct of romantic love within many contemporary cultures.

Whatever one's feelings about 'love', it is certainly the case that those of us living in the contemporary West are bombarded by cultural representations of romance. Consider the thousands of texts, images, concepts and phrases evoked by Box 11.1: each tries, in its way, to say a little about what 'love' is like. But even with this vast cultural resource at hand, it can be hard to adequately describe and explain a particular feeling or mood. Putting 'love', or any other feeling, into words is profoundly difficult: perhaps impossible to do – so Aoki (2004) argues in Box 11.1 – without falling back on a repertoire of words, texts, phrases and clichés that can feel inadequate, stilted and lifeless in comparison.

So thinking about a feeling like 'love' directs attention to two phenomena. On the one hand, there exists a shared cultural notion of what 'love' is like, and a shared vocabulary and set of norms relating to 'it', all of which are present and reproduced in a vast array of cultural media and discourses. On the other hand, there are the small incidents, interrelations and bodily sensations – the blushing cheeks, hands held, involuntary sighs, butterflies in the stomach, and so forth – that are ultimately what 'love' is, and what it feels like, but which may be impossible to adequately put into words. Human geographers increasingly recognise the difference between these two things: between shared cultural descriptions and concepts relating to feelings (e.g. the word 'love') and the often unsaid or unsayable physicality and *sensation* of particular feelings at particular moments (e.g. the complex feelings and relationships that the word 'love' attempts to describe). To make this distinction, the term 'affect' is increasingly used to describe the latter of these phenomena (the physical and sensuous stuff of feelings as they happen), whereas the term 'emotion' is reserved for the former (feelings translated into shareable social constructs). Box 11.2 summarises some key distinctions between these two concepts.

Emotions and affects are not easily separated in 'real life' contexts. Indeed, they are invariably mutually constituted (Ahmed 2004): for example, socially constructed emotions frequently spark bodily affects (a trite love song can move one to tears), whereas affective sensations lead to the social construction of emotions (being moved to tears can prompt the writing of trite love songs). Crucially, *both* emotions and affects are important for cultural geographers (and, as the following sections outline, matter in a wide range of important social-cultural contexts, far beyond the writing of love songs).

Box 11.2
Comparing 'emotion' and 'affect'

'Emotion' . . .	*'Affect' . . .*
Refers to those aspects of 'feelings' that can be expressed, reflected upon and communicated to others	Draws attention to those aspects of 'feelings' that cannot be expressed, and which make conventional representations of 'emotion' feel lifeless or inadequate
Refers to the **discourses** and social constructs that result from the "socio-linguistic fixing of intensity" (Anderson 2009b: 9): i.e. the 'translation' or 'packaging' of bodily feelings into neat, shareable words, ideas and representations (Greco and Stenner 2008: 10)	Is often unconscious, irreducible to words, and prior to cognition or vocalisation
Is implicitly understood to 'belong' to an individual: hence 'I feel *x*', 'you feel *y*', 'she feels *z*'	Is **embodied**: e.g. "the blush of a body shamed . . . , the heat of a body angered . . . the restless visceral tension of body bored" (Anderson 2006: 736)
Draws attention to the supposed universality of some feelings: how some emotions are experienced similarly by different individuals and across different cultures and times/spaces	Critiques understandings of 'emotion' as singularly residing in individuals, by emphasising how feelings occur and circulate in practice, through complex interactions between people, things, places and events: i.e. how they are "distribute[d] . . . throughout the world, through, for example, the nervous system, hormones, hands, love letters, screens, crowds, money" (Anderson and Harrison 2006: 334)
Is a concept that is intuitively understood by a popular audience	Recognises the contingency, subtlety and complexity of specific affective events, encounters and 'intensities': e.g. how feelings can be prompted, in a particular moment, by "the push of the terrain upon the . . . body . . . , the spiritualised pull or uplift of a chord of music, and the stillness struck by the colour of paint" (Dewsbury *et al.* 2002: 459)
Is sometimes inadequately defined or theorised by human geographers	Is a concept that is preferred by many cultural theorists, but is obscure to most other people
	Is sometimes complexly defined, with reference to theoretical literatures that many find impenetrable

11.5 Geographies as emotional-affective

Why, then, should emotions and affects be considered important for the work of cultural geographers? Collectively, the following points amount to an argument that emotions and affects are interrelated in human geographies, that human geographies should be understood as *inherently* emotional-affective, and that emotional-affective characteristics are key to understanding geographical issues.

Point 1: Being human is always emotional-affective

The sociologist Simon Williams reminds us that, as human beings, "we are never . . . devoid of an emotional stance on the world" (Williams 2001: vii). We are always feeling something (even if it is unremarkable, or difficult to describe). Whoever, wherever and whenever we are, we are always in the midst of some kind of affective state (whether or not we are able to put a finger on, or explain, the emotion we feel).

Moreover, affective experiences, like trauma, care, fear, repression, anxieties, shocks, and desires, shape our **identities** (see Chapters 3 and 8), and are complexly "inscribed in . . . the **body**" (Attfield 2000: 241) (Chapters 7 and 12), in ways that we may not fully understand (indeed, in ways that even leading-edge psychologists may not fully understand).

If all human experience is affective – and can often be described as emotional – it follows that all human *geographies* are emotional-affective. Each of the following points attests to the emotional-affective nature of all social and cultural geographies, and should draw attention to two characteristics of human being. First, it should be clear that affects and emotions are fundamental to our experience of the world: they "affect the way we sense the substance of our past, present and future; all can seem bright, dull or darkened by our emotional outlook" (Bondi *et al.* 2005: 1). This 'outlook' can be taken for granted and pass unnoticed for long stretches of time; but other times/spaces are characterised by much greater affective intensity. For, second, certain geographies or moments are experienced as more emotional than others: "at particular times and particular places, there are moments where lives are so explicitly lived through pain, bereavement, elation, anger, love and so on that the power of emotional relations cannot be ignored" (Anderson and Smith 2001: 7).

Point 2: Geographical issues have emotional-affective consequences

Central to work within geography's 'emotional turn' has been a realisation that all human geographies are affective and have potentially emotional consequences. This is invariably true, even of geographical issues that one may be accustomed to thinking of as devoid of emotion. Thus Anderson and Smith (2001: 8) describe how even issues that may intuitively seem somewhat 'grey' and emotionless are, on closer inspection, "shot through with emotional relations": for instance, they demonstrate how the theories and mathematical models of economic geography should be understood as disguising complex personal stories, "emotionally present intricacies" and relationships within particular everyday spaces of economic activity. Broadly, then, it can be argued that *any* issue you will have considered while studying human geography has emotional consequences to some extent (although one may not guess this from the ways in which geographers have tended to write or theorise such issues).

As noted previously, feminist social scientists have been especially influential in uncovering the (often previously overlooked, and often gendered) emotional-affective consequences of social and cultural issues. As just one example of work in this vein, see Box 11.3: consider how, in this particular setting, a large-scale political/historical conflict has had deeply personal and emotional consequences for women's everyday lives. A wide range of work by feminist social scientists has made similar points in relation to a wide range of other large-scale geographical issues, highlighting the everyday emotional-affective impacts of globalisation, economic development, armed conflict, natural disasters (see Box 12.4), social exclusion, poverty and pandemics in diverse contexts.

Point 3: Emotions and affects shape social relations

Almost self-evidently, how we feel (about ourselves, others, issues, spaces and situations) is important in shaping our interactions with other human beings. Affects and emotions are therefore key to the constitutions and shaping of all social relations and geographies. For, on the one hand, our encounters with other people and other social groups will, to an important extent, be coloured by our extant feelings: we are likely to act and interact differently depending on whether we feel love, hatred, empathy or indifference towards them. Moreover, on the other hand, different kinds of social interaction afford different affective/emotional responses: in all likelihood, we shall feel and think differently about another person or social group if they grant us continuous care or if they treat us with unremitting cruelty, and so forth.

These truisms should make plain the capacity of affects/emotions to "align some subjects with some others and against some others" (Ahmed 2004: 25): a realisation that has prompted many social and cultural geographers to investigate the small, personal, everyday feelings, anxieties and tensions that are ultimately at the heart of much broader forms of social exclusion and inequality. For example, in his key text *Geographies of Exclusion*, the social geographer David Sibley (1995) urges social scientists studying large-scale issues of social and spatial exclusion to closely examine (often hitherto overlooked) small-scale and personal emotional experiences in these contexts. That is, Sibley (1995: 3)

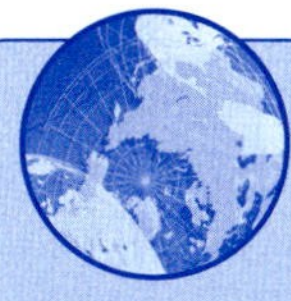

Box 11.3

Affective consequences of a social-historical issue

Photograph 11.2 Material markers of social-cultural division on a Belfast street.
Source: Alamy Images/Richard Baker.

Since the seventeenth century, Unionist (largely Protestant) and Republican (largely Roman Catholic) communities have lived in close proximity in Northern Irish towns/cities. Centuries-old histories of political discord (relating to the deeply contested relationship of Ireland to the United Kingdom) and pan-European religious sectarianism have been manifest as tensions, animosity and violence between these communities. In particular, a period spanning the late 1960s to the late 1990s – popularly referred to as 'the Troubles' – witnessed significant acts of political and sectarian violence, resulting in the deaths of more than 3,000 members of paramilitary groups, police and armed forces, as well as civilians in Ireland, England and mainland Europe (see Boal 2002).

Although a 'ceasefire' was signed in 1998, the affective aftermath of 'the Troubles' has been significant. For example, research by the feminist psychologist Alice McIntyre (2002) illustrates how this social-historical context continues to be *felt* in the everyday spaces and routines of women living on a Belfast street. The following extracts from her interviews exemplify two key sets of feelings described by women living in this context. These affects pervade these women's everyday lives and spaces.

- *Normal abnormality* – "a state of being/living where people come to anticipate living with multiple forms of sanctioned and unsanctioned violence, marginalisation and oppression, all of which inform and shape the dailyness of their lives" (McIntyre 2002: 394)

  I was 11 in 1969 when the Troubles were at their worst. We had the army inhabiting half our school . . . That was normal. I seen gunmen in the

Box 11.3 continued

> streets. That was normal. I seen soldiers shot. That was normal. To us, that was what we grew up with. I didn't know any different.
>
> (p.398)

> The time the [local pub] was blown up in 1974, I was at school and I came home and I was seeing those people lying on the pavement.
>
> (p.396)

- *Frozen watchfulness* – "a constant sense of impending doom about what might happen to them, their families and, in particular, their children" (McIntyre 2002: 394)

> We have a security gate on our stairs. It gets locked at night, to stop unwanted visitors . . . from harming my family when we are in bed. Most families, 90% of the families . . . in the community have some sort of security in their homes. The children in the community are used to this. Mine have never asked why we have this and we will never remove this gate as I do not trust enough yet.
>
> (p.399)

> I am scared to go in my kitchen. I have a terrible fear about going into my kitchen when there's no ceasefire. Because all there is behind me is a railway. And [one time] . . . the [unionist paramilitaries] came up the back and lit all the bushes and the shrubbery and they burned it all. And I always have a fear in my mind. I'd always be scared of them coming up the back of the houses and maybe shooting in.
>
> (p.397)

suggests that large-scale geographical issues might be better understood by:

> start[ing] by considering people's feelings about others because of the importance of feelings in . . . social interaction, particularly in instances of racism and related forms of oppression. If, for example, we consider the question of residential segregation, which is one of the most widely investigated issues in urban geography, it could be argued that the resistance to a different sort of person moving into a neighbourhood stems from feelings of anxiety, nervousness or fear . . . It is often the case that this kind of hostility to others is articulated as a concern about property values, but certain kinds of difference, as they are culturally constructed, trigger anxieties, and a wish on the part of those who feel threatened to distance themselves from others.

These personal feelings of difference, anxiety or hostility are often importantly shaped, (re)produced and reinforced via contemporary cultural circulations of discourses, representations and norms. So, in particular

Box 11.4
The production of emotions in mass media reporting

In May 2004, eight East European countries (Czech Republic, Estonia, Hungary, Latvia, Lithuania, Poland, Slovakia, Slovenia) joined the European Union (EU). Three EU states (UK, Ireland, Sweden) granted relatively free employment, movement and citizenship rights to migrants from these countries. Consequently, around 630,000 East European workers migrated to the UK during 2004–07. As an example of the ways in which this population group has been socially constructed, consider the following front-page headlines carried by the UK's two bestselling tabloid newspapers during a twelve-month period: "Migrant rush hits services hard: ROCKETING immigration has left schools, hospitals, police and housing at full stretch, watchdogs warned yesterday. Towns have been hit by racial tension, street crime and binge drinking as foreign arrivals flood in. And roads are more dangerous than ever because of poor driving by some newcomers." (*The Sun*, 31 January 2007); "Signs warn Polish anglers to 'stop eating all the carp'" (*Daily Mail*, 5 March 2007); "Polish street signs 'are a waste of funds'" (*Daily Mail*, 15 June 2007); "Migrants to get AIDS test" (*The Sun*, 16 June 2007); "British workers denied jobs 'because they can't speak Polish'" (*Daily Mail*, 19 June 2007); "Polish immigrants take £1bn out of the UK economy" (*Daily Mail*, 28 June 2007); "Eastern Europeans 'cause 15 per cent of fatal accidents on rural roads'" (*Daily Mail*, 1 August 2007); "Migrant tax hike fears: COUNCIL tax will have to rise above the rate of inflation to pay for the cost of immigration - town hall chiefs have warned." (*The Sun*, 8 August 2007); "Poland yobs get their kicks here: POLISH football hooligans are teaming up with British yob firms to cause mayhem at matches this season. Eastern European thugs who have headed here to find work are making friends with home-grown hooligans. Now police fear the immigrants could spark a massive increase in violence at and around Premiership grounds." (*The Sun*, 21 August 2007); "Migrant rapist is allowed to stay" (*The Sun*, 1 September 2007); "Migrants 'put Brits on dole'" (*The Sun*, 18 September 2007); "£1m of child benefit paid out a month - to mothers in Poland" (*Daily Mail*, 21 September 2007); "Migrants' 'huge rise' in crashes: East European motorists' attitude to drink-driving and inability to understand road signs is fuelling a massive rise in crashes, police said yesterday." (*The Sun*, 25 September 2007); "Beer prices set to rise as Polish lager sales soar 250% in Britain" (*Daily Mail*, 7 November 2007)

Indentify the diverse forms of anger and anxiety inherent in these mass-mediated reports. What do these reports *not* tell us about these migrants?

times and spaces, emotional responses to 'others' are, to an important extent, anticipated and scripted by contemporary **hegemonic** popular norms and discourses (Chapters 2 and 6). Consider Box 11.4: note the production of multiple forms of anxiety and anger in mass media reportage about a particular population group in this case study. Geographers have investigated similar mass-mediated circulations of anxieties in relation to other socially excluded groups, including asylum seekers, homeless people, young people, people with mental health conditions, and 'deprived' communities.

Point 4: Emotions and affects are fundamental to experiences of any space

Just as they are key to social relations, emotions and affects also "contribute to . . . the avoidance of certain places . . . or, conversely, attraction to particular places and social [settings]" (Sibley 1999: 116). For instance, the experience of passing through a particular space may prompt, or be affected by,

> feelings about others, people marked as different . . . , [n]ervousness about walking down a street in a district which has been labelled dangerous, nauseousness associated with particular smells or, conversely, exhilaration or a feeling of calm . . . , [r]epulsion and desire, fear and attraction, attach[ed] to both people and places in complex ways.
>
> (Sibley 1995: 3–4)

These feelings, and the everyday geographies of avoidance or attachment they engender, are complex in origin. They may be logical, or somewhat personal and quirky, or essentially inexplicable. They may reflect local or site-specific psycho-geographies of the space in question: the idiosyncratic and locally understood feelings, emotions, narratives, histories and associations attached to particular landscapes by "the people who inhabit and create the place" (Nold 2009: 37).

Often, these personal and local emotions are structured and criss-crossed by broader inequitable **cultural politics,** and social geographies of **identity,** inclusion and exclusion (Chapters 3 and 8). For example, feminist social scientists have explored how emotional experiences of particular spaces are often strongly gendered, with women's everyday geographies in diverse settings affected by gendered forms of fear

(Koskela and Pain 2000). Thus, for example, Valentine (1989: 386) highlights how "many [women's] apparently 'taken-for-granted' choices of routes and destinations are in fact the product of 'coping strategies' women adopt to keep safe", such is their latent fear and anticipation of (especially male) violence in many public spaces. Elsewhere, parallel work by social geographers has explored similarly fearful geographies – and consequently restricted usage of public spaces – of migrants, minority ethnic and religious groups, gay and lesbian communities, elderly people and disabled people.

Moreover, these kinds of emotional-affective experience are structured and patterned by contemporary cultural norms and panics. For example, research by children's geographers has explored how children's and young people's experiences of public spaces in the UK are structured, limited and affected by norms about parenting, contextualised by historically/culturally specific panics that imagine unaccompanied younger people in public spaces as invariably either vulnerable (to 'stranger danger', bullying and many other hazards) or 'out of place' and threatening (as a potential source of diverse forms of 'antisocial behaviour') (Valentine 1996). In practice, these norms and anxieties are often experienced (and reproduced) by children and parents/carers as a sense of fearfulness in, and avoidance of, particular spaces, scenarios and social interactions in public spaces. In practice, the above forms of affective spatial experience interrelate and combine in complex ways in specific spaces: see Box 11.5 for one example. These examples illustrate the broader significance of emotions and affects in all everyday spaces, routines and practices (see Stewart 2007).

Point 5: Emotions and affects are fundamental to culture

It follows, from the preceding points, that emotions and affects must be understood as significant in any of the cultural practices, politics, forms and genres introduced throughout this book. Indeed, the cultural theorist Raymond Williams (1979: 48) suggested that culture (or *a* culture) might best be understood by exploring its "structure of *feeling*": the "living result of all the elements in the general organisation" of the culture in a particular time/space. Later cultural theorists have understood this concept as calling attention to the complex, affective characteristics "always 'present', 'moving', 'active', 'formative', 'in process', 'in solution'" in particular spatial and cultural contexts (Thrift 1994: 193). Following a similar line of argument, Thrift (2004: 57) suggests that cities – and, implicitly, 'their' cultural lives and spaces – must be understood as complex, processual,

> maelstroms of affect. Particular affects such as anger, fear, happiness and joy are continually on the boil, rising here or subsiding there . . . either at a grand scale or simply as part of continuing everyday life . . . [whether] the deafening roar from a sports stadium when a crucial point is scored . . . , the mundane emotional labour of the workplace, the frustrated shouts of road rage, the delighted laughter of children as they tour a theme park, or the tears of a suspected felon undergoing police interrogation, [and so on].

Moreover, many cultural, and especially consumption, practices are often valued, and explicitly and deliberately used, for their emotional qualities. The quotes from Anderson's research on the consumption of popular music (see Box 3.2) provide examples of how popular cultural texts may be used within personal strategies to affect one's own mood.

As noted earlier, popular cultural texts and media are deployed in all manner of similar, affecting ways through countless **consumption** practices (Chapter 3). Indeed, some cultural theorists suggest that Western popular cultural practices are *increasingly* geared towards such activity. They identify a so-called 'emotionalisation' of twenty-first-century culture evidenced by, for example, increasingly 'emotionally saturated' content in screen and print media (e.g. the many popular 'confessional' talk, discussion, docu-soap and 'reality' television shows now broadcast in Europe and North America), and the proliferation of commodities and services explicitly addressing the emotional lives of consumers. For example, Williams (2001: 9–10) points towards countless "commoditised and commercialised forms of emotional experience and expression", including the commercialisation of pharmaceuticals and forms of therapy, the marketing of products on the basis that they will make us *feel*

Box 11.5
Complex emotional geographies of a public space

Photograph 11.3 A playground in north-east London, UK.
Source: John Horton.

Photograph 11.3 shows a children's playground in a public park in north-east London. The authors' research with 750 5-13-year-olds living in this area revealed a diverse range of meanings and emotions associated with this small space. Examples of children's anxieties about this space are extracted below. Note how these feelings relate to a complex mixture of personal anecdote, local urban myths and broader social geographies.

Not allowed here - road is too busy.
Hate it - too many dogs ☹
Too many teenagers and bullies.
Swings always broke by vandals.
Gangs from [name of particular street] hang out.
I came here once and hurt my toe.
Graffiti and perverts.
Nasty place - not safe. Too dark.
Old kids always skateboard here.
There was a murder here [apparently not true, but widely believed].
Not allowed here - mum says it's rough.
Older kids always break the swing.
Many Asians around here.
People put razors on the slide.
Dodgy men might jump out at you.
Might get attacked by Islam.
Rats and litter all around.
Dog poo everywhere.

better, the preponderance of self-help books and websites, and so on.

Point 6: Doing cultural geography is itself emotional-affective

It is argued that geography's 'emotional turn' – typically involving exploration of feelings of *other* people, *elsewhere* – must be accompanied by an acknowledgement of *our own* emotional experiences as geographers (Widdowfield 2000). Indeed, it follows, from the preceding points, that *doing* any piece of geographical research is invariably an emotional-affective experience and process. This is evident from the accounts of geographers who have reflected upon the emotional nature of their work. For example, emotions are frequently important prompts to do particular kinds of research. They are often, fundamentally, what motivates geographers to do, and care about, what they do. For example, as Askins (2009: 10) – whose quotation opened this chapter – notes:

> I'm motivated to be an academic in the first place by emotion. I feel passionately about social and environmental issues, which intensifies an array of emotions embedded in a range of experiences, and how I make sense of them.

And these emotions and motivations are complexly woven into geographers' work and everyday lives:

> I feel compelled to be involved . . . , being involved precipitates new/more emotions and, crucially, these feelings cross over into all other aspects of my work and everyday life, and emotions work back again into my research.
>
> (Askins 2009: 3)

Moreover, doing a piece of geographical research can often place the researcher in affecting situations. For example, consider Askins' (2009: 8) description of her work with asylum-seeking families at the community centre mentioned at the beginning of this chapter:

> There's a lot of sadness . . . Tears in fear of asylum claims being denied and being forcibly removed; crying in frustration when claims aren't processed month after month; upset and anger at the too-often experienced racism; frustration at the helplessness of not knowing what is happening to loved ones; grieving for absent family and friends, who may be in immediate danger – or on hearing the worst news. There's a lot of happiness . . . Screams of joy when refugee status is gained by someone in the group; giggles at a child's comedic behaviour; smiling compliments paid when anyone turns up in a new hairstyle/coat/dress/bag; chuckles and amusement at cultural and language misunderstandings . . . ; contented 'mmms' attend every meal cooked and served . . . Tears in someone's eyes . . . quickly beget tears in many eyes; smiles spread on recognition and circulation of happiness.

These kinds of emotional/-affective experience are central to many geographers' experiences of investigating social/cultural issues and problems (see Cloke *et al.* 2000; Horton 2008). Yet tears and laughter – and the countless other bodily and affective experiences of doing research – are normally written out of geographers' research publications. The realisation that many past and present geographers have "appear[ed] to remain wary of acknowledgement and expressions of emotion" (Widdowfield 2000: 200) has prompted numerous calls for geographers to recognise the emotional nature of their work and subject matters (Laurier and Parr 2000; Bondi 2005), as well as the importance of empathy, care and emotion in conducting effective research into social and cultural issues (Jupp 2007, 2008).

11.6 Affective practices

In addition to this general recognition of the importance of emotions and affects for geographical experiences, issues and research, geographers have also begun to explore the role of specific practices that target affects within contemporary societies. This work draws attention to the many, often covert ways in which affects are manipulated, to various ends, in diverse social and cultural contexts. These kinds of practice have a long history. Those exercising forms of political and/or cultural power have long, and almost by definition, sought to affect the feelings and behaviour of their subjects. Thus, for example, Thrift (2004: 64) describes "the marshalling of aggression through various forms of military training such as drill": that is, the way in which a centuries-old tradition of repetitive training procedures can be used to manipulate and 'damp down' the emotions of soldiers, thus rendering them more capable of carrying out orders in the interests of their state (see also Box 12.3). He notes that:

> From the seventeenth century onwards these kinds of training have become more and more sophisticated . . . These trainings were used to condition soldiers and other combatants to kill, even though it seems highly unlikely that this would be the normal behaviour of most people on the battlefield. These trainings involved bodily conditionings which involved fear to be controlled. They allowed anger and other aggressive emotions to be channelled into particular situations. They damped down revenge killings during outbursts of rage, and they resulted in particular effects (e.g. increased firing rates and higher kill ratios) which the military had not been achieving heretofore.
>
> (Thrift 2004: 64)

Elsewhere, social scientists have chronicled similar kinds of affective training or manipulation in other contexts: for example in the subtle strategies used to maintain calm and 'good behaviour' in schools and other institutions, to discourage or suppress dissent in cities, to afford particular consumption habits, or to foster feelings of loyalty, deference or patriotism in different times and spaces.

The remainder of this chapter outlines some examples of this kind of affective practice that are relevant to cultural geographers. These diverse practices share several characteristics: they often go unsaid or unnoticed in contemporary cultural contexts; they often entail sophisticated manipulation of small-scale bodily and everyday spaces; they are often deployed "knowingly [and] . . . politically (mainly but not only by the rich and powerful"; they are increasingly prevalent, perhaps especially in Euro-American cultures (Thrift 2004: 58).

As a first example of this kind of affective practice, consider an act of consumption that is routine and taken for granted in many contemporary cultural contexts: a visit to a supermarket. How does your local supermarket make you feel? To what extent do supermarket spaces affect our behaviour? Retail corporations routinely deploy an ingenious and complex array of spatial tactics in supermarkets, which seek to influence consumer behaviour by subtly affecting shoppers' bodies and feelings. Some examples of tactics that are allegedly commonly used in UK supermarkets are listed in Box 11.6. Whether or not these strategies actually work is questionable, especially in the light of understandings of **consumer** agency (see Chapter 3). What *is* certain is that substantial quantities of time, effort and capital are – quite deliberately – expended on tactics and spaces that *may* affect consumer behaviour, produce feelings and desires conducive to shopping, and ultimately increase profit margins. These numerous spaces and tactics of retail – and the panoply of retail psychologists, brand technologists, consultants and marketeers devoted to them – are frequently encountered, but often unnoticed, within everyday cultural geographies.

The example of the supermarket directs our attention to countless other examples of affective strategies designed into, and enacted through, the **production** of contemporary built and cultural spaces (see Chapter 2). For example, as noted in Chapter 3, spaces designed for consumption are frequently purpose-built to be conducive to behaviour and emotions that serve particular ends or ideals. More broadly, cultural geographers have explored interactions between **buildings,** feelings and behaviour in diverse contexts (see Chapter 4). Two notable examples are spaces for education, and spaces for care. For, on the one hand, geographers have chronicled how past and present school buildings have been explicitly designed to encourage particular, limited forms of pupil behaviour that correspond to contemporary norms about education and discipline. Box 11.7 includes one example from the authors' own research on school design.

The features listed in Box 11.7 represent only a small selection of the features, logics and technologies designed into contemporary school buildings. In each of the listed examples, the use of spatial features to target small-scale practices (and thus foster emotions and 'good behaviour' believed to be conducive to effective learning) was evident. Similar attention to details and practices is evident in geographical research on the creation of spaces of/for care. Here, geographers have investigated the material and social constituents of spaces that are used for the provision of diverse forms of therapy, counselling, support and medical care. That is, they have sought to understand what makes a space 'caring', or indeed how spaces might be made more 'caring'. Small practices relating to spatial design and layout are invariably crucial to the constitution of such care. Box 11.8 includes some examples from the authors' research in a space designed for the care of preschool children in the English Midlands.

Research in spaces of care, such as in Box 11.8, reveals the forms of careful, patient 'emotional labour' that is done by practitioners in such spaces (see Conradson 2003a, 2003b; Milligan and Wiles 2010): that is, the toolkit of techniques and strategies that are employed by practitioners to create 'care' itself by affecting others within these spaces. In the contexts of spaces of/for care, these techniques typically entail subtle bodily practices, conducive to the gradual development of relationships of trust and openness. This kind of subtlety is evident in McCormack's (2003) account of attending dance music therapy

Box 11.6

Supermarket design: affective management in a consumption space?

Photograph 11.4 How does the design of supermarkets create particular moods and behaviours?

Source: Alamy Images/Aardvark.

Activist groups such as Spacehijackers (2009) have critiqued the manipulative tactics and design features that are, allegedly, increasingly being built into UK supermarkets. Some examples are listed below. As you read, think about the spaces in which you shop: in what ways do their design, layout and features encourage you to spend?

Some "retail tricks to make you shop" purportedly include (after Spacehijackers 2009):

- placement of 'essential purchases' (e.g. bread, milk) at disparate points at the back of the store to maximise the time shoppers spend in-store, and the number of aisles and offers they have to pass;
- positioning of key offers and eye-catching products at visible points on major 'routes' between aisles in stores;
- circulation of baby powder or similar gentle fragrances via air conditioning (supposed to relax adults by evoking memories of childhood);
- handing out shopping baskets on arrival (supposed to lead to multiple purchases, as shoppers feel awkward about taking a basket with a single item to the checkout);
- circulating canned appetising smells (baked bread, cinnamon) to instil hunger;
- the use of floor textures, and contrasts between them, to steer shoppers to particular points or routes in stores;
- arranging smaller floor tiles, or noisier underfoot textures, in aisles with more expensive goods or 'treats' (the increased tempo of clicking as shopping trolleys pass over the smaller tiles supposedly leads shoppers to slow down and spend more time in the aisle);
- the choice of background music, with mid-tempo instrumental tunes supposed to instil a relaxed state and optimal steady tempo of shopping;
- the placement of 'impulse' purchases (sweets, magazines) beside queues at checkouts, to tempt distracted shoppers and their children.

sessions (see also Box 7.3). This particular therapeutic practice seeks

> to use movement experimentation to explore new ways of being and feeling, and to gain access to feelings that cannot be verbalised . . . [T]he dance music therapist . . . is non-directive, and neither prescribes the emotions the client is to express, nor instructs the clients on how to move.
>
> (Stanton-Jones 1992: 3)

McCormack (2003: 497) describes a number of different kinds of dance music therapy activities intended

Box 11.7
School building design: behavioural and affective control?

Photograph 11.5 How does school design create particular moods and behaviours?
Source: Peter Kraftl.

In the authors' research on school building design, the following design features were found to have been carefully built in to one newly designed school in the English East Midlands (Photograph 11.5; see den Besten *et al.* 2011). As you read, think about the ways in which institutional spaces such as schools, universities, hospitals and prisons are designed to foster particular moods, atmospheres and behaviours (see also Photograph 6.2 and Chapter 12).

- Wide, curvilinear corridors to prevent 'corridor jams' of pupils during busy periods, thereby reducing restlessness, collisions and heated tempers. The corridor had been designed through computer modelling of crowd patterns through the corridors.
- Walls, carpets and fabrics coloured in soft pastel shades, supposed to be relaxing, and conducive to contemplation rather than excitability.
- Removal of spaces associated with unauthorised activities or congregation, to reduce forms of antisocial behaviour believed to be associated with pupil inactivity and 'hanging out'. In particular, dead ends or isolated spaces (believed to be associated with the fear of bullying for many pupils) were designed out of the school building.
- Ensuring that all spaces where pupils can congregate can be overlooked or easily seen by school staff, thus supposedly encouraging orderly, civil behaviour via the possibility of surveillance. This sense of surveillance was extended through the installation of CCTV cameras.
- Installing electronic locking systems on all interior and exterior doors, thus ensuring that pupils' access and mobility could be controlled (and if necessary, the school could be 'locked down').
- 'Inspirational' architectural spaces within the school (e.g. a spacious, glass-paned atrium) to instil feelings of pride, identity and aspiration.
- Design features intended to evoke aspects of the school's local community (e.g. through the use of locally quarried stone), to instil community-mindedness.
- Aspects of design that can be determined or customised by pupils, in order to foster feelings of 'ownership' in the school building.
- Careful attention to the design, dimensions and layout of seating in classrooms in order to optimise pupils' concentration, behaviour and work ethic in relation to their learning in class.

Box 11.8

Creating a caring space for preschool children

Photograph 11.6 A children's centre in the English Midlands.
Source: John Horton.

In the authors' research in a space designed for the care of preschool children in the English Midlands (Photograph 11.6; see Horton and Kraftl 2011), we noted a wide range of practices where staff worked to make the space more 'caring', 'friendly', 'trusted', 'nice' and 'fun' for the children in their care. Some examples included:

- removing a padlock from a cupboard (as it had upset a child who had had traumatic experiences of being locked in a room by an abusive family member);
- brushing cobwebs and scuff marks from a doorstep (as they felt unpleasant to children crawling over them);
- mowing a small patch of grass (to improve the texture for outdoor play with scooters);
- rearranging boxes of toys to spread different groups of children throughout the space, and reduce arguments over popular playthings;
- using air freshener to disguise smells of nappy-changing (as older children found this unpleasant, and refused to play with toys in the same room);
- using a different brand of antiseptic hand-cleaner (as children found the previous brand unpleasant in terms of its smell, texture and drying effect on hands).

to foster this kind of access to 'feelings that cannot be verbalised', such as the following:

> Everyone in the room is asked to get into pairs. One person of each pair must lead the other around the room. The person being guided must keep their eyes closed and the exercise is to be nonverbal. With my right hand I take my partner's left hand and hold it a little out in front of me, experimenting with different ways of holding. I find a position that is loose, yet hopefully influential, and very slowly, begin to guide this hand around the room . . . As the movement continues, the surface of contact between hands becomes the most important place in the room.

McCormack thus draws attention to the small activities through which practitioners of this, and many other kinds of, therapeutic practice seek to foster an atmosphere and relationships of trust, support and care within a particular space. Through such work, spaces that may not be purpose-built as spaces of/for care can be temporarily transformed, at least in terms of the way in which they affect those within them. Crucially, to return to the distinction outlined earlier in this chapter, such practices should often be understood as affective, rather than emotional. For as McCormack (2006: 331) reflects, "Throughout . . . 18 months of attending drop-in [dance music therapy] sessions, I was never once asked how I felt, nor was anyone else. And I never asked anyone else how they felt."

Elsewhere, geographers have investigated affective practices and emotional labour in spaces as diverse as airports, workplaces, churches, monuments, museums and tourist destinations. Moreover, as noted in Chapter 2, the production of particular emotions and interactions in/through larger civic spaces, and the generation and marketing of different kinds of 'buzz' or 'atmosphere' in specific urban districts and events, have also extended our understandings of the importance of affective practices in contemporary cultural lives.

Through these diverse research contexts, geographers have begun to explore the centrality of emotions/affects in broader (geo)political issues and practices. Thrift (2004) argues that emotions/affects should be understood as integral to forms of political power and persuasion in several senses. For example:

- in the arguably heightened attention to details such as bodily gestures, appearance and performance by politicians, in order to construct a particular image, identity and emotional appeal;
- likewise, in the arguably increasing use of event management, design, lighting and careful mediatised staging for optimising the emotional appeal and effectiveness of politicians' public appearances;
- in politicians' use of forms of rhetoric and discourses that deliberately appeal to the emotions;
- in the increased centrality of 'micro-biopolitics' in large-scale political acts (that is, the targeting and involvement of small-scale bodily and affective spaces to further larger political ends).

For example, many social scientists have charted the role of particular 'fear events' in contemporary geopolitics, especially in the aftermath of terrorist attacks in the West and the so-called 'War on Terror' (Pain 2009: 471). It is argued that representations of fear-ful events, and discourses and practices in their aftermath, have a heightened political and cultural importance in this context. Such 'fear events' are noted to have some recurrent characteristics.

- References to fear-ful events are used within politicians' rhetoric to galvanise feelings of outrage, patriotism, self-righteousness and injustice.
- Discourses of fear circulate widely through diverse cultural media.
- These discourses typically portray particular regions, individuals or social or ethnic groups as 'other' and irredeemably villainous, thus sparking or heightening feelings of hostility towards them.
- These discourses fuel a kind of 'affective contagion' whereby everyday feelings, experiences and social interactions become haunted, at least temporarily, by anxieties.
- The threat of future fear-ful events is used as a justification for political actions that might be unacceptable in other circumstances (indeed, political and media discourses can appear to actively 'stir up' anxiety, thus making such actions more palatable).
- 'Decisive' discourses, symbolic gestures and iconic images are widely used in political and media discourses to assert a sense that situations are under control.
- A specific lexicon of terms (e.g. the term '9/11'), symbols and representations relating to particular 'fear events' becomes prominent in contemporary popular culture.
- Forms of bodily training (e.g. fire drills, air raid drills, disaster-preparedness training) can be used by authorities to produce a sense of calm and faith in the preparedness of one's community, city or nation.

Consider how these characteristics have been played out in one specific example of a 'fear event': the so-called 'War on Terror' (Box 11.9). Note that although social scientists interested in the politics of affect have often been drawn to this kind of event (where affective practices can be seen to have been deployed purposefully and cynically by powerful political agents), it is also the case that other kinds of affective practice are important in more countercultural, subversive or activist forms of identity, practice and dissent (as illustrated in Chapters 3 and 8).

Box 11.9

Constituting the 'War on Terror'

Photograph 11.7 George W. Bush's address to Congress on 20 September 2001.

Source: Getty Images.

The following quotations are taken from President George W. Bush's address to Congress, simultaneously broadcast on US television networks on 20 September 2001, nine days after terrorist attacks had caused the deaths of 2,973 people in New York, Washington DC and Pennsylvania (a sequence of events popularly termed '9/11'). All quotations are from DHS (2001: unpaginated). In this speech, President Bush proposed a "war on terror . . . [which] will not end until every terrorist group of global reach has been found, stopped and defeated". Consider how these quotations construct an emotive 'fear event'.

> In the normal course of events, presidents come to this chamber to report on the state of the union. Tonight, no such report is needed; it has already been delivered by the American people. We have seen it in the courage of passengers who rushed terrorists to save others on the ground . . . We have seen the state of our union in the endurance of rescuers working past exhaustion. We've seen the unfurling of flags, the lighting of candles, the giving of blood, the saying of prayers in English, Hebrew and Arabic. We have seen the decency of a loving and giving people who have made the grief of strangers their own. My fellow citizens, for the last nine days, the entire world has seen for itself the state of union, and it is strong. (APPLAUSE)
>
> . . .
>
> Our nation has been put on notice, we're not immune from attack. We will take defensive measures against terrorism to protect Americans . . . Many will be involved in this effort, from FBI agents, to intelligence operatives, to the reservists we have called to active duty . . . And tonight a few miles from the damaged Pentagon, I have a message for our military: Be ready. I have called the armed forces to alert, and there is a reason. The hour is coming when

America will act, and you will make us proud (APPLAUSE). This is not, however, just America's fight . . . This is civilization's fight. This is the fight of all who believe in progress and pluralism, tolerance and freedom . . . The civilized world is rallying to America's side. They understand that if this terror goes unpunished, their own cities, their own citizens may be next. Terror unanswered can not only bring down buildings, it can threaten the stability of legitimate governments. And you know what? We're not going to allow it. (APPLAUSE)

. . .

Americans are asking, "What is expected of us?" I ask you to live your lives and hug your children. I know many citizens have fears tonight, and I ask you to be calm and resolute, even in the face of a continuing threat. I ask you to uphold the values of America and remember why so many have come here . . . I ask your continued participation and confidence in the American economy . . . And finally, please continue praying for the victims of terror and their families, for those in uniform and for our great country. Prayer has comforted us in sorrow and will help strengthen us for the journey ahead . . . Great harm has been done to us. We have suffered great loss. And in our grief and anger we have found our mission and our moment . . . Our nation, this generation, will lift the dark threat of violence from our people and our future. We will rally the world to this cause by our efforts, by our courage. We will not tire, we will not falter and we will not fail. (APPLAUSE)

. . .

Each of us will remember what happened that day and to whom it happened. We will remember the moment the news came, where we were and what we were doing. Some will remember an image of a fire or story or rescue. Some will carry memories of a face and a voice gone forever. And I will carry this. It is the police shield of a man named George Howard who died at the World Trade Center trying to save others. It was given to me by his mom, Arlene, as a proud memorial to her son. It is my reminder of lives that ended and a task that does not end. I will not forget the wound to our country and those who inflicted it. I will not yield, I will not rest, I will not relent in waging this struggle for freedom and security for the American people. The course of this conflict is not known, yet its outcome is certain. Freedom and fear, justice and cruelty, have always been at war, and we know that God is not neutral between them. (APPLAUSE) Fellow citizens, we'll meet violence with patient justice, assured of the rightness of our cause and confident of the victories to come. (APPLAUSE)

Summary

- Historically, human geographers have tended to underestimate or overlook the importance of 'emotions' in geographical issues.
- Human geography, like many other social sciences, experienced something of an 'emotional turn' in the wake of the cultural turn. The diverse work of humanistic, behaviouralist, feminist, non-representational and medical geographers has outlined the importance of 'emotion' for geographical research.
- Many geographers distinguish between 'affects' (complex, bodily, often unsaid or unsayable sensations) and 'emotions' (the cultural, shareable representations, labels and concepts we use to make sense of affects).
- Recent geographical research has begun to explore how: human geographies are always emotional-affective; geographical issues have emotional-effective consequences; emotions and affects shape social relations, and are fundamental to experiences of space; emotions and affects are integral to 'culture' and doing cultural geography.
- Geographers have also sought to understand how affects are targeted and manipulated via strategies, and spatial practices. Examples include: the production of desires in supermarkets; the production of good behaviour in schools; the creation of spaces of care; the targeting of affects through political rhetoric (e.g. popular/politicised representations of 'fear events').

Some key readings

Anderson, B. and Harrison, P. (2006) Questioning affect and emotion. *Area,* **38,** 333–335.

A clear-sighted statement on the need for rigorous concepts and understandings of emotion and affect in human geography.

Butler, R. and Parr, H. (eds) (2004) *Mind and Body Spaces: Geographies of illness, impairment and disability,* Routledge, London.

As we noted in Section 11.3, geographical work on illness and disability was important in geography's 'emotional turn'. This collection includes geographical research on diverse – and diversely emotional – forms of illness, impairment and disability. For example, chapters by Parr and Milligan reflect upon how discourses and medical practices around mental health represent particular emotional-affective states as problematic, and seek to regulate and control emotions and affects.

Davidson, J., Bondi, L. and Smith, M. (eds) (2005) *Emotional Geographies,* Ashgate, Aldershot.

An important publication in geography's 'emotional turn'. Bondi *et al.*'s chapter provides a useful overview of the place of emotions in geographical research. Subsequent chapters feature geographical research on a wide range of emotions and emotive topics.

Greco, M. and Stenner, P. (eds) (2008) *Emotions: A social science reader,* Routledge, London.

A very useful compendium of key social scientific writings, past and present, dealing with emotions and affects. Could be read in conjunction with Williams' (2001) *Emotion and Social Theory,* which provides contextual discussion of the history of social scientific work relating to emotions and affects.

Pain, R. and Smith, S. (eds) (2008) *Fear: Critical geopolitics and everyday life,* Ashgate, Farnham.

An excellent collection of essays about geographies of fear in different contexts, and at different scales. Includes chapters on the 'war on terror', 'pandemic anxiety', and the production and regulation of fear in diverse political, social and cultural spaces.

www.softhook.com

The website of the London-based artist Christian Nold. Nold's work has often involved using digital and GPS technologies to produce 'emotion maps' of different locales. These maps, and the accompanying discussions, are a fascinating resource for anyone interested in emotional and affective geographies.

Thrift, N. (2004) Intensities of feeling: towards a spatial politics of affect. *Geografiska Annaler B,* **86,** 57–78.

A key theorisation of the political and geographical importance of affects. Can be read in conjunction with Stewart's (2007) *Ordinary Affects,* which considers how politicised affects are part and parcel of everyday lives, spaces and routines.

Bodily geographies

Before you read this chapter . . .

- Look at your body. Consider how you feel about it. Do you find you are more aware of your body in particular places or situations?
- Think about somebody (perhaps a celebrity) who has an 'attractive' body. What is attractive about it?
- Notice how your body *feels*. What is going on within your body?

Chapter Map

- Introduction: working out
- Acknowledging bodies
- Representing bodies
- Bodily practices
- Embodiment

12.1 Introduction: working out

Like more than 50 million other people in the USA and UK, the authors occasionally work out in a gym. By its nature, the gym is a space that prompts reflection upon bodies, and particularly one's own body.

For example, when working out, we might become acutely aware of:

- anxieties about our own body (Do you go to the gym 'enough'? Do you feel at ease in the locker room?);
- our body's capacity or inability to perform certain tasks (How much weight can you lift? How far can you run?);
- our body's physical workings (heartbeat, sweat, straining muscles, lactic acid);
- the enormous range of multimillion-dollar industries and commodities (training shoes, apparel, energy drinks, gym membership) devoted to bodies in many contemporary societies.

The gym therefore draws our attention to some of the ways in which bodies matter in many contemporary cultural contexts. Of course, working out in a gym is just one example of an activity and space where bodies matter. In this chapter, we explain how bodies are, in fact, central to all human geographies (indeed, it should become clear that all of the chapters of this book have been about bodies, all along). The chapter introduces a turn within cultural geography wherein bodies have become an important focus of research

Photograph 12.1 The gym: a space that prompts reflection upon the body.
Source: Shutterstock.com/Vasaleki.

and theory, and we provide some examples of geographical work that has focused on bodily practices and embodiment.

12.2 Acknowledging bodies

Let us begin with a truism: "There is an obvious and prominent fact about human beings: they have bodies and they are bodies" (Turner 1984: 1). If all human beings have bodies and *are* bodies, it follows that human geographies must always, by definition, involve bodies. Everything that you will ever learn about while studying human geography – or indeed sociology, history, politics, psychology, cultural studies, economics – will essentially involve human bodies, acting in relation to one another and their environments. It seems obvious: all humans are bodies; *therefore* all human geographies involve bodies.

However, during the **cultural turn** (see Chapter 1), it became apparent that the discipline of human geography had often, curiously, had very little to say about the bodies that are fundamental to all human geographies. It was argued that much classic work in human geography overlooked, or shied away from, what the American poet Adrienne Rich (1986: 212) famously called "the geography closest in – the body". For example, it was noted that bodies had been overwhelmingly absent from the research interests, maps, models and debates that traditionally made up the discipline of human geography. This neglect of the 'geography closest in' can be understood as part of a broader, historical tendency for social scientific research to "silence . . . the body and emotional life" (Seidler 1994: 18).

Even now, it may seem odd, or perhaps startling, to be asked to think about bodies in the context of a human geography textbook (even though we *know* that human beings have bodies, and so human geographies must be about bodies).

Why, then, have bodies been so neglected by many geographers and other social scientists? An extensive set of answers to this question has been developed within a so-called 'sociology of the body'. Here, sociologists such as Brian Turner (1984, 2008) (whose quotation opened this section) have argued that the body's absence from classic social scientific work was a result of four related factors (see also Chapter 11).

- The foundational concerns of social scientific disciplines were largely related to major contemporary social change: industrialisation, imperial expansion, ideological revolution, the futures of institutions such as states and religions. All these concerns appeared to require 'big' theories, models and projects (Shilling 2003). A focus on individuals, let alone on individuals' bodies, would arguably have seemed irrelevant and esoteric in this context.

- The social sciences are characterised by a 'Cartesian' legacy – after the philosopher René Descartes (1596–1650). Descartes is perhaps best known for his phrase *cogito, ergo sum* ('I think, therefore I am'), which expressed the idea that conscious, rational thought (and *not* the body) is the distinctive, defining characteristic of human existence. Sociologists of the body argue that this kind of 'Cartesian' thinking has been central to the foundation and progress of Western academia. Subsequently, academic work regarding human beings has overwhelmingly prioritised their consciousness, intellect and ideas. Meanwhile, human bodies have overwhelmingly been neglected. For – after *cogito, ergo sum* – humans' *thoughts*, not bodies, were and are believed to define 'personhood' (Howson 2004: 3).
- Efforts to establish social sciences as 'reputable' subjects within the Western academic establishment involved positioning them as radically different from longer-established disciplines (e.g. the 'natural sciences'). Arguably, this had two effects: (i) social sciences came to focus on subject matters that were clearly *not* covered by the natural sciences; (ii) social sciences disciplines developed a critique of, and distanced themselves from, the approaches and concerns of natural sciences. Since the workings of bodies were seen to 'belong' to the natural sciences, bodies came to be absent from social scientific disciplines (Shilling 2007: 3).
- Until relatively recently, the vast majority of prominent academic researchers have been privileged, Western males. Classic concerns of social scientific disciplines have reflected this imbalance, betraying a fascination with large-scale – perhaps masculine – issues such as politics, economics, imperialism, statecraft and civic spaces. It is argued that, in this context, many aspects of the world were dismissed as 'trivial' and 'feminine' (e.g. childcare, domesticity, 'small-scale' and private spaces). In particular, a range of contemporary discourses understood women as closely (supposedly 'dangerously', 'irrationally') linked to their bodies and nature, via experiences of menstruation and childbirth (Valentine, 2001). Thus, within academic social sciences, bodies came to be: (i) defined as 'obviously' not relevant to large-scale 'proper' academic concerns; (ii) avoided, as inherently 'feminine' and 'irrational'.

In the wake of the cultural turn, and the consequent critique of traditional academic norms and approaches, many social scientists began to consider the importance of bodies for their disciplines, often in the course of a turn towards **everydayness and emotions** (see Chapters 9 and 11). Sociologists of the body suggest that several broad, contemporary Western social-historical shifts lay behind this turn to 'the body'.

- Many social scientists identify a so-called 'individualism' of society. For whereas, previously, individuals defined themselves in relation to traditional markers of collective identity (e.g. religion, customary morals, primary industries, social movements), social-cultural changes have caused the disintegration of these identities. Consequently, contemporary societies have come to be characterised by individuals who are much more self-regarding and self-reliant. In this context, individuals arguably have greater freedom to make their own cultures and **identities** (see Chapters 3 and 8), in line with their own goals, desires, opinions, interests and consumption preferences. In practice, bodies are central to many of these projects of identity formation. For example, individuals work to an unprecedented extent on their body's appearance and fitness, as an expression of their personality and identity. Consequently, it is argued that there is "a tendency for people . . . to place ever more importance on the body as constitutive of the self" (Shilling 2003: 3). Attention to this constitutive role of bodies has therefore become key to understanding contemporary societies.
- Many social scientists recognise a deepening politicisation of the body during the twentieth century. This is evident in the formation of a diverse range of radical and activist groups concerned to critique contemporary understandings of the body.
    - First, the work of 'second wave' feminists has been instrumental in placing bodies on the agenda within the social sciences. In the 1970s and 1980s, feminists developed diverse, important critiques of popular contemporary representations of the female body: for example, in relation to

'gender difference', 'beauty', 'glamour' and 'slimness'. More generally, feminists' work underlies virtually all of the concepts and critiques outlined in this chapter.

- A second important line of politicisation has emerged from gay and lesbian rights movements, and critiques of heteronormativity and homophobia. For example, accounts of the politics of drag, transgenderism or transsexuality have highlighted the body's "changeable, malleable and contingent characteristics" (Turner 1996: 5), in marked contrast to received cultural notions about 'normal' bodies.
- Third, an important line of critique has emerged from disability rights activism, contesting cultural representations of the diverse mind–body–emotional differences that are termed 'disability' in contemporary societies (Holt 2007).
- Finally, sociologists of the body describe a proliferation of countercultures concerned to cultivate politicised bodily experiences through, for instance, consciousness-altering substances, performance, or practices such as meditation and yoga (Shilling 2007).

- Recent decades have witnessed a proliferation of multimillion-dollar industries and economies devoted to bodies. On the one hand, there exist many consumer industries and products that specifically address consumers' bodies (see Section 12.4, or think of the introductory example of the gym). In turn, these industries have afforded complex consumer cultures where bodies, and bodily practices (like going to the gym), become explicitly central to consumers' lifestyles and identities. On the other, hand, there exist numerous economies where bodies, or parts thereof, have themselves become commodified. Examples include people trafficking, the legal and illegal exchange of organs for transplant, and the exchange of bodily traces in the form of biometric databases.
- A range of new and emergent technologies and techniques – such as neuroscience, genetics, body imaging, microsurgery, cybernetics, organ transplant, in-vitro fertilisation – have effectively constituted "a weakening of the boundaries between science, technology and bodies" (Shilling 2007: 8). In so doing, these innovations have led to substantial popular, media and academic debate about relationships between science, technology, societies and bodies (see also Section 10.4).
- Demographic change in many Western societies is resulting in ageing populations. Arguably, this has meant that ageing bodies, and 'non-ideal' and/or debilitating bodily conditions of old age, are significantly more visible within these social/cultural contexts. This has prompted contemporary debate regarding care for ageing bodies, and ensured that societies are confronted daily with the unsettling "brute facts of . . . thickening waistline, sagging flesh and inevitable death" (Shilling 2003: 7).

Collectively, these trends have prompted diverse social scientific work regarding bodies. This is perhaps most evident from a succession of books published by social scientists, predominantly during the 1990s, such as *The Body* (Featherstone *et al.* 1991), *The Body and Social Theory* (Shilling 1993), *The Body and Society* (Turner 1984), *The Body in Society* (Howson 2004), *The Body and the City* (Pile 1996), *The Body in Everyday Life* (Nettleton and Watson 1998), *Body Matters* (Scott and Morgan 1993), *The Lived Body* (Williams and Bendelow 1998), and so on. Each of these books – and many similar texts – makes a detailed case for social scientists to explore the importance of bodies to their field of interest. This, in turn, has prompted a substantial range of research and theory regarding bodies in very diverse contexts.

Reflecting on this extensive body of work, Moss and Dyck (2003) identify three key areas of interest:

- the popular representation or social construction of bodies in particular contexts;
- the importance of bodily practices in relation to particular issues and spaces;
- the implications of bodies – and their complex materiality – for understandings of society and culture.

These lines of work have been important for many cultural geographers. This chapter explores each of these three points in turn. Section 12.3 introduces research regarding the cultural representations and constructions of bodies, and their geographical implications. Section 12.4 introduces the 'rediscovery' of bodily practices in diverse geographical contexts. Section 12.5 outlines some questions posed by 'embodiment' for understandings of the 'human' in human geography.

12.3 Representing bodies

One of the many important lines of work to emerge from second wave feminism was a detailed critique of the ways in which bodies were/are represented and socially constructed in contemporary popular media and **discourses** (see Chapters 2 and 6). The feminist philosopher Susan Bordo (1993: 16) reflects on this critique in the following way:

> We have learned to read all of the various texts of Western culture – literary works, philosophical works, artworks, medical texts, film, fashion, soap operas – less naïvely, . . . attuned to the historically pervasive presence of gender-, class- and race-coded dualities, alert to their continued embeddedness in the most mundane, seemingly innocent representations . . . [F]ew representations – from high religious art to depictions of life at the cellular level – can claim innocence.

Bordo uses the term 'duality' to refer to the social construction of a perceived difference between self/other or us/them. She therefore suggests that a key outcome of feminist theory has been an unveiling of the 'pervasive presence' of such dualities – based upon stereotypes of gender, class and race – in all kinds of **cultural products** and **representations**, past and present, (see Chapter 2). Central to this project was a critique of the ways in which *bodies* have been represented in these contexts. For it is apparent that:

> Discourses in the media, fashion industry, medicine and consumer culture map our bodily needs, pleasures and possibilities to create geographical and historically specific bodily 'norms' . . . In the late twentieth century, we are surrounded by images of ideal bodies against which we are invited to measure our own.
>
> (Valentine 1999: 349)

This section will introduce some key aspects of this critique, which has been important in prompting the social scientific turn towards bodies. Such is the importance of feminist scholarship here, that most key examples cited here refer specifically to the representation of gender, and especially female bodies. However, the underlying point – that attention to contemporary representations of bodies is important for cultural geographers – should prompt reflection upon the popular representation of *all* bodies.

Let us take one case study, which will illustrate some important ways in which representations of bodies matter. Much research and critique by feminist social scientists has focused upon the many representations of 'normal' and 'ideal' female bodies that are circulated via contemporary popular culture. Lupton (1996) refers to 'the food/health/beauty triplex' to describe three major, related sets of cultural industries, spaces and discourses that produce many of these representations in Western societies. The first part of this triplex draws attention to some pervasive, and contradictory, assumptions about 'ideal' relationships between female bodies and food in Western societies. For on the one hand, in contemporary Western societies there is an expectation – passed down through many generations and normalised via many kinds of discourse and domestic spatial practice – that females should be responsible for food preparation and 'at home' in the kitchen. On the other hand, as Bordo (1993) outlines in her moving reflections on women's relationships to body size, Western popular culture abounds with representations that idealise females' regulation of their food consumption, in pursuit of a slender, 'healthy', 'beautiful' body.

Second, related to this, the presence of 'health' at the heart of Lupton's triplex draws attention to the importance of medicalised discourses in contemporary representations of the 'ideal' female body. That is, in Western societies, an 'ideal' female body is very widely represented as a 'healthy' body, with a 'healthy' diet and lifestyle. The nature of 'healthiness' here is frequently defined in terms of a range of scientific and medical standards and apparatuses, which effectively – and apparently 'factually' and definitively – define bodies as 'healthy' or 'unhealthy', 'normal' or 'abnormal'. An example is the usage of 'body mass index', height–weight charts or 'weight-for-age' tables with the majority of children born in the UK and USA, to determine which of their bodies are 'obese' or 'underweight' (see Figure 12.1). As

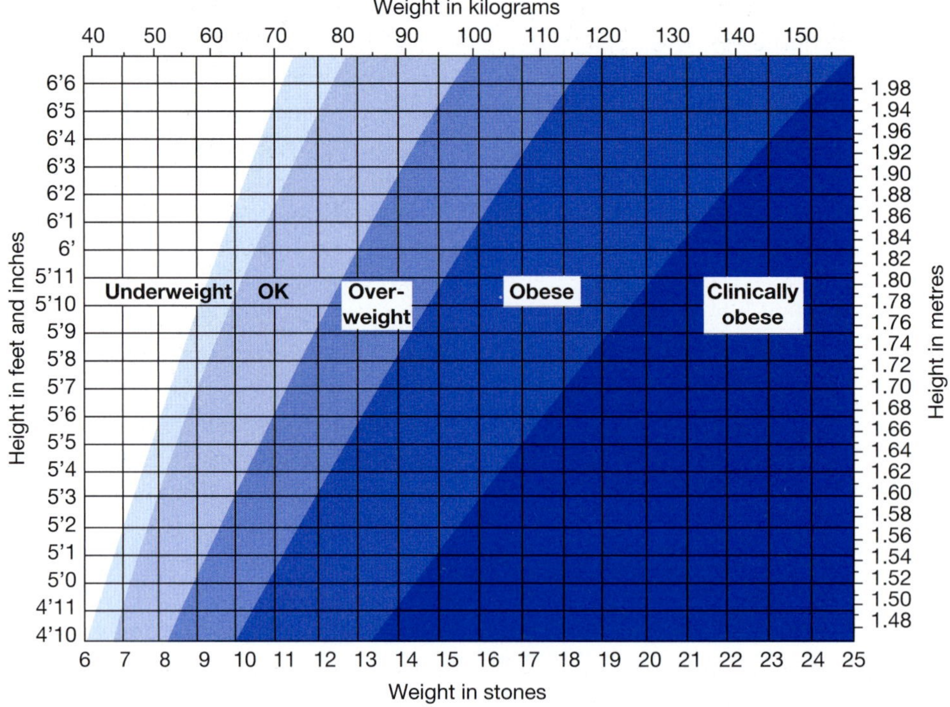

Figure 12.1 A standard height-weight chart: a widely used instrument for judging the healthiness of human bodies.

Evans and Colls (2009) identify, this medical construct has been important in normalising political and popular (and frequently exclusionary and gendered) ideas about obesity that profoundly affect many individuals' lives and geographies, by fostering norms and value judgements about body size.

The third term in Lupton's food/health/beauty triplex draws attention to the complex range of representations of female beauty and glamour that circulate in contemporary popular cultures. For:

> we are presented everywhere with images of perfect female beauty - at the drugstore cosmetics display, the supermarket magazine counter, on television. [For women] [t]hese images remind us constantly that we fail to measure up. Whose nose is the right shape, after all, whose hips are not too wide - or too narrow?
>
> (Bartky 1990: 40)

Here, then, we should note the importance of systems of celebrity and **cultural production** (see Chapter 2 and Photograph 12.2) in (re)producing contemporary notions of attractiveness, and the vast array of commodities and services that exist to support consumers in attaining a 'beautiful' face and body. The notion of beauty inherent in these representations and commodities may be taken for granted, and seem 'natural', at the time (although note that norms about 'beauty' can and do differ, often dramatically, over time and space).

Feminist critics argue that representations of bodies perpetuated via these kinds of discourse are important for social scientists, in three senses. First, because popular representations of bodies contain recurring 'codes' that reveal broader social/cultural norms and ideals (see McRobbie 2000). For example, feminist analyses of representations of women in popular Western magazines, soap operas and television commercials during the 1970s–1990s identified the following, recurring 'codes' that were widely circulated in contemporary popular culture.

- Women who are presented as 'attractive' in popular media representations are usually slim, tanned, toned, 'fashionably' dressed, heterosexual; "a visual object, 'attractive' and preferably blond and white skinned" (Frost 2001: 75).

Photograph 12.2 Magazines devoted to fashion and celebrity gossip: a widely circulated source of normative notions of female beauty.

Source: Alamy Images/Alex Segre.

- It is represented as the norm for females to strive to have an 'attractive' body and face, to dress 'fashionably', to eat and live 'healthily', and to pursue romantic heterosexual love. Thus "the female body is revealed as a task, an object in need of transformation", in pursuit of these ideals (Bartky 1990: 40). It follows that it is represented as 'acceptable' for females to be judged on the state of their bodies in relation to these ideals: that is, the extent to which they are visibly successful at these 'tasks'.
- It is represented as the norm for females to be imagined as 'inferiorised' (Frost 2001): that is, to exist in roles and positions that are inferior or submissive to male contemporaries.
- It is represented as the norm for women to be visually attractive objects to be gazed at by males. In turn, it is deemed 'normal' for women to critically scrutinise the 'attractiveness' of themselves and other women. Berger (1972: 47) describes this process of 'objectification' in the following way: "[m]en look at women. Women watch themselves being looked at . . . Thus [a woman] turns herself into an object – and most particularly an object of vision: a sight."

Second, these popular representations of bodies are important because they constitute a large part of the contemporary 'cultural imaginary' about bodies (Gatens 1983): that is, the store of representations and ideas about bodies that are available within a particular cultural/historical context. It is argued that the cultural imaginary in contemporary Western societies provides a limited and problematic range of representations of bodies. Thus it is argued that certain bodies, and ways of representing bodies, are over-represented and normalised by popular culture. Meanwhile many other bodies, and alternative ways of representing bodies, are under-represented and thereby treated as 'abnormal'. For example, feminist critics like Frost (2001) argue that contemporary Western popular culture provides only a narrow and problematic range of representations of teenage females' bodies.

On the one hand, it is argued that some bodily features, such as menstruation, are treated as taboo and are seldom represented or openly addressed in popular culture. Frost argues that where menstruation is acknowledged in popular culture – in television commercials, magazine advice columns, for instance – the emphasis is on the need for concealment and cleanliness; moreover, these messages are typically delivered in a neutral, 'medicalised' tone. The underlying 'code', therefore, is that menstruation is understood as an embarrassing bodily problem.

On the other hand, Frost argues that certain aspects of young female bodies are over-represented as (usually sexualised) objects within Western popular culture. For example, she argues that the popular cultural concealment of menstruation is in marked contrast to the excessive and sexualised presence of young females' breasts in popular cultural representations: for example in glossy magazines (Photograph 12.2), pop videos, fashion and celebrity culture, and pornography. Through such representations, it is argued that female breasts have come to be "highly fetishised by contemporary western cultures, seemingly almost the defining feature of whether a woman will be seen as sexually attractive or not" (Frost 2001: 74). The underlying 'code' here, then, is that this particular bodily feature is, 'naturally', desirable, sexually attractive, and available for scrutiny and/or appraisal by others.

It is argued that, in these senses, young Western females are presented with a limited, problematic repertoire of popular cultural representations with which to come to terms with their (developing) bodies. They are surrounded by popular cultural discourses that variously position their bodies as embarrassing and taboo (e.g. menstruation) or on display and subject to scrutiny (e.g. the fetishisation of breasts). Alternative discourses, such as some feminists' attempts to celebrate female coming of age and fertility, are overwhelmingly marginalised.

Third, and crucially, popular representations of bodies are important because they (re)produce social and spatial relationships, at diverse scales. At an individual level, it is apparent that constant bombardment by popular representations of 'ideal' bodies can often be an important causal factor for self-image and self-esteem issues. Since the way in which we see ourselves, and imagine we are seen by others, importantly shapes our everyday social encounters and experiences, the 'codes' of popular culture thus come to have a "direct grip . . . through the practices and bodily habits of everyday life" (Bordo 1993: 16). For example, there is abundant evidence that contemporary representations of teenage female bodies often render *individual* young females intensely self-conscious about their bodies: via embarrassment, inadequacy, the scrutiny of others, or else a newfound awareness of sexual power.

This self-consciousness subsequently permeates and affects practically all personal bodily practices, as well as social interactions. For, at a broader, societal level, it is argued that the popular cultural representations are central to the production of shared norms about the body. For example, it is argued that glossy, widely circulated magazines devoted to beauty, glamour or celebrity fuel local talk about bodies among particular communities (at a high school, or in a workplace, for example). This kind of process fuels collective understandings and norms about what constitutes an 'ideal' or 'normal' body. By definition, this also produces a set of exclusions: wherever there exist shared understandings of 'ideal' and 'normal' bodies, there will also typically be understandings of 'non-ideal' or 'ab-normal' bodies. As many geographers have explored, such understandings produce tangible geographies of social exclusion based on bodily differences: for example in relation to disabilities, skin colour, body size or bodily adornments (for a range of examples, see Valentine 2001).

12.4 Bodily practices

Since the **cultural turn**, and particularly via the work of second wave feminism, social scientists have begun to explore the importance of diverse bodily practices. This can be broadly defined as any of the things done with bodies, to bodies, between bodies or for bodies. It should be noted that all human life involves some form of bodily practice. Wherever we are, whatever we are doing, our body is always doing *something*. Whether we are working out in the gym, or sitting quietly reading a book, or even sleeping, we are *always* engaged in a bodily practice. So, to extend the truism that opened this chapter, we can say that all human geographies involve bodily practices. Anything you will ever learn about human geography – or any other social science subject – will, fundamentally, be about some *body* (or bodies) engaged in some kind of practice, in relation to their environments, and other bodies within it. The realisation (or reminder) that all human life involves bodily practices may seem banal and obvious. However, many social scientists have found this realisation helpful in developing and reinvigorating their concepts of social/cultural life. Drawing together a range of this work, we might say that bodily practices are especially important for cultural geographers for four key reasons.

Photograph 12.3 Mealtime: a bodily practice that can be a key site for family relationships and tensions.
Source: Alamy Images/Photo Alto.

Reason 1: Bodily practices create spaces and routines

We may take for granted the extent to which "[b]odily practices (eating, sleeping, washing, and presenting ourselves to others) dominate our everyday lives" (Valentine 1999: 329). However, consider the extent to which these kinds of bodily practice matter in your everyday life. Certainly, in contemporary Western cultures it is commonplace for dedicated spaces and times to be set aside specifically for bodily practices. Think of how most homes built in Europe and North America now have specific rooms for particular bodily practices: 'dining rooms', 'bedrooms', 'bathrooms'. Likewise, think of how many lives in these contexts are structured around routines devoted to bodily practices: 'dinnertime', 'bedtime', 'bath time' and so on. Those of us living in contemporary Western cultures may often take such everyday spatial arrangements and routines for granted. However, they do structure many everyday lives. Moreover, the spaces and times associated with bodily practices are often important to the maintenance of social relationships and structures.

For example, 'mealtimes' can be a site for daily family togetherness, or daily family arguments. Similarly, 'bedtime' can be a time of great intimacy between a couple (perhaps at the beginning of a relationship), or a time of great sadness (towards the end). In each of these examples, a bodily practice, and its space/time, is visibly central to the nature of particular social relations (between particular families, or between particular couples) and, as such, of more generic social structures and norms ('the family', 'the couple'). The proliferation of commodities and spaces devoted to specific bodily practices (e.g. gyms, or the examples of urban **consumption** spaces discussed in Chapters 2 and 3) also provides numerous case studies of this centrality.

Reason 2: Bodily practices are fundamental to major social/cultural issues

To reiterate, all human life involves some kind of bodily practice: therefore bodily practices must be at the heart of all human geographies. Although apparently simple, this observation becomes revelatory in the contexts of academic research and theory that have too often overlooked the 'geography closest in'.

One outcome of the social scientific turn to bodies has been the realisation of how apparently small bodily practices are central to major social/cultural issues and norms. Feminists' critiques of the social construction of 'gender' are important here. In particular, the philosopher Judith Butler's work on gender has been widely used by social scientists to understand the importance of bodily practices in the (re)production of taken-for-granted social/cultural norms. Butler's work emerged from an observation that was key to much second wave feminism: that biological differences between men and women ('sex differences') do not account for the substantial differences in how males and females are treated within cultures and societies ('gender differences'). For Butler (1990a: 411), this latter kind of difference must be understood as 'performative': "quite simply . . . it is real only to the extent that it is performed."

So, although gender inequalities are taken for granted and 'normal' across diverse social/cultural contexts, there is nothing 'natural' about this; indeed, gender differences are typically out of all proportion to the difference made by physical, biological differences between men and women. Instead, gender differences are reproduced through "repeated acts . . . [which] congeal over time to produce the appearance of a natural sort of being" (Butler 1990b: 33). Butler argues that small-scale, everyday bodily practices are central to the production, regulation and normalisation of gender differences: "gender is instituted through the stylization of the body and, hence, must be understood as the mundane way in which bodily gestures, movements, and enactments of various kinds constitute the illusion of an abiding gendered self" (Butler 1990b: 402).

As an example of this kind of mundane 'stylisation', consider how newborn babies are conventionally treated in contemporary Western cultures.

> From the moment they are born (and sometimes before), human beings are treated differently because of their sex. In contemporary Western society, for example, one of the first questions people ask of a newborn baby is 'is it a boy or a girl?', and a look at any baby care shop will show you how clothes for newborns are designed to answer this question visually when the immediate physical appearance of the child may be more indeterminate.
>
> (Gregson *et al.* 1997: 50)

As the preceding quote suggests, male and female infants may be physically very similar. Nevertheless, in Western cultures they are conventionally dressed, addressed, held and named differently, and supplied with very different sets of toys, texts, spaces and **material culture** (see Chapter 10): that is, a significant gender difference is *performed*, through bodily practices, that is disproportionate to the actual difference between any given pair of male and female infants.

The realisation that social/cultural norms are (re)-produced, 'performatively', through mundane bodily practices is widely applicable. Indeed, thinking along these lines, many social scientists have begun to explore the bodily practices that are important to issues that have long concerned academic researchers and theorists. In addition, there have been efforts to reveal a kind of "secret history of the body in social theory" (Turner 1991: 12): that is, how bodies and bodily practices have been central to the concerns and theories of many classic social scientific theorists, even though their work may not refer directly to bodies.

An example of this kind of approach can be found in recent, retrospective reflections on the importance of bodies and bodily practices in Marxian critiques of class-based inequalities (see Box 12.1). Here, it is argued that an attention to bodily practices enables understandings of social/cultural issues to be extended and developed. An attention to bodies can thus enhance understandings of how, in (bodily) practice, key social/cultural issues happen and matter (see also Chapter 11 on **emotions and affects**). Indeed, bodily practices should be understood as a key 'medium' of all social/cultural norms and issues:

> The body – what we eat, how we dress, the daily rituals through which we attend to the body – is a medium of culture . . . Banally, through table manners and toilet habits, through seemingly trivial routines, rules and practices, culture is 'made body'; . . . – converted into automatic, habitual activity.
>
> (Bordo 1993: 165)

This realisation – the bodily and performative nature of all identities – has been very important for geographical work on **identities** and cultural practices (see Chapters 3 and 8).

Reason 3: Bodily practices are key sites of power and resistance

The example of 'mealtime' (see also Reason 1) serves to illustrate this point. Family mealtimes are frequently a site where forms of adult power ('eat those vegetables!') can be resisted by children's agency (refusal to

Box 12.1
Williams and Bendelow (1998) on bodily practices and Marxism

It is argued by many sociologists of the body that bodily practices were an 'absent presence' within classic works of social theory: i.e. bodily practices were always important to these works of theory, but not mentioned explicitly. Consequently, there have been numerous re-readings of works of classical social theory, with the aim of 'rediscovering' bodily practices therein. For example, the work of Karl Marx has been re-read in this light. Particular significance has been attached to Marx's notion of the 'alienated' working class labourer. Williams and Bendelow (1998: 9–16) ask us to consider the following extract.

> What constitutes the alienation of labour?
>
> Firstly, the fact that labour is *external* to the worker – i.e., does not belong to his [sic] essential being; that he, therefore, does not confirm himself in his work, but denies himself, feels miserable and not happy, does not develop free mental and physical energy, but mortifies his flesh and ruins his mind. Hence, the worker feels himself only when he is not working; when he is working, he does not feel himself. He is at home when he is not working, and not at home when he is working. His labour is, therefore, not voluntary but forced, it is *forced labour*.
>
> (Marx 1959[1844], n.p.)

Marx's notion of 'alienated labour' refers to work that "does not belong to [a labourer's] essential being". In short, Marx observed that capitalist labour relations typically require workers to complete tasks that are basically alienated from workers' own interests, desires and needs. For example, alienated labour can refer to labour that is repetitive (and boring) and exploitative (advancing the interests of capitalists at the expense of workers themselves). More broadly, Marx noted that capitalism relies on the reproduction of a workforce (labouring bodies), and typically results in workers (and their bodies) being viewed primarily as objects and means of production. In all this, sociologists of the body have found Marx's observation that alienated labour 'mortifies flesh and ruins minds' especially significant. Clearly, here, there is a gesture to how Marx's influential critiques are intimately linked to bodily practices. Sociologists of the body have developed this observation with reference to contemporary first-hand observations of working class life, such as those made by Marx's collaborator Friedrich Engels:

> A pretty list of diseases engendered purely by the hateful greed of the manufacturers! Women made unfit for childbearing, children deformed, men enfeebled, limbs crushed, whole generations wrecked, afflicted with disease and infirmity, purely to fill the purses of the bourgeoisie.
>
> (Engels 1987[1845]: 184–185)

Through this kind of re-reading Williams and Bendelow demonstrate that Marxian critiques of capitalism must be understood as a thoroughly *bodily* business.

eat vegetables). This is a banal, but telling, example of ways in which bodily practices (such as eating, or not) can be central to the workings of social power relations. This observation can be usefully scaled up. Indeed, it should be apparent that, if bodily practices are central to all human geographies, they must also be at the heart of all geographies of social-cultural power and resistance (on which, see Chapters 2 and 3).

Many social scientists have found the work of the philosopher Michel Foucault useful in understanding relationships between social/cultural power and bodily practices. Foucault sought to explore how powerful institutions (e.g. states, the military, religion, medicine, legal systems) exert power through practices, rules and instruments that target individual bodies. Thus these institutions 'discipline' bodies and render them 'docile': that is, receptive to these forms of institutional power, to the extent that their norms and expected behaviours are reproduced with minimal resistance. Box 12.2 provides extracts from one of Foucault's accounts of this kind of power: on bodily practices of state-sponsored execution and imprisonment.

Foucault's work was principally historical: his case studies relate mostly to the social/cultural context of eighteenth-century France. However, his work has influenced diverse explorations of bodily discipline and its relation to social/cultural power, past and present: there is a wide range of geographical and social scientific work that applies Foucault's ideas to the forms of bodily disciplining that characterise present-day

Box 12.2
Foucault on bodily practices of punishment

Nobody who has read Foucault's (1991[1975]) *Discipline and Punish* is likely to forget the lurid scenes of violence in its opening pages. The book begins with archival material describing the punishment meted out to Parisian prisoners in the eighteenth and nineteenth centuries. Two short extracts will give you a flavour of this material.

A public execution, 1757

The prisoner was condemned to be:

> taken and conveyed in a cart, wearing nothing but a shirt, holding a torch of burning wax weighing two pounds . . . [to] a scaffold [where] . . . the flesh will be torn from his breasts, arms, thighs and calves with red-hot pincers, his right hand, holding the knife with which he committed the [murder] . . . burnt with sulphur and on those places where the flesh will be torn away, poured molten lead, boiling oil, burning resin, wax and sulphur melted together and then his body drawn and quartered by four horses and his limbs and body consumed by fire, reduced to ashes and . . . thrown to the winds.
>
> (p.3)

A prison timetable, 1837

> *Rising.* At the first drum-roll, the prisoners must rise and dress in silence, as the supervisor opens the cell doors. At the second drum-roll they must be dressed and make up their beds. At the third, they must line up and proceed to the chapel for morning prayer . . .
>
> *Work.* At a quarter to six in the summer, a quarter to seven in winter, the prisoners go down into the courtyard where they must wash their hands and faces, and receive their first ration of bread. Immediately afterwards, they form into work-teams and go off to work, which must begin at six in summer and seven in winter.
>
> (p.6).

Foucault draws attention to the ways in which states and institutions develop forms of punishment and correction that work by impacting (sometimes horribly) upon individuals' bodies. He also notes that these bodily practices vary significantly over space and time. For example, he identifies a shift (roughly at the end of the eighteenth century) in the forms of punishment enacted by European states, with the disappearance of punishment as a spectacle (as in the first extract above) and the emergence of more subtle, hidden forms of disciplining, order and correction (as in the routines and drills in the second extract).

schools, institutions, the policing of cities, forms of military training, and so on. Consider an example of this kind of research, exploring the US Army's training of new officers (Box 12.3).

If bodies are key targets for forms of social/cultural power, it follows that certain bodily practices can also resist these powers. Certainly, all of the forms of resistance, activism and **subcultural** agency described in Chapters 3 and 8 involve some form of bodily practice. Bodily forms of resistance can vary in intent, ambition and poignancy. At one end of the spectrum, small bodily gestures (growing one's hair, getting a tattoo or piercing, refusing to eat vegetables) can feel rebellious and important within the parameters of one's identity and self-image: perhaps giving a small "sense of control over one's body, that one has somehow taken possession of one's body" (Featherstone 2000: 2). Elsewhere, however, individuals' bodily practices can represent important moments within major social/cultural and political contexts. For:

> Historically, bodies have been a powerful political tool in activist work. They have been used as an instrument or object in campaigns where bodies might be chained to trees, placed in front of bulldozers, camped outside of nuclear bases or marched down to Parliament House. Bodies have also been put at risk in actions: zapped by radioactive waves, beaten by opposing forces, or deprived of adequate food, water and medical supplies.
>
> (Parviainen 2010: 312)

Forms of activism, protest and subcultural subversion frequently involve some form of purposeful and resistive bodily practice (Photograph 12.5).

Think, for example of Rosa Parks' decision to *sit down* on a particular bus in Montgomery, Alabama, in

Box 12.3
Bodily discipline and US Army officer training

Photograph 12.4 Parade ground drills: a disciplining practice to produce combat-ready bodies.
Source: Getty Images/AFP.

In the extracts below, Lande (2007) observes the US Army's training of new officers on a parade ground (Photograph 12.4) and shooting range. Curiously, in both settings, becoming an officer is fundamentally about learning to breathe 'correctly'.

> Two platoons are lined up in front of . . . a young . . . cadet who has just been rotated into a leadership position. The [platoons] chat loudly, standing 'at ease' . . . [The cadet] says 'ATTENtion'. It comes out sounding more like a question than a command. The platoons come to attention, but do not do so with the sense of urgency and assurance that is customary when a sharp, loud and correctly intoned command is given by one of the . . . officers. One of the [instructors] runs to the front of the formation and tells the [cadet] 'Don't speak from your throat. No-one will hear what you say as a command. Breathe from your stomach.'
>
> (p.95)

> The Instructor says 'OK, we are going to work on trigger squeeze and breathing. When you shoot, you want to reduce your movement as much as possible. Even a little movement can send a bullet off in the wrong direction' . . . [the training continues] 'Squeeze slowly and don't jerk it' . . . 'Use the meaty part of your finger' . . . 'Hold [your breath] at the bottom of the exhale . . . [T]ake a breath and just pause when you have emptied your lungs of air . . . That is when you want to shoot.'
>
> (p.103)

It becomes clear that the usually taken-for-granted bodily rhythm of breathing is crucial to becoming an effective military leader. The consequences of breathing 'incorrectly' are potentially grave: a misplaced breath might result in misdirected bullets, or the failure to command a body of soldiers. A key aspect of officer training is therefore a programme of physical endurance and breathing exercises, designed to discipline trainees' bodies until they can breathe reliably without effort: that is, in Foucault's terms, until their bodies are 'docile', such that they conform to their superiors' demands without resistance. Thus, to reiterate, in this example the interests of the state (to reproduce an army of combat-ready soldiers, capable of exerting state power) are enacted through disciplining practices that target – and ultimately render 'docile' – individual bodies.

December 1955: a small bodily act that, in defying contemporary social/cultural rules and norms about racial segregation, became an iconic, galvanising catalyst for North American civil rights movements. Of course, very few bodily practices have this kind of political power. However, the examples juxtaposed above

Photograph 12.5 Using the body in activist protest practices.
Source: Getty Images/AFP.

should remind us of how different forms of social/cultural power are (often) reproduced and (sometimes) resisted through bodily practices. This notion of resistance is explored in more detail in the discussion of **subcultural** identities in Chapters 2 and 8.

Reason 4: Research is a bodily practice

It follows, from the observation that all human geographies involve bodily practice, that doing human geography must involve bodily practices too. Research, theory-making, teaching and learning are all bodily practices. On reflection, academic activities such as fieldwork are unquestionably bodily practices (Okely 2007). For example, geography fieldwork frequently involves physical exertion (climbing, hiking, digging, measuring, recording) or interactions between human bodies (interviewing, discussing, observing). Perhaps less manifestly, "the grander façades of . . . disciplinary schools of thought"– like the discipline of 'cultural geography'– are made up of all manner of bodily encounters with the world, and habitual disciplinary bodily practices (Dewsbury and Naylor 2002: 254). However, a great deal of this bodily practice gets written out of the canon of published accounts of academic practice. We are rarely told about the physical exertions and bodily practices that are central to fieldwork, for example. When reading academic research papers and monographs, one seldom gets any impression of the "meaningful and lively practices" that must be central to the production of knowledge (Lorimer 2003: 213).

This realisation might prompt two anxieties. First: what forms of bodily labour go unjustly unacknowledged in the knowledge production process? For instance, postcolonial historians have begun to outline the extent to which early geographical fieldwork was possible only as a result of the substantial, but unacknowledged, physical exertions of native servants and slaves supporting (sometimes literally) explorers (see Pratt 2007). Second: how do bodies matter in research? For example, there is now a large body of literature critiquing the masculinist and ableist nature of geographical fieldwork, where strong, fit, able-bodied explorers of the landscape have historically been normalised and valorised (to the exclusion of diverse bodies and identities) (Rose 1993, and Photograph 12.6).

Alternatively, consider how different kinds of bodily appearance, size, gesture, fitness, behaviour and dress might make a difference to doing research with other people. For example, adults who conduct research with children are acutely aware of the problems for research posed by differences in bodily size (how might one get down to a child's 'level' to engage in an interview?). Similarly, social scientists working in this context have become acutely aware of potential ethical problems associated with bodily contact: certainly in the UK, it

Photograph 12.6 Geographical fieldwork: a masculinist, exclusionary bodily practice?

Source: Alamy Images/Robert Harding Picture Library (tl); Dr Faith Tucker (tr); Alamy Images/Kevin Britland (b).

would at present be considered deeply inappropriate for an adult researcher to hug or smack a child with whom they were working. Indeed, social scientists conducting research with children find themselves engaged in a project of *avoidance* of bodily contact between themselves and their research participants.

12.5 Embodiment

The term 'embodiment' refers to the complex material, messy, processual *stuff* that makes up – that *is* – our body (some of which becomes evident if we stop to think about how our body feels, here and now). Increasingly, it is acknowledged that many social scientific accounts of the world – even those that focus on 'the body'– tend to overlook, oversimplify, or cover up this embodiment:

> The fluid, volatile flesh, . . . the messy surfaces of bodies, depths of bodies, their insecure boundaries, the fluids that seep and leak from them . . . [and how the body] breaks its boundaries – urinates, bleeds, vomits, farts, engulfs tampons, ejaculates and gives birth.
>
> (Longhurst 2000: 23)

It is argued that this complex bodily materiality has important implications for social scientists. For instance, we might say that embodiment is important for cultural geographers for four key reasons.

Reason 1: Embodiment is who we are

Another truism: the materiality of our bodies – for instance, in terms of our bodily shape, size, characteristics, capabilities, habits and needs – is a significant part of who we are. The stuff of our bodies *is* – to a great extent – us. For example, each of us has our recognisable bodily habits and quirks: a particular "way of walking, a tilt of the head, facial expressions, ways of sitting and of using implements", for example (Bourdieu 1977: 87).

The feminist theorist Iris Marion Young (1990) calls this 'bodily comportment': the characteristic way in which we each act and 'hold' our bodies. These (largely unconscious) habits attend everything we do, every social interaction, every encounter with the world (see Reason 2). Bodily comportment can be interpreted as 'body language': certainly, we might recognise types of bodily comportment that characterise somebody who is shy, angry, outgoing, afraid and so on. Many aspects of bodily comportment are genetically and socially inherited: think of the ways in which fathers and sons typically have uncannily similar ways of sitting, standing, moving, behaving and so on. Our bodily materiality can also reveal life histories: our body's scars, flinches and phobic reactions can be read like a **text** (see Chapter 6). For Young, however, bodily comportment is important, as it bears witness to how our bodies have been nurtured, socialised and disciplined, within contemporary social/cultural contexts. Young (1990: 146) takes the example of throwing and catching a ball. In particular, she reflects upon differences in male and female bodily comportment during this kind of playful activity:

> Men more often move out towards a ball in flight and confront it with their own countermotion. Women tend to wait and then *react* to its approach, rather than going forth to meet it. [Women] frequently respond to the motion of a ball coming *at* us, and our immediate bodily impulse is to flee, duck or otherwise protect ourself from its flight.

'Throwing like a girl', then, is a bodily comportment characterised by hesitancy, nervousness, self-consciousness, inhibition, using *part* of one's body rather than throwing the *whole* body into a situation. Young argues that this is a microcosm of many females' bodily comportment in public spaces: reflecting contemporary differences in the training of males and females to encounter public space, and a lifetime of being subjected to the gazes of others. Here then, broad social/cultural norms (relating to 'gender') are manifest in individuals' bodily materiality and habits ('throwing like a girl') that constitute a particular bodily comportment which, to a large extent, defines a person (how they act, how they see themselves, how they are seen by others). In this way, broad social/cultural issues and phenomena are intimately linked to individual embodiment.

Reason 2: Embodiment is our encounter with the world

Yet another truism: "Our embodiment is implicated in everything we see or say" (Harrison 2000: 497). Our entire experience of the world happens through our bodies; indeed, we might say that our embodiment *is* our experience of the world. One way to appreciate this is to pause to think about how you are experiencing the world *now*. Imagine, now, how your experience of wherever you are could be very different, if your bodily materiality was somewhat different. How might your experience be different:

> if you were bigger or smaller; if you were more short-sighted; if you had a headache or toothache; if you had more difficulty getting around; if you were hung-over or High; if you felt cold, or had a Cold; if you were really tired or really stressed-out or really calm; if you'd just smacked your head on a low doorframe; if you were really hungry or full-up; if you were feeling depressed, or if you were chuckling away at something?"
>
> (Horton and Kraftl 2006a: 76)

This exercise might prompt three reflections. First, our encounter with the world is always embodied and multisensuous. Our bodies are open to the world, and open to being 'touched by the world', in all kinds of ways, which ultimately relate to our embodiment (Paterson 2005). Second, even very small changes in our embodiment can fundamentally affect how we encounter the world. For example, a few millilitres of caffeine or alcohol in the bloodstream alters our reactions and perceptions; ingesting a microscopic bacterium can cause sickness, debilitation or death. Third, experiences of **places, spaces** and **landscapes** (see Chapters 5 and 13) must be understood as encounters between the materialities of one's body and that space, there and then.

Thus it is argued that human geographers must understand everyday life as a "multisensual . . . relationship to places" (Rodaway 1994: 4). That is, we should acknowledge the multiple kinds of sensory stimulus that constitute a space. For example, Chapter 5 reflects on the importance of **landscape** for cultural geographers, and the changing ways in which geographers have attempted to register the importance of bodies and emotions in encounters with landscapes. Increasingly, it is apparent that many chief accounts of landscape overwhelmingly prioritised the *visual* representation of landscapes, and paid much less attention to the multisensual experience of particular events of being-in-landscape. Thus social scientists such as Urry (2007) call for understandings of landscape to be expanded to also encompass: 'soundscapes' (sounds, noise, silence and auditory phenomena in the landscape), 'smellscapes' (smells, odours, tastes), 'haptic geographies' (textures, touch, physical sensations) and 'psychogeographies' (emotions, moods, associations evoked by/in landscape).

Reason 3: All social/cultural issues have embodied consequences

If all human geographies are embodied, it follows that the social/cultural phenomena that have long concerned human geographers can frequently have embodied consequences. Box 12.4 presents just one example: the consequences of natural disasters for young children's embodiment. Note, here, how a major, pan-continental geographical event (e.g. a disaster or war) has poignant consequences at the level of an individual's embodiment. For further examples of this kind of embodied **affect,** see Chapter 11. Consider, too, how case studies elsewhere in this book, relating to **consumption** and **emotions** (Chapters 3 and 11) are also embodied in nature, and have embodied consequences.

Reason 4: Embodiment raises questions about the 'human' in human geography

It is argued that reflection upon embodiment requires a fundamental questioning of some basic assumptions that underlie much social scientific work and practice. For example, it is commonplace in social scientific accounts to see human beings represented, neatly, as bounded individuals. However, theorists such as Donna

Box 12.4

Geographical events, bodily affects: impacts of disasters upon children

Research by charities and aid agencies in the aftermath of natural disasters and armed conflict documents some everyday, bodily consequences of major geographical events. For example, Ansell (2005: 197; after Jabry 2002) collates evidence regarding impacts of natural disasters and armed conflict upon toddlers, children and adolescents. These impacts are fundamentally personal, *bodily* and affective, with significant consequences for everyday life.

Impacts upon toddlers

Regression in behaviour; decreased appetite; nightmares; muteness; clinging; irritability; exaggerated startle response.

Impacts upon school-age children

Marked reactions of fear and anxiety; increased hostility with siblings; somatic complaints (e.g. stomach aches); sleep disorders; school problems; decreased interest in peers, hobbies; social withdrawal; apathy; re-enactment via play; post-traumatic stress disorder.

Impacts upon adolescents

Decreased interest in social activities, peers, hobbies, school; inability to feel pleasure; decline in responsible behaviours; rebellion and behaviour problems; somatic complaints; eating disorders; change in physical activities (both increase and decrease); confusion; lack of concentration; risk-taking behaviours; post-traumatic stress disorder.

Haraway insist that attention to the fleshy, messy materiality of bodies should unsettle this assumption. As we noted in Chapter 10, Haraway directs attention to the many ways in which human embodiment is augmented and extended by non-human materials. Think of spectacles or contact lenses, clothes, dental repairs, hearing aids, prosthetic limbs. All of these material technologies can be incorporated into one's body, and extend its capabilities. Similarly, material commodities can extend our bodily capabilities in particular contexts: returning to the example that opened this chapter, we might feel that sportswear, sports equipment, energy drinks, nutrient supplements or portable music devices enhance our ability to work out at a gym. In instances such as these, embodiment is supplemented by non-human **material objects** (see Chapter 10). Haraway (1991) defines these kinds of assemblage, linking human embodiment and non-human materials, as 'cyborgs'. For Haraway, 'cyborgs' raise difficult questions for social scientists: after all, in a cyborg assemblage where does the boundary between human embodiment and non-human materiality begin?

A second assumption that is commonplace in many social scientific accounts is that human beings can reliably be understood as conscious agents: we (mostly) think before we act. However, again, an attention to embodiment, especially the embodiment of senses and cognition, unsettles this assumption. For example, Dewsbury (2000) juxtaposes the following quotations, which raise questions about the relationship between embodiment and consciousness.

> The subjects were asked to flex a finger at a moment of their choosing, and to note the time of their decision on a clock. The flexes came 0.2 seconds after they clocked the decision. But the EEG machine registered significant brain activity 0.3 seconds *before* the decision . . . a half second lapse between the beginning of a bodily event and its completion in an outwardly directed, active expression.
>
> (Massumi 1996: 222–223)

> A finger turns in light circles across your toes and the pads of your feet. Is this irritating? ticklish? erotic? relaxing?
>
> (Brusseau, 1998: 10)

Consider the first quotation: consider, especially, that 'half-second lapse'. Significantly, the experiment cited here identified a 0.3 second gap between 'the beginning of a bodily event' and consciousness of that bodily event. We are left to wonder: although we *feel* as though we think before acting, is this truly the case? What of that 0.3 second period that, the experiment suggests, precedes conscious recognition that one is acting?

Now consider the second quotation: how *would* that finger turning in light circles feel? We might answer: it depends on who the finger belongs to (a lover? a masseur? a child? a bully?) or where this scenario is taking place (in bed? in a shoe shop?). Or timing may be key: this sensation would probably be unwelcome when one was sleeping deeply, or during a period of illness. However, all of these provisos draw attention to another kind of 'gap': between an embodied sensation and our ability to consciously 'make sense' of it. Conceivably, for example, a ticklish sensation experienced in a shoe shop could be momentarily experienced as pleasant, before one consciously 'catches oneself' and suppresses this as 'inappropriate'. Conversely, the experienced and expensive touch of a masseur might be experienced as ticklish, requiring the suppression of laughter. The embodied sensation in all of these possible cases remains the same; yet the experience of that sensation – after a few seconds of conscious interpretation – might differ greatly, depending on context and, then, cultural norms and expectations within that context. Again, we might wonder: what are the implications of this kind of 'gap' – between embodied sensation and consciousness – for the social scientific characterisation of humans as conscious beings?

Finally, an attention to embodiment reveals the extent to which bodies are in flux: they change, develop, age. It is easy to take our bodies, and their capabilities, for granted. However, the complicated stuff of bodies is, of course, finite, vulnerable and subject to change, sooner or later. For instance, studies of ageing and chronically ill, pained bodies reveal the extent to which embodiment matters for everyday geographies. Here, chronically pained embodiment is experienced as "a profound disruption of . . . biographies, selves and the taken-for-granted structures of the world upon which they rest" (Williams and Bendelow 1998: 159). Pained embodiment reveals how "the fidelity of our bodies is so basic that we never think of it – it is the grounds of our daily experience" (Kleinman 1988: 45); when this fidelity breaks down, social and spatial consequences are numerous (see Box 12.5).

Box 12.5

Some social, cultural and spatial consequence of arthritis

The term 'arthritis' encompasses around 200 types of debilitating conditions of muscular and connective tissues and joints. Around 10 million people in the UK, and 46 million in the USA, suffer from an arthritic condition (Arthritis Care, 2012). Research by the sociologists Williams and Barlow (1998) explores the consequences of this embodied condition for sufferers' everyday lives. A selection of extracts from their interviews with people recently diagnosed with arthritis is shown below. Reflect on how these factors might impact on the social-cultural lives of people with arthritis.

A sense of 'falling out with one's body'

> My concern is with my swollen fingers, my swollen feet, whichever bit that decides to swell. My fingers are all twisted and you get this sort of gargoyle appearance . . . I don't like my body, not the way it is. I mean who would? It is disfigured.
>
> (p.128)

Self-consciousness and shifting social relations

> You don't feel so attractive at all . . . You want to still feel attractive and still feel that he fancies you . . . You hope that they think of you as the same as you were, before all of this happened, but you don't feel as if they can . . . You feel self-conscious of your body, you want to disguise it as much as you can.
>
> (p.134)

Self-consciousness and avoidance of social contact

> You feel very self-conscious, you try to hide your body and the swellings. You avoid things that take you into contact with people, you avoid places where you think you are going to meet friends . . . you avoid contact of any kind like the plague.
>
> (p.134)

Inability to complete physical activities (e.g. inability to wear fashionable clothes)

> I don't wear pretty blouses because of the cuffs, collars and buttons up the front. I wear looser clothes so that should I get the swellings, the clothes don't restrict it. You have to choose your clothes carefully, you choose clothes that you can cope with.
>
> (p.132)

Depression and self-doubt

> I get depression a lot . . . I often feel annoyed and angry with myself. It is a self-destructive anger . . . tears streaming down your face and thinking 'I wish I wasn't alive, I wish I was dead' . . . [I]f you have got arthritis it happens a lot because when you are in pain, terrific pain . . . severe body pain, you doubt yourself constantly.
>
> (p.134)

Summary

- Human geographers, like most social scientists, have historically shied away from what we can think of as 'the geography closest in – the body'. This is problematic: for all human life is *bodily*.
- Representations of 'normal' and 'ideal' bodies abound in contemporary popular culture. These kinds of representation have tangible implications for individuals' self-image and broader forms of social exclusion.
- All human activity involves some kind of bodily practice. Acknowledging bodily practices can enable more careful understandings of the nature of major social/cultural issues. For example, bodies are key sites for the reproduction and resistance of forms of power, at diverse scales.
- The messy, material stuff and processes that make up bodies pose difficult questions for social scientists. The term 'embodiment' is used to denote this messy, processual stuff. In particular, a range of assumptions often taken for granted by social scientists become problematic when attention is paid to embodiment.

Some key readings

Environment and Planning D **18** (4) (2000) – Theme issue on *Spaces of performance*

A collection of pioneering, theoretically orientated papers by human geographers. All of the papers are relevant to the discussion of bodies and embodiment in this chapter. For example, see: the paper by Gregson and Rose for detailed discussion of Judith Butler and the bodily performativity of identities; the paper by Longhurst for a case study of activist practices and bodily difference (in this case relating to the pregnant body); papers by Harrison, and Thrift and Dewsbury, for discussion of the importance of embodied performativity for human geographers.

Hörschelmann, K. and Colls, R. (eds) (2009) *Contested Bodies of Childhood and Youth,* Palgrave Macmillan, Basingstoke.

A collection of chapters by geographers interested in childhood and youth. Chapters apply many of the concepts discussed in this chapter to specific research contexts, and provide empirical examples of representations of bodies, bodily practices and embodied geographies.

Longhurst, R. (2001) *Bodies: Exploring fluid boundaries,* Routledge, London.

A very readable call for geographers to focus on bodily practices and especially the complexities of embodiment. In particular, many of our students have found this book a clearly written 'way in' to feminist debates and concepts in this context.

Rodaway, P. (1994) *Sensuous Geographies: Body, sense, and place,* Routledge, London.

A useful source for discussion of how successive waves of human geographers have tackled issues relating to the senses, emotions and the body.

Shilling, C. (2003) *The Body and Social Theory,* Sage, London.

You may have noticed that a great deal of the discussion in this chapter has drawn on the work of sociologists, rather than geographers. This is because sociologists of the body provided some of the earliest and most compelling arguments that social scientists should take bodies seriously in their work. Shilling's book provides an excellent synthesis of concepts and debates from this body of work.

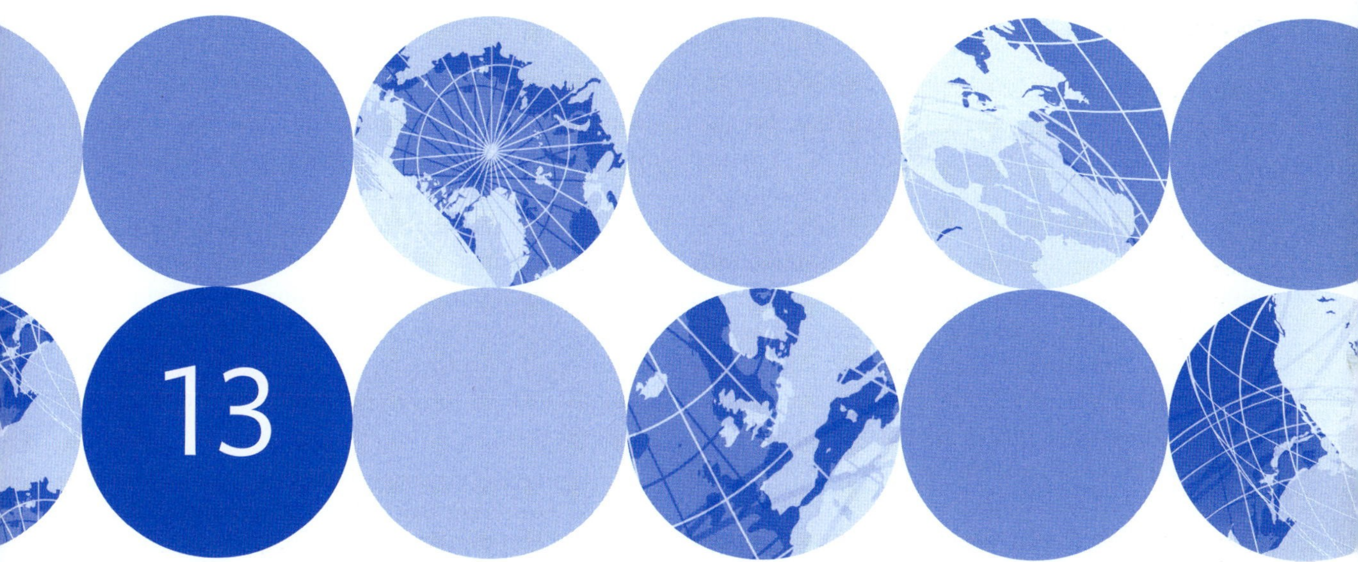

Space and place

Before you read this chapter . . .

Consider the many and varied ways in which the terms 'space' and 'place' are used in everyday English. For instance (to name but three):

- 'David is in a *bad place* right now.'
- 'I need *space to think*.'
- 'It's not really *my place* to comment on your fashion sense.'

You can probably think of many other examples. In each case, think carefully about what 'space' or 'place' mean – literally or, in these cases, perhaps more metaphorically. Do they describe a person? A thing? An emotional or embodied state? An action? An identity?

Chapter Map

- Introduction: geography, space and place
- Space and place: definitions and debates
- Key motif 1: Place (and space) as socially constructed
- Key motif 2: Scale
- Key motif 3: Flow/mobility
- Discussion: where next for space and place?

13.1 Introduction: geography, space and place

At some point or another, the vast majority of geographers (students, lecturers, teachers, researchers) have been asked what geography is. Perhaps they have been asked why geography is *not* sociology; or cultural studies; or anthropology. Perhaps they have been asked to summarise what geographers *do*. Or, perhaps, they have been asked what the particular approaches are – conceptual, methodological, political – that characterise geographical endeavour.

Ultimately, many geographers' answers tend to involve reference to one of two key words: 'space' and 'place'. Each term has been, and continues to be, hotly contested. For this reason, one could write entire books about them; indeed, the annotated bibliography at the end of this chapter lists what we consider some of the best texts whose sole purpose is precisely that. Another difficulty is that each term is so central to human geography that one could equally easily end up writing about any and all philosophical approaches to the discipline. They are all, somehow, about space, or place, or both. And again, there have been many excellent texts to which we would refer you if you want to read about these approaches (e.g. Hubbard *et al.* 2002;

Johnston and Sidaway 2004; Clifford *et al.* 2009; Nayak and Jeffrey 2011). Thus it is a daunting task to either write, or read, about space and place. But it is also important to acknowledge the complexity and magnitude of geographical writings on these topics.

In this chapter, though, we do something a little different, and much more modest. We encourage you to think about how space and place matter to cultural geography. The chapter looks 'beneath' some of the -isms (such as Marxism, feminism and poststructuralism), which are often cited as 'the conceptual approach' taken to a piece of research, in order to foreground what is being said about *space* and/or *place*. This means that the present chapter would most helpfully be read alongside some or all of the others (especially Parts 1 and 2), because it teases out a selection of ideas about space and place that have been *deployed* in empirical work by cultural geographers. For this reason, we make plenty of cross-references to examples and case studies in Parts 1 and 2, and we would encourage you to follow some of these up to read about ideas of space and place 'in action'. This will also give you an idea of how *we* understand the use of space and place in cultural geography – as diverse as the many definitions of the subdiscipline (Chapter 1) but orientated around a series of key cultural practices and politics (Part 1) and cultural materials, texts and media (Part 2). Otherwise, this chapter may seem a little 'abstract' and hard to grasp in places, although we do include several new examples to help explain key ideas.

This chapter begins with a section that offers brief definitions of 'space' and 'place'; but it emphasises that there are some productive tensions and overlaps between these definitions. It then charts a series of ideas collected under three very broad 'key motifs' that extend notions of space and/or place in sometimes more particular, sometimes more politicised, and sometimes more theoretically contentious ways.

13.2 Space and place: definitions and debates

Look again at the familiar English colloquialisms that began this chapter. It is noticeable that they share two features. On the one hand, none of them refers to a 'real', geographically referenced space or place. They do not indicate named features on a map; nor do they evoke points that could be given grid coordinates. Second, each colloquialism involves, in some sense, some kind of *life*. In all of the cases listed, these are human (sometimes termed 'social') lives. In its own, quiet way, each phrase indicates how space and place *can never* exist without these lives. As we argue at the beginning of Chapter 8, space and society are always co-implicated. We agree wholeheartedly with Doreen Massey that:

> [an] approach to the understanding of the social, the individual, the political, itself *implies and requires* both a strong dimension of spatiality and the conceptualisation of that spatiality in a particular way. At one level this is to rehearse again the fact that any notion of sociability . . . is to imply a dimension of spatiality. This is obvious, but since it usually remains implicit . . . its implications are rarely drawn out.
>
> (Massey 2005: 189)

Massey is arguing that the human life cannot exist without space, and vice versa. Moreover, at the end of the quotation, she is suggesting that the implications of this (seemingly obvious) observation are rarely considered. This seems a very odd point for a geographer – writing in 2005 – to make. Surely geographers have spent years considering how society and space interact?

In some ways, the answer to the above question is yes. But the point is that particular assumptions about *space* and *place* have tended to dominate, as we show in the two subsections that follow. Remember that these are *assumptions,* not necessarily the 'correct' definitions of each term. Try to spend a little time thinking about whether you agree or disagree with each point – perhaps based on your own ideas about space and place, perhaps based on classes you have sat in, or perhaps based on your reading of other chapters in this book. The remainder of this section outlines some of the key debates and problems surrounding these definitions.

Dominant assumptions about space: quantitative or 'scientific' approaches

1. Most fundamentally, definitions of space refer to an *abstract* realm. Space is 'out there', not in here; it is beyond the human mind and body. It is

extensive: at its greatest scale, it covers the globe (or refers to its surface).

2. Space is either a *surface* or a *container*. According to the German philosopher Gottfried Leibniz, space is simply an expression of the relationships between things on the world's surface. Space is an abstraction of the relationships between those things. But according to the English mathematician Isaac Newton, space exists independently of matter; matter is contained within space.
3. In each of the above formulations, space is *measurable*. Space can be given *x* and *y* coordinates to measure distance; a *z* coordinate could be added to measure depth and volume. One could also add a *t* dimension to show changes over time. In all dimensions, space (and time) are finite resources and closed realms in which individuals and societies operate (Haegerstrand 1978).
4. Following the above, space can be *represented*. Once measured, space can be mapped; even the globe can be projected onto a flat page and made subject to a grid. Space can also be represented in three dimensions by the diagrammatic elevations used by architects, and especially in computer-aided design (CAD).
5. Space is politically neutral. Space does not move. Space is closed (a map, surface, container or globe has an edge).
6. Because space is mappable, it can be subject to mastery by society. In particular, space can be subject to mastery by dominant groups who can view spaces as 'empty' and ready to be 'planned' or 'colonised' (see Chapter 5 on 'seeing' **landscape**).
7. To generalise enormously, assumptions 1–4 formed the basis of most research in human and physical geography before 1970. This approach is known variously as the quantitative or positivist approach to geography, because it was based on apparently objective, scientific and rational rules about the location of phenomena in mappable space (Hubbard *et al.* 2002). More fundamentally, it was the most common expression of geography's core task: to investigate what Richard Hartshorne (1959) – drawing on the cultural geographer Carl Sauer – termed 'the study of areal differentiation'. In this way, as you may well know from earlier studies, geographers produced a series of (what appear now as rather dry, abstract) models that schematised and attempted to predict land use in cities and their hinterlands – from the urban land-use models of the 'Chicago School' to von Thünen's rural model (Box 13.1).
8. To be a little contentious, assumptions 5 and 6 were, more or less, ignored by geographers until the 1970s. However, we should recognise (as we shall discuss throughout this chapter) that, just as cultural geographers *do* actually write about space, some quantitative geographers have wrestled with ideas of how to think about place in ways that overlap with the seemingly livelier, if sometimes seemingly esoteric, concerns of cultural geography (see Box 13.3).

Dominant assumptions about place

1. Most fundamentally, place refers to a *point* in space. Place is a point on the earth's surface.
2. It follows that 'a place' is *also mappable, representable and measurable*. A place can have specific coordinates in the *x*, *y*, *z* and *t* dimensions.
3. Whereas space is abstract, place is *concrete* (Massey 2005: 184). Places have material bits and pieces that make them unique. This was the core assumption of the idiographic approach taken by the **Berkeley School** in their research on **landscape** (see Chapters 1 and 5). Places tend to have names: 'Beijing'; 'Flagstaff'; 'Caracas'; 'Mount Everest'.
4. Space can be mapped at the local scale, but refers ultimately to an abstract, global system. But places are *local*, because they are named points on the earth's surface that are 'here' and not elsewhere. What makes a place concrete and unique also makes it local, and vice versa (i.e. a self-fulfilling prophecy!).
5. Places *mean something* to people; people must find them meaningful to make them places. A space without meaning is not 'fully' a place, just a point on the map that can be 'place-d'. People can be attached to places to differing degrees. Edward Relph (1976) called this 'insideness': the degree of association of involvement that someone has with a place.

Box 13.1

von Thünen's (1826) model of rural land use: an isotropic plane from the quantitative era of geography

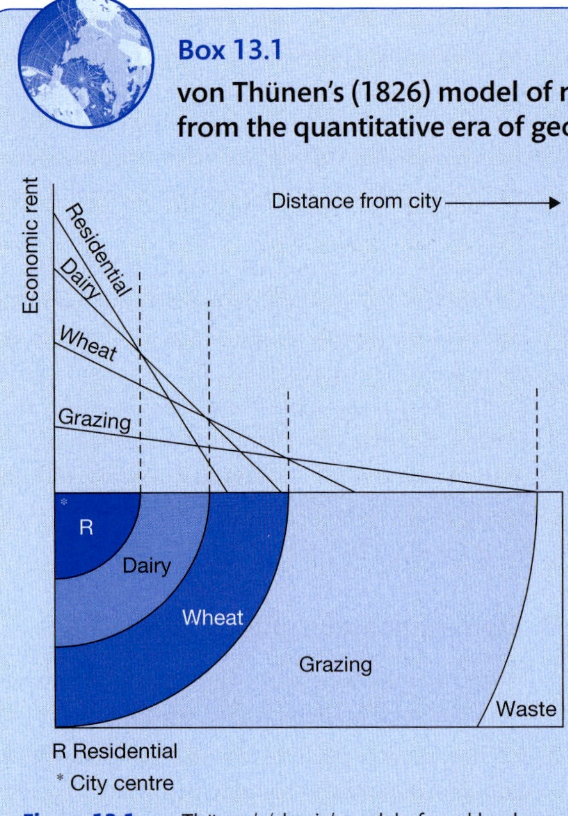

Figure 13.1 von Thünen's 'classic' model of rural land use.

Source: from *Der Isolierte Staat in Beziehung auf Landwirtschaft und Nationalökonomie*, Pergamon Press, New York (Von Thünen, J. 1826), Von Thünen Model.

You may well have seen the model in Figure 13.1 before. It is one of the most famous images to come from the discipline of geography. It shows how the intensity and kind of land use changes with distance from a city centre, dependent upon rental costs and transportation. It is situated on an 'isotrophic' plane – meaning that there are no human- or physical-geographic variations in any direction that could disrupt the model (it is 'flat', in all senses of the term). Of course, this is intended as a *model*, but it provides a pretty good sense of the ways in which space was conceived during the quantitative era (the 1950s and 1960s) – as something flat, measurable, subject to modelling, and actually quite dry and devoid of any (cultural) meaning. At the same time, we must recognise that models such as this are not neutral at all (even if they purported to be objective) – they were often used to prop up particular political or economic demands. Moreover, they represent the cultural beliefs of the time – in the power of scientific models and experimentation to understand and change the world.

6. *Places are associated with diverse human traits.* According to humanistic geographers (the most famous proponents of the term), places may be characterised by dense nodes of human activity, they may be sites of fond individual or collective memory (and nostalgia), or they may simply be locales with which people have a strong emotional bond (Relph 1976; Tuan 1977; Cresswell 2004: Chapter 2). One's home or place of birth may be a particularly important place.
7. Places may involve particular spatial traits (and here there is an overlap with notions of space). The internal volumes of a building – doorways, arches, atria, corridors, décor – may be arranged in such a way as to create an emotional or affective attachment to place (Tuan, 1977; Kraftl and Adey 2008; Box 13.2).
8. Cultural geography is about place, not space. In particular, cultural geography is about the **emotions,** meanings, **texts,** images and **performances** that constitute the meanings we give to place (see Chapters 4 to 8 and Chapter 11). Thus much cultural geography is influenced by humanistic understandings of place, rather than an understanding of places as 'points' in space.

Debates about space and place in cultural geography

What begins as undifferentiated space becomes place as we get to know it better and endow it with value . . . the ideas 'space' and 'place' require each other for definition. From the security and stability of place we are aware of the openness, freedom and threat of space,

Box 13.2
Forest schools

Photograph 13.1 A forest school in Cornwall, England.
Source: Peter Kraftl.

Originating in Denmark, forest schools attempt to use local, 'natural' environments to educate children outside the classroom. Before the children arrive – as in this case – Forest School practitioners try to create a 'base' or 'home'. In this case, they do so with some simple chairs and a colourful 'parachute' that provides and is meant to symbolise 'shelter'. This area acts as a focal point for children, and helps them to feel secure in a large and complex space that could otherwise be confusing. We could see this as a relatively rudimentary, but contemporary – and culturally specific – way in which humans try to create a feeling of belonging by using simple structures. We can also see this as an example of what humanistic geographers meant when they looked at how architectural volumes (arches, doorways, rooms) could make people feel attached to places.

> and vice versa. Furthermore, if we think of space as that which allows movement, then place is pause; each pause in movement makes it possible for location to be transformed into place.
>
> (Tuan 1977: 6)

The quotation above (from the humanistic geographer Yi-Fu Tuan) neatly captures the key differences between space and place in the lengthy definitions we have provided. Space *becomes* place through its association with meaning and (usually long-term) inhabitation – with what the German philosopher Martin Heidegger called *dwelling* (see also Ingold 2000). We discuss humanistic conceptions of **landscape** in Chapter 5, and, relatedly, the ways in which poetic, humanist notions of place are expressed in relation to fictional **texts** in Chapter 6. But the point here is that a place needs space – a realm, volume, surface or locale that can be named and provide the physical and/or imaginative space for dwelling to occur. Those people who cannot have a space to make place may be 'out of place', 'dis-placed', 'home-less' – and so on (Cresswell 1996; see also Chapters 3 and 8 on **identities**).

So far so good. But the two definitions we have provided in the previous two subsections are lengthy, complicated and, above all, hotly contested by cultural geographers. We cannot simply reduce them to

Tuan's dualistic formulation (and nor did Tuan in his own work retain such a simple separation between the terms). There have therefore been many critiques of the definitions listed above, and the dualistic sense in which space and place are often written about. Below, we focus on just five.

First, there are many ways to think about space beyond picturing it as a 'surface' or 'container'. Some examples of other (albeit still largely visual) ways of figuring space are considered in Section 13.5 on *flow/mobility*.

Second, certain features of space (and, indeed, place) resist mapping, and are hard to pinpoint through grid coordinates. Consider, quite simply, how difficult it is to delineate the 'edge' of a city, or where a mountain 'is' (Box 13.3). Places exceed the very space (the pinpointable place) where we think they 'are'. We extend this initially rather counter-intuitive idea in Section 13.3 on *place as social construction* and Section 13.4 on *scale*.

Third, using a mathematical/cartographic system to organise and visualise space has many advantages. Indeed, many people rely on these systems (from walking maps to GPS car devices; see Photograph 13.2). But we should be concerned with the ways in which those systems have come to dominate – both as ways of seeing and knowing the world, and through their complicity in colonial or oppressive acts (see Chapter 5, on **landscape**). Mathematical/cartographic space is never politically neutral, and nor is any other way of conceiving space. We consider this issue in Section 13.5 on *flow/mobility*.

Fourth, space has often been seen as the opposite of time (Massey 2005). Where time is open-ended, unpredictable and brimming with infinite possibility, space has been figured as static, finite, closed and limiting. In part, space has been viewed in this way precisely because of the mathematical/cartographic systems that have dominated societal understandings of the term. But many theorists now argue that there is no space without time, and vice versa. We expand our discussion of space–time later on in this chapter.

Fifth, there is a danger that locales must contain certain ingredients in order for them to qualify as places (at least, in some readings of humanistic geography). Humanistic geographers were criticised for characterising particular sites as 'placeless' or 'non-places' – shopping malls, airports, motorway service stations – because they seemed not to contain the deep attachments and long-term inhabitation that, say, a 1,000-year-old village in rural France might (Relph 1976; Augé 1995; see Chapter 9 for a fuller discussion

Photograph 13.2 A satellite navigation device. Arguably, many car users and hikers now experience places through the 'lens' or 'frame' of such devices, which can encourage them to ignore all but the immediate context of their route.
Source: Shutterstock.com/Pincasso.

Box 13.3
Where is a mountain? 'Seeing' landscapes, GIS and 'fuzzy logic'

Both humanistic and Marxian approaches have been key to the development of cultural geography. They share (in different ways) a critique of quantitative geography, characterised by mathematical/cartographic notions of space. You may be surprised, then, that this box contains reference to GIS, which shares with quantitative geography a basis in mathematical/cartographic space.

Despite their many differences, however, some GIS theorists agree with cultural geographers that the nature and significance of places exceed the explanatory logic of a cartographic grid. Simply knowing the coordinates of a place is rarely - if ever - enough. Take the seemingly innocent question "Where is Mount Everest?" A high school pupil will be able to reply that it is in Asia, on the Nepalese/Chinese border. They may be able to pinpoint its latitude and longitude. But both cultural geographers and GIS theorists will agree that the matter is not quite so simple, and that we need to tell other stories about the spaces and places in which Mt Everest is located.

Cultural geographers will - if you read Chapter 5 (on **landscape**) - cite the many natural and cultural processes that mean that the world's highest mountain is 'located' in many different senses, in different spaces and places. Today, it is revered as an object of stupendous size and splendour - but only as a result of a change in the ways mountains were valued and represented during the Romantic era in European art. It holds a key imaginary role in the history of (Western) exploration and (masculine) sporting endeavour. It is a familiar sight, the subject of countless photographs, films and fables. Its location in geometric space is cross-cut by the criss-crossing of countless images, practices and texts at many spatial scales (see Section 13.3).

However, Mt Everest is still 'there', after all, and not 'elsewhere'. Or is it? Writing about Helvellyn (a mountain in the English Lake District), Fisher *et al.* (2004) show that the task of defining a mountain's location is more complex than might seem.

- Is a mountain 'the summit', the 'highest point'? In terms of a map (with a little triangle symbol), perhaps; but the word 'summit' is rarely a synonym for 'mountain'.
- "For most people therefore, the mountain is not a point, it is a region which to some degree they can visit" (Fisher *et al.* 2004: 106). But if they do not get near the summit, will others agree that they have visited the mountain?
- A mountain has a spatial extent - but to where does that extend: the beginning of the next mountain; its foothills; the nearest river or valley?

In one sense, Fisher *et al.* (2004) are discussing 'fuzzy logic': that is, the logic that all of the above definitions of the location of a mountain can be 'true' - but since there is more than one definition, each of which is different, there is fuzziness about any absolute (single) truth. And this is where GIS theories around fuzzy logic - remember, emerging directly out of abstract notions of space - begin to coincide with cultural geographers' notions of space, place and landscape. Fisher *et al.* (2004: 106–107) - who are GIS specialists, and not self-defined 'cultural geographers' - put it well:

> The mountain is not a real (bona fide) feature; it can be interpreted as simply a region in the continuum of variation of the surface of the earth, which people choose to identify and name. It can be interpreted as a flat object . . . which only exists in the human understanding and division of the landscape.

Ultimately, Fisher *et al.* (2004) use a novel form of quantitative multi-scale analysis to identify 'peakness' (*not* mountains or summits). They identify significant statistical relationships with named peaks (toponyms) whose inclusion on a database of Lake District landforms results from a long social and scientific history of place-naming, mapping and the intergenerational transmission of local knowledge. In both cases, the identification of place and the delineation of space are processes involving technical *and* cultural practices. Significantly, this is still a 'scientific' argument, but one that recognises the utter complexity of geographers' notions of space and place, and the centrality of perception and cultural values to space and place. Indeed, it also highlights that simply choosing one 'story' about space over another (e.g. the cultural over the geometric) is often a mistake.

This example demonstrates the value of a generous approach (in this instance to landscape) that makes room for approaches that reach beyond cultural-geographical understandings of space and place.

of 'non-places'; also the discussion of phenomenology in Section 13.3. So we need to critically question how and why some spaces become places – and what the real distinction is between them, which we do in Section 13.3 on *place as social construction* and Section 13.5 on *flow/mobility*.

In sum, then, we began this section by setting up a rather 'false divide' between definitions of space and place. Space and place have (certainly under the auspices of Anglo-American science and, it has to be said, human geography) become associated with key assumptions: space with the abstract and global scale, with speed and movement and with uncontrollable, unfathomable unfamiliarity; place with the local scale, with slowness, with tradition, dwelling and familiarity and, even, with stasis. But these associations are far from natural, and we have also begun to show that these associations and seeming divisions between the two terms can be intensely, politically, problematic.

From these arguments, it is perhaps most appropriate to argue that the distinction between space and place is at best blurred. Therefore in the rest of this chapter we explore – through three 'key motifs' – some recent, productive ways in which the tensions between definitions of space and place have been explored by cultural geographers. Remember that, as we argued in Chapter 1, it is very difficult to generalise about any 'one' cultural geography, and we therefore pick out some important 'moments' in this book (especially from Parts 1 and 2) where theories about space and place have been particularly important to research in cultural geography. This enables us to tease out diverse ideas about space and place elsewhere in the book, which we have sometimes left a little implicit in order to deal with them in this chapter.

13.3 Key motif 1: Place (and space) as socially constructed

As we suggested in the previous section, it could be argued that much of cultural geography is concerned with 'place'. For instance, the **Berkeley School's** approach to **landscape** concentrated on what made *regional* places – culture regions – unique. Work inspired by humanistic, Marxian and non-representational theories has, broadly, explored the **discourses, performances, everyday routines, materials** and **emotions** (Chapters 2, 7, 9, 10 and 11) through which *local* places are constructed and experienced. Those same cultural geographers have, often, also attended to the *national* and *global* processes that may frame, flow, be resisted or reconstituted through local places.

Two related points follow from these initial observations about place. Both points relate to the idea of *social construction*. That is, the idea that place (and, as, we will come to, space) is not natural or given, and is instead always the *result* of social processes (Harvey 1996). The first idea relates directly to the social construction of place; the second to what is known as the 'production of space'.

The social construction of place

Places may be socially constructed in various ways. Quite literally, a place may have been constructed through physical acts of design, building, planning and regulation. This seems obvious; places are marked out by boundaries (walls, fences), and by physical features (buildings, trees, pavements, street furniture) that have been put there to serve the needs of society. To read about detailed examples of how cultural geographers have explored the physical construction (or **production**) of places in this way, please turn to Chapter 2 and look at the examples of the design features and boundaries governing public spaces, Victorian slum clearance, and spectacular architecture in Dubai and Las Vegas. The crucial point about *place* here is that cultural geographers have shown that different norms, values and meanings can be invested into places and become part of their social construction.

The place of 'home' is a significant case in point. Home is not only the material or regulatory space that corresponds with a 'house'; rather, those built, legal frameworks intersect with a complex and culturally varied set of emotions and meanings that are associated with 'home' (Jacobs and Smith 2008). Thus, the social construction of the place we call home is a result of material, legal, moral, embodied and emotional processes that vary over time and space.

Finally, as we have shown elsewhere in this book, places can be critical to the construction of

identity – and vice versa. Particular places may become hugely significant in the representation of national or ethnic identities and/or political struggles – think of Trafalgar Square in London; or of Tiananmen Square in Beijing (Box 13.4; also Chapter 8 on **identities**). Places can be celebratory or reconciliatory, haunting or even painful, precisely because they are subject to social constructions that may privilege one identity group to the detriment or exclusion of another (Sibley 1995; Cresswell 1996).

Not all cultural geographers agree that places are *only* the result of *social* construction. Some, drawing on **actor-network theory** (see Chapter 10), argue that places are the result of ongoing interaction between human and non-human agents – the latter including animals, plants, minerals and the many tools and technologies that keep places running (Whatmore 2006; Hinchliffe 2008). Elsewhere in the book we suggested that this was a key way to understand **landscapes** as material phenomena – something that has been a central concern for cultural geographers (Chapter 5). Specifically, we argued that landscapes acquire meaning (as 'places') through the relationships between the human *and* non-human agents that constitute them. So the migration of birds, the invasion of a plant species, or the erosion of sedimentary rocks over countless millennia can be as important in constituting what a landscape means as how 'we' (humans who think and feel from a particular vantage point) experience them (see Box 5.2).

Elsewhere, Cresswell (2004) shows how geographers inspired by work on phenomenology – revising earlier work by humanistic geographers – seek to reinstate a more essentialist understanding of place. For Sack (1997), places are "the bedrock of human meanings and social relations" (Cresswell 2004: 32). While

Box 13.4
Tiananmen Square

Photograph 13.3 Tiananmen Square in Beijing.
Source: Getty Images/Chen Hanquan.

In April–June 1989, the Square was the site of popular protests against the ruling communist party, particularly among students who were demanding political liberalisation. The protests ended on 4 June, when an unknown number of protestors were killed by the military. Also the site of several other important events in Chinese history (such as the invasion of British and French troops in the 1860s), the Square is a central, (in)famous and contested part of Chinese cultural identities.

they agree that places can be socially constructed, they take a step back, in a sense – to argue that without place, we would not exist. Places come first; places are the primary facet of human existence; to be human is, first, to be in the world. As Cresswell rightly points out, this fact (for it is often viewed as such in a philosophical sense) does not explain many of the conflicts, **identities**, **texts** and practices in which cultural geographers are interested, and which form the vast majority of empirical examples cited in this book. But it does remind us that human experience is utterly inconceivable without place. Thus an essentialist understanding of place is both local (to the buildings, settlements or social spaces that mean most to people and *without which* they could not make meaning) and global (in the straightforward sense that the 'place' of humankind is planet earth).

The production of space: Henri Lefebvre

A different version of the social construction of space can be found in the work of French philosopher Henri Lefebvre (1991), which has been influential for cultural geographers interested in contestations over urban and especially public spaces. Lefebvre sought to understand the role of space and especially city spaces in the perpetuation of capitalist economies (see also Chapter 9 on **everydayness**). He also explored how, throughout history, the everyday lives of 'ordinary' citizens were played out in capitalist spaces (Gardiner 2004). Lefebvre formalised some of his ideas through an explicit theory of the 'production of space' that brought together some of the ideas detailed in Section 13.2. He sought to expose, decode and read space – a little like the way Marxian-inspired geographers sought to read **landscape** (Chapter 5). In the process, he suggested that the production of space took place in three interrelated ways – what has become known as a 'trialectics of space' (Soja 1996).

- *Representations of space*. These are spaces conceived, planned, drawn and written about by professionals, planners and politicians. Such spaces include jargon, codes and "objectified representations" that make spaces appear abstract (Merrifield 1993: 523; see also Section 13.2). In any society, representations of space dominate, because they perpetuate ruling class (i.e. capitalist) interests, and dictate how material spaces look. Key examples include historical studies of inter-war Britain, which have explored how planners like Clough Williams-Ellis tried to restrict suburban development in order to create a clear, neat divide between the city and the country (see Chapter 2 for more details).
- *Representational space*. This is the space of everyday life – the experience of particular places. As everyday inhabitants of a city move around its public spaces and buildings, they make meanings (see Lees 2001, and Chapter 4 on **architectural geographies**). Representational space is the 'lived' space that is dominated by and takes place within *representations of space*.
- *Spatial practices*. These are the practices that structure the realities of urban life – daily rhythms, commutes, and comings-together. Sometimes, spatial practices may be resistant to both *representations of space* and the *representational spaces* they control. Some authors (Soja 1996) refer to *spatial practices* as 'third spaces'; in the work of Matthews *et al.* (2000), third space is used to describe the ways in which children find collective spaces or almost hidden 'niches' in ordinary streetscapes that elude adult surveillance and enable them to perform their own identities in relative comfort. We discuss these kinds of 'third space' in the example of young people's **consumption** of shopping mall spaces – which can be understood as subversive – in Chapter 3.

Lefebvre's work has been widely used in critical and cultural geography (e.g. Gregory 1994; Shields 1999). It has formed one inspiration for cultural geographers who have theorised how resistant and subversive practices take place in and produce spaces that challenge dominant social and political assumptions (see Chapter 3 on **subcultures**). It also forms part of a critique of notions of space that privilege the 'abstract' and visual above the lived and **embodied** ways in which space is experienced (see Chapter 12). Lefebvre in fact argued that the abstract and the bodily were inextricable parts of the production of space.

While influential, Lefebvre's work has been subject to stinging critiques. Lefebvre's gender-blindness and privileging of heterosexuality have been critiqued by feminist scholars (Blum and Nast 1996), and make it of limited use to geographies of sexualities (see Chapter 8 on **identities**). As interest in Marxian dialectics has waned somewhat in geography, more diverse and complex ways of theorising difference have been taken up that are less involved with social class and the workings of capitalism. Moreover, the opaque style and frequent generalisations in his work make his work frustrating for some readers.

In this section, we have considered different ways in which place and space can be said to be socially constructed. There is a sense that space and place are never completely 'natural' or 'neutral' containers for social action, and that individuals and societies intervene in shaping spaces for their needs. Moreover, Lefebvre's work on the production of space is especially instrumental in showing how many spaces tend to represent the needs of dominant groups, and especially professional elites. Space is, then, something processual, lively and ongoing: spaces and places are always and already works-in-progress.

13.4 Key motif 2: Scale

As identified above, 'place' is usually associated with particular spatial scales – most commonly, the local (i.e. village or neighbourhood). The local is understood as being at the 'bottom' of a 'hierarchy' of scales, which span from local to global (see Box 13.5). At first glance, this hierarchy seems natural. It seems sensible that the 'local' is somehow 'smaller' than or even opposed to the 'global'. Indeed, that is how many experiences of local protest are figured by the media and even those participating – as local direct action *against* the global. We routinely understand that certain decisions (say, United Nations Conventions on human rights) are made at a global scale, ratified by national governments, and implemented by practitioners working at regional or local scales. A common basis for the organisation of a national election or census is that chunks of a country's land surface can be broken down into ever smaller segments, allowing ever greater degrees of analytical sophistication (and, ideally, political representation, as Box 13.5 shows).

Scale matters, then. But just as cultural geographers disagree on the extent to which place is socially constructed, they also disagree on the extent to which *scale* is socially constructed. In turn, this has profound implications for how cultural geographers understand the very processes they write about and the political position they take in writing about them (most notably in writings about globalisation). It also has implications for understanding the extent to which the 'local' really is 'smaller' than, or in opposition to, the 'global' – and, indeed, the extent to which the local is actually local at all.

The following represent four important ways in which geographers have revised the apparently logical scales depicted in Box 13.5. To confuse things further, not all of the following authors explicitly write about 'the social construction of scale', although we are assuming that to an extent they all agree that scale is not a completely natural register of spatial extent.

First, most geographers – especially cultural geographers – now question the idea that scale is a pre-given platform for society to operate upon. Instead, scale is constructed. In his work on the world economy, for instance, Taylor (1982) argues for the *emergence* of three scales – the urban, national and global – as an *effect* of the expansion of global capitalism, first through colonialism and then through global business. Marston *et al.* (2005) show how this has led to a 'politics of scale', whereby boundaries and flows (for instance of commodities) are regulated – as per Box 13.5.

A second way to understand the social construction of scale is to see scales as the *outcome* of cultural, economic and political processes that are then made seemingly 'stable' by telling particular stories about them. Gibson-Graham's (2006) influential work on 'diverse economies' is particularly significant in developing this previous point. In Gibson-Graham's view, a particularly pressing task for geographers is to disassemble the idea that 'globalisation' is a process that is simply 'happening' to the world, at a global scale. They argue *against* trying to 'understand' globalisation and its cultural, economic or political effects, because this just reinforces the idea that globalisation 'exists', independent of human intervention. Instead, they adopt an "anti-essentialist approach [to] theorizing the contingency of social outcomes rather than the unfolding of structural logics" (Gibson-Graham 2008: 615).

Box 13.5
A 'nested' hierarchy of scale

Common understandings of scale (in both academic geography and many societies) draw on a seemingly intuitive sense of 'size'. That is, the idea that the 'global' is bigger, and ultimately more important, than the national, and so on. A little less commonly, the following schema is underpinned by notions that: the local and global represent two ends of a self-contained spectrum; that the local and global are somehow opposed; that the local is more 'authentic' than the global; that ideas, regulation, materials and cultural objects somehow 'flow' from the global to the local. The examples to illustrate the following schema are taken from Austria.

Scale (biggest first)	*Example*
Global	United Nations (UN). Responsible for, among other things, a series of Conventions on human rights.
Continental	European Union (EU). Responsible for, among other things, legislation on trade within and between the EU.
National	Austrian national government. Responsible for making and enforcing federal law on health, education, etc. Also responsible for upholding and ratifying UN and EU laws.
Regional	Provinces (roughly equivalent to counties in Britain, states in the USA, Germany, Australia and Canada). An example is Steiermark (Styria). A province has power over planning and nature protection, but is also responsible for some aspects of taxation and implementing Federal laws.
Sub-regional	Bezirk (equivalent to counties in the United States). In Austria, there are 84 Bezirke. In Austria, this scale is most notable for its designation of place-based vehicle licence plates – for instance 'JO' for Sankt Johann. As in other countries, this designation may have day-to-day significance for a person's place-based identity.
Local	An individual town or village. Some aspects of local planning and cultural events (such as local festivals) are organised at this scale.

Ultimately, this means that globalisation (and the global scale of this process) is viewed by many cultural geographers as an *outcome* of human endeavour, shaped by regulation, creativity and a measure of risk-taking, as the 2008 'global financial crisis' showed. This is why, in practice, seemingly 'global' industries are actually located in quite specific, local places. As we showed in Box 2.6, various **creative industries** in the UK that purport to being 'global' are located in particular parts of the country – architectural firms in the North-East, designer fashion in the East Midlands, and so on. It is just that they tell particularly powerful stories about the fact that they are 'global industries' as part of their branding. To cut against the grain of these pervasive stories about globalisation, geographers like Gibson-Graham have documented a series of alternative cultural-economic practices. These include unpaid labour and care in households, consumer cooperatives, and movements caring for the environment, all of which may cut across and create their own spatial scales, and are not necessarily 'local' spaces in opposition to 'global' capitalism.

One notable example of such 'movements' is the 'No Tesco' protest in Stokes Croft (see Box 3.11), which, although emphasising local food networks, sought to destabilise the power of a pre-eminent supermarket chain that wanted to locate a new superstore in Bristol, England. A second, and very different example that we include elsewhere in the book is Khoj's 'Negotiating byways' performance art/dance installation (see Box 7.4). The installation works so well as a piece of political critique because, in drawing attention to road-building projects across India, it brings together people *across* local and national scales – and, with an Internet presence, has attracted increasingly global attention (see Chapter 7 on

performed geographies for more details; also Brown 2009, on diverse gay and lesbian economies).

A third way to think about scale is to move away from notions of 'local' or 'global' to think about what we generally conceive to be 'big' or 'small'. Again, although it may sound strange, those categories are not pre-given, but the result of (often) geographically and culturally specific processes. We discussed one example of this, in relation to GIS representations of mountains, in Box 13.3. Recent cultural geographies of architecture have also grappled with these questions of size (Jacobs 2006; Rose *et al.* 2010). For instance, Jane Jacobs' work is concerned in part with the telling of 'global' stories, this time about high-rise buildings (for details, see Chapter 4 on **architectural geographies**). She argues against an approach that simply follows how global architectural styles are 'diffused' around the world – in other words, the top-down, one-way flow shown in Box 13.5. Rather, she argues that cultural geographers should explore 'the processes by which certain things cohere', with the result that something identifiable as a 'building' or 'high-rise' emerges. She deliberately *avoids* the term 'building', however, because – as with Gibson-Graham's critique of globalisation – this refers to an apparently stable set of architectural stories, practices and forms of regulation that seem above scrutiny. To avoid these stories, she calls the residential high-rise a 'big thing', which can be looked at with fresh eyes through a form of 'surface accounting'. This means that:

> [t]hrough this technique it is possible to see that a seemingly minor thing like a 'toy rabbit' lies on the same level as 'advanced capitalism' . . . An interest in surface does not mean a disinterest in the wider systems in which a 'thing' is entangled . . . Rather, it is to bring into view how the coherent given-ness of this seemingly self-evident 'thing' [the high-rise] is variously made or unmade. Further to that, I seek to demonstrate the part played by the *constitution of scale* in relation to stabilizing this big thing called the high-rise.
>
> (Jacobs 2006: 3, our emphasis)

In other words, Jacobs does not assume that the high-rise is a 'global' architectural form; nor is it simply a 'big' thing by virtue of its physical size. Rather, the high-rise has *become* a big thing because it has been involved in a series of big, important stories (e.g. the careers of internationally famous architects like Le Corbusier), which in turn have helped to *create,* not simply reinforce, what has come to be called 'globalisation'. We could say that a high-rise is deliberately reduced to the level of a commodity – an object like any other (hence the 'toy rabbit') – and can thereby be understood in exactly the same way as the commodity chains and commodity circuits that are discussed in Chapter 2. It only *gains* its powerful meaning because of the ways in which it is positioned in those chains and circuits – gaining power from governmental discourses, from its comparatively great financial worth, or from its emotional meaning to the hundreds of people who live in it or visit it (see also Kraftl 2009). This, then, is a wonderful example of the social (and material) construction of scale.

Fourth, in a significant and provocative paper, Marston *et al.* (2005) query to what extent one could imagine human geography (and cultural geography, for that matter) *without scale.* On the face of it, this might seem like a difficult, if not impossible, position to take: surely scale is fundamental to human geography? They cite a number of critiques of scale, but perhaps most pertinent here is the problem that there is a danger that scale becomes simply a 'conceptual given' (Marston *et al.* 2005: 422). Once articulated, empirical work simply relates a series of social or cultural practices that 'fit' those scales:

> Research projects often assume a [scalar] hierarchy *in advance* . . . whereby objects, events and processes come pre-sorted, ready to be inserted into the scalar apparatus at hand.
>
> (Marston *et al.* 2005: 422)

Thus Marston *et al.* (2005) question what geography might look like *without* scale. They propose instead a 'site ontology', which means inventing new spatial terms, as follows.

- Focusing not on scales but on *sites,* which emerge through socio-spatial relations and not the other way around.
- Sites emerge through *events.* So socio-spatial relations are the outcome of events, and of relationships between events. Events involve both space *and* time. So, for instance, rather than a political demonstration happening 'at' a local place, against a 'global' issue, that demonstration *produces* a site

all of its own. It is an event that creates all kinds of effects – socio-spatial outcomes. We start with the event (the protest), *not* the scale.

- Socio-spatial outcomes may be more or less *ordered,* of a greater or lesser *duration,* and may be more or less constrained by *boundaries.* The argument is not (as some critics have mistakenly understood poststructuralist theory) that the world is a space of borderless flows. Instead, events create, forge, fold and break relations (Doel 1999: 16).
- The *effects* of 'event relations' (as Marston *et al.* 2005: 424 term them), *may,* then, be felt variably: as borders (delineating temporary no-go zones, out of the reach of protestors); as channels (routes taken through city streets); or as images that travel vast distances (on CNN or Fox News, relayed at points around the world).
- Thus, sites – of whatever size, constitution or duration – are the effects of event relations that endure for different lengths of time. They are always spatial *and* temporal. They are emergent. And they may bleed into or relate with *other* emergent sites in all sorts of ways – some quite predictable, some planned, some completely surprising. Marston *et al.* (2005: 425) should have the last word here, as they provide an explanation and example of what they mean in terms of events in the built environment:

> [This] strategy avoids misrepresenting the world as utterly chaotic and retains the capacity to explain those orders that produce effects upon localized practices. Thus, for example, a site ontology provides the explanatory power to account for the ways that the layout of the built environment – a relatively slow-moving collection of objects – can come to function as an ordering force in relation to the practices of humans arranged in conjunction with it. Particular movements and practices in social sites are both enabled and delimited by orderings.

Marston *et al.*'s conception of scale has supporters and detractors, and we explore a couple of critiques later in the chapter (Section 13.6). For now, in summarising this section, we want to emphasise that the four examples really do represent the tip of the iceberg. We have suggested that most cultural geographers agree that scale is to some extent a social construction, and that this notion is at least an implicit guide for their work. We have also broached the idea that human geography might 'go beyond' scale, however contentious this suggestion has proven. But this is not the end of the story about scale; thinking about scale implies other kinds of spatial metaphors and motifs. With this in mind, we turn to spaces of 'flows'.

13.5 Key motif 3: Flow/mobility

We are used to hearing about 'flows': of commodities; of migrating peoples, by choice or force; of electronic transactions; of emails, letters, faxes and telephone calls; of aeroplanes and boats (see Chapters 2 and 10 on **cultural production** and **material things**). We are also used to hearing about how *global* flows affect, interact with and are re-interpreted at 'local' places (see Box 7.1 on hybrid musical **performances** in France and Bangalore, for example). Thus the notion of 'flow' has considerable importance for cultural geographers; we concentrate on three quite different kinds of flow in this section. The first relates directly to questions of scale. The second relates to material things. The third relates to the idea of 'mobility'.

Flow and questions of scale

We suggested earlier in the chapter that traditional definitions articulate place as an identifiable, concrete and locally scaled realm (see Section 13.2). Our key point in this section is that local places are not simply local. Furthermore, to name this or that practice (say, protest against climate change) 'local' is in many senses a way to dismiss it as marginal or irrelevant. Doreen Massey (1991) forcefully made this point against David Harvey's (1989) charge that while feminist struggle was 'local', the more general (and by implication more pressing) struggle was a global one over class inequality.

Massey has perhaps done more than any other geographer to expand – even explode – earlier notions of place articulated by cultural geographers, principally through the idea that places are created through flows. As Callard (2004: 224) puts it:

> Massey's emphasis on mobility, openness, flow and differential power relations has contributed to those terms becoming guiding principles in human geography.

Since the early 1990s, Massey has developed the idea of a 'global sense of place' (Massey 1991: 29). Her work foregrounds the differently scaled flows that *constitute* a 'local place'. Just – as we argued above – as globalisation is created and not given, so too local places are not given. Cities, towns, villages, even homes are rarely, if ever, simply the result of local processes. All sorts of people, things and ideas from all sorts of places make them what they are. Some of those things *may* be locally sourced; some may come from the next county or state; others may come from the other side of the world.

Think of your own home, for instance. If you live in a bungalow (a one-storey dwelling), you may or may not know that the term 'bungalow' was initially used to denote a form of housing used first by British sailors in seventeenth-century India and, later, to describe the spacious lodgings of the British colonial elite in the same country. The bungalow has since appeared around the world in many different guises (King 1984). Or take a moment to look around at the objects that surround you. Where did that apple, that banana, that papaya come from (Cook 2004; see also Chapters 2 and 10. All of these spatial processes may be thought of as *flows* that operate at different scales *and* which constitute those scales. This sounds like a kind of self-fulfilling logic (and in the case of globalisation it often is). But what it means for 'local' places is that they are, in effect, always *translocal* (Brickell and Datta, 2011). Rather than an introspective sense of place, then, this is the idea of 'place beyond place' (Massey 2005): a town is always formed by, and in relation to, those things that appear to be 'beyond' its apparent boundaries. To read more about the implications of Massey's work, and some of the many empirical studies it has inspired, please turn to Box 13.6.

Flow, fluidity and material things

Here, the work of John Law – a key exponent of actor-network theory – is significant. Law and Mol (2001) contrast different ways of thinking about objects in space. They initially take the example of a ship crossing an ocean. They argue that, in Euclidean space (i.e. the geometric space of the map), when a ship is in port, it does not move; as soon as it leaves port, it "displaces itself" (Law and Mol 2001: 611). It moves. This is the commonsense understanding of 'movement', with which most readers will identify.

But we can also think of space as relational, as actor-network theorists insist. In this understanding of space, we are interested simply in whether the relations that constitute an object (or a system) are *themselves* changing. This is a different understanding of movement (or of fluidity). It is not about whether an object has moved in three-dimensional space, but about whether something about that object has changed such that it changes shape – for instance, it no longer can be called a 'ship'. As Law and Mol put it:

> Is there no change in the working relations between the hull, the spars, the sails, the sailors, and all the rest? If this is the case then the ship is immutable . . . It does not move in relation to network space.
>
> (Law and Mol 2001: 612)

Although it seems somewhat counter-intuitive, this argument makes perfect sense. The ship is *still* a ship; it has not sunk, and nor has it become, say, a train. Thus, in relational terms, we can still call it a ship. The very fact that it moves, in fact, reinforces this fact: a ship is hardly useful if it simply floats, tethered to a quay, or is rested on dry land. This is, then, a relational notion of space. Note the similarity to the previous points about flow and scale: these are also relational conceptions of space, place, the local and global that follow how relations constitute places (not vice versa).

Arguably, Law and Mol (2001) are discussing *flows* in their example of the ship. However, the idea of 'flow' or 'fluid' space can be pushed further. Moser and Law (2006) critically analyse plans to implement electronic record-keeping ('health informatics') in the UK and Norway. The argument for health informatics is that better, more efficient 'flows' of information are, in the long run, better for the safety and well-being of patients. Moser and Law (2006) do not dispute this directly. Rather, they argue that flow also involves change. When any information is transferred – say, from one medical practice to another – there is no guarantee that that information will be presented in the same format, or be interpreted in the same way as originally intended. And, as Moser and Law (2006: 68)

Box 13.6

Doreen Massey's work on place, and examples of cultural geographies inspired by her work

Massey provides an extended discussion of space and place in her book *For Space* (Massey, 2005). In it she makes many propositions about space and place. Most notably, she argues that space has been marginalised by philosophers in preference of time. This is because, as we suggest earlier in the chapter, time appears to be more open-ended, and to allow for creativity, whereas space has traditionally been assumed to be static. Furthermore, she suggests that place is often seen as the site of problematic assumptions and practices – from forms of 'authentic' identity that exclude on the basis of race or heritage, to forms of nationalism that have led to physical violence against those who do not 'belong' (see also Chapter 8). *Instead*, she argues for a conjoined view of space and time, where places are the result of ongoing processes of creation and meeting, and are never fixed. The following represent four of the implications of her work, along with examples of cultural geographers who have taken up her ideas.

- Space is "a product of interrelations" (Massey, 2005: 10; see also Box 2.4). As we have highlighted in this chapter, many people, ideas, things and practices make space something that is **produced** (also Lefebvre, 1991). For Massey (2005: 10), an implication of this is that seemingly 'already-constituted identities' are far from that. Instead, **identities** are constructed (see Chapter 8), and there is no such thing as an authentic, local, place-based identity.
- Massey's writing has implications for ideas about responsibility. In her own work, Massey claims that the intimate interrelations between places, coupled with the possibilities of mass communication (in some contexts), mean that the urge to care for those closest-to is now problematic. In a provocative closing chapter, Massey (2005) questions why charity should 'begin at home', as the old adage goes. In an abstract sense, we should care about what happens on the other side of the world – not simply because this or that natural disaster has wreaked havoc for thousands of 'distant' strangers, but because those strangers are actually closer to us than we think. In simple terms, they may have made the shoes we are wearing; their uncle may live next door; they are part of the very same 'global' processes as we are.
- Writing about migrant labourers in London, May *et al.* (2007) show that a disproportionate number of workers in low-wage jobs are foreign born. After outlining the conditions under which they work, May *et al.* conclude that, as the previous point suggests, *all* Londoners (policymakers and public alike) have ethical responsibilities for ensuring the well-being and fair treatment of these seeming 'strangers'. While residing in London, these workers are, as Massey suggests, representative of the "distant places and people who make London what it is [and] the influence London exerts on those [distant] places" (May *et al.* 2007: 163). This leaves a difficult question in practice, however. On the one hand, some Londoners (many of whom themselves were migrants, or are descended from migrants) have taken up this challenge, helping with groceries or working as labour activists on behalf of newer migrants. On the other hand, May *et al.* (2007: 164) ask pertinently, "how to address the needs of those previously distant others now 'here', without undermining the equally pressing needs of other Londoners". This is a question for which Londoners – and Massey herself – have no easy answer.
- Massey's work has been particularly influential for studies wishing to explore the political implications of 'going beyond' local places. Nicholls (2009) explores a range of social movements, originating in places as diverse as France and South Korea, that 'string' together activists from different locales. He argues that the process of creating transnational solidarities along class, ethnic or religious lines – what he calls 'social movement spaces' – is not a natural one, and requires considerable work. He explores the mechanisms that connect local activists to distant allies – from continental meetings like the European Social Forum to communication technologies such as online social networking sites. He also argues that some activists are more able than others to overcome effects of 'distance' – effects that include the ability to travel, to communicate in another language, and to adapt to new environments. Hence the stringing together of social movement spaces is highly variable, and the result of complex social, cultural, financial and technological processes that transcend the seemingly inherent localism of activist work.

put it: "electronic information will always intersect with information in other material forms that are complex and heterogeneous." As soon as information enters a new space it enters into new sets of relations, combines with other forms of information and the expertise of different medical professionals, and is put into practice in different ways. The information is, in one sense, 'the same'; but it is also always different as soon as it moves – as soon as it flows. Indeed, that information may 'fail' in certain contexts; or it may become simply a component part of another piece of information.

In the context of their example, they caution against rigid informatics systems that try to anticipate every way in which information might be used. But in the broader conceptual sense we are interested in here, the implication is that fluid objects – and fluid spaces – are those that change shape in different contexts. They are adaptable. They may be unrecognisable, but they are still, in some sense, 'this' piece of information, and not another. With reference to medical practices (such as the measurement of blood pressure), Moreira (2004: 62) summarises the argument in this way:

> Instead of a common understanding of the position and role of components (as in regional [or geometric] space) or a single definition of those components (as in network space), in the space of fluids entities can be performed in more than one way.

Hence – and thinking back to previous examples in this chapter and the book – pieces of information are *not* like 'the ship' that retains its essential 'ship-ness' in network space. But they *are* like the many versions of 'ship-ness' that have followed the invention of the first ship, back in the mists of time, and which have become yachts, dinghies, trawlers, liners and so on.

On the face of it, few cultural geographers explicitly write about 'fluid' or 'flow' spaces – but these *ideas,* expressed most clearly by John Law, are key to understanding ideas such as hybridity. Or, to cite earlier examples, they help us to understand the 'high-rise', which has travelled around the globe, but has been changed in so many ways that we cannot simply suggest that all high-rises are alike (Jacobs 2006). And again, these ideas recall the multiple ways in which commodities travel around the globe (Chapter 2) and are subject to forms of **production** and **consumption** that alter their meaning (and perhaps their shape) but do not change the fact that a papaya is a papaya (Cook 2004; see also Chapter 10).

Flow and mobility

Studies of mobility are concerned with forms of movement, of migration, of travel, of trajectory, of communication and/or of connection. They are indicative of stories about contemporary globalisation that figure people as differently mobile – from the jet-setting businessman flying between global cities for meetings, to the family seeking asylum and taking up low-paid work in London (Box 13.6), to the perambulation of well-to-do gentlemen as they **consumed** the arcades of nineteenth-century cities (Chapter 3). They are indicative, if you like, of an increasingly accepted dictum that the world is mobile in a sense never experienced before; that the forces of time–space compression (Harvey 1989), driven by cheap air travel and electronic communication, have effectively shrunk the globe; that, at least in more privileged contexts, being mobile is not simply a cultural good but a popular expectation. In Europe, the advent of the budget airline means that in some countries a majority of the populace *expects* to take one non-domestic holiday per year, even if they cannot afford it.

There is not space here to do justice to cultural geographers' writings on mobility; for excellent introductions, see Cresswell (2006) and Adey (2009). We also provide other examples of mobility elsewhere in the book (see Chapters 2 and 10, on commodity chains, for instance, or Chapter 12.4, on different kinds of **bodily** movement, disciplining and resistance). And, in a sense, all of the examples in this section of the chapter nod to the idea of mobility. However, the concept of mobility has certain, further, implications for how we think about space and place.

First, humanistic geographers have argued that more mobility – and increasingly fast forms of communication – have diminished the meaning of places. For instance, Relph (1976) argues that the increased mobility of homeowners in the USA, changing home frequently, means that the very notion of 'dwelling' in a place loses meaning. Elsewhere, Urry (1990) argues (albeit in a general sense) that mass tourism leads to a 'popular tourist gaze' (see Chapter 3) whereby places are reduced to a short visit, a couple of photographs and a souvenir, before moving to the next stop on a tour. Cresswell (2004) highlights a range of similar work that suggests how contemporary transport infrastructures both give rise to large, institutional, ostensibly faceless (and placeless) buildings, and increase the very speed at which places are experienced, further reducing their meaning. We deal with these arguments in more depth in our discussion of **everyday geographies** (Chapter 9).

Second, Thrift (1994) suggests that mobility is a 'structure of feeling' that accompanied the rise of modernity. With faster travel and communication, the world is 'in-between'; places become strategic locations that simply capture flows of people, goods and transport (Thrift, 1994). In this world of speed, the *image* becomes king; so, in order to carve out a place for themselves, places (such as cities) must constantly (re-)invent and market themselves as *the* latest tourist destination, *the* attractive place to live, or *the* place to invest in real estate (Hall and Hubbard 1998). Berlin, in Germany, is an example of such reinvention (Box 13.7).

Box 13.7
Berlin: The Sony Center

Photograph 13.4 The Sony Center at the Potsdamer Platz in Berlin, Germany.

Source: Shuttestock.com/Dainis Derics.

According to a German tourist website,[1] Berlin is a city reinventing itself. It is branded as 'The creative capital of Europe: from classical to cool', and is apparently at the forefront of avant-garde **cultural production** (Chapter 2) – from new architectural forms, such as the one in this image, to over 20,000 visual art installations in public spaces around the city. Thus the very material spaces of the city itself are 'reinvented' as images to be circulated around the globe, which, it is anticipated, will attract millions of tourist every year.

[1] http://www.germany.travel/en/towns-cities-culture/towns-cities/berlin.html

Third, Adey *et al.* (2007) draw attention to the multiple, social and especially *cultural* spaces implicated in 'aero-mobility' – in travel by plane. They seek to move beyond simply describing different kinds of mobility to "detail the social and technical practices that produce them" (Adey *et al.* 2007: 776). For instance, they show that flying (and, indeed, airline companies) may promote new, glamorised senses of 'global citizenship', where seeing the earth from miles up, and travelling over an undifferentiated 'earth surface', removes a sense of the different political territories and borders below. Very differently, they move to the scale of the **body** (see Chapter 12), to suggest that, far from being meaningless, placeless sites, airports are meaningful places for the frequent flyers, baggage handlers, refugees and air traffic controllers who are some of the many people who routinely inhabit the key nodes of a mobile world (Adey *et al.* 2007: 779). They call for greater attention to the **emotional** connections that people forge *through* mobilities at and through places like airports.

Fourth, a range of work in cultural geography has also explored the forms of *im-mobility* – of stasis, control and boundary-making – that are an increasingly important element of mobile societies. Much of this aligns with cultural geographers' interest in how human **bodies** (Chapter 12) and **emotions/affects** (Chapter 11) are regulated by governments and other agencies, especially with respect to possible risks (like terrorist threats). For example, Amoore (2006) focuses on what she terms 'biometric borders' at US border controls, especially airports. There, forms of risk-profiling separate 'legitimate' mobilities (travel for work or leisure) from those that pose a potential threat to US security (illegal immigration or terrorism). A range of techniques are used to determine the threat posed by a traveller – from biometric data to credit card records. The intention here is to (attempt to) predict the future: to determine whether, in the interest of US national security, a person's journey should continue, whether they should be turned back, or whether, in extreme cases, they should be detained for interrogation. Thus mobilities do not simply produce spaces of flows, but involve stutters, rests and, on occasion, longer-term periods of detention, each of which casts places as *relatively* im-mobile.

In this section, we have suggested that the notion of 'flow' characterises some very different kinds of work in cultural geography. Flows and interrelationships between places are the very stuff that creates those places. Indeed, local places are always *translocal* – they are constituted by flows of people, materials and ideas that come from near and far, in ever-changing constellations. Fluid spaces may also have implications for how information and objects are transported around, and received at, different places. Fluid notions of space are implicit in cultural geographers' work on identity, material cultures and built spaces. Finally, we have suggested that much work in cultural geography is turned by an interest in mobility. In all cases, as Doreen Massey perhaps best reminds us, the production of flows and mobilities involves political decisions and kinds of responsibility that almost always work to the advantage of some social groups and to the disadvantage (and, sometimes, im-mobility) of others.

13.6 Discussion: where next for space and place?

Reflect again on the popular uses of the terms 'space' and 'place' listed in the box that opens this chapter. Do those uses of the term evoke anything different now that you have read the chapter? For instance, what does it mean to have 'space to think'? Which spatial motifs – which spatialities – does this term evoke? Perhaps having a place to call one's own, in the phenomenological tradition, or a similar sense of belonging that affords the comfort and conditions to spend time ruminating this or that issue. Perhaps a 'spatial representation' – an everyday experience of a quiet café or park bench, designed and built by another architect or city planner. Or perhaps the phrase relates to something substantive – space to think about your responsibilities to distant others.

The question is – do any of the notions of space and place above do justice to, or help you to explain, the many ways in which we talk about, feel, move through, imagine, plan and represent space and place? Hopefully, they do, to some extent. But maybe not completely. So: where next for space and place in cultural geography? There are, it could be suggested, three ways of breaking down this question.

First, by not being too quick to dismiss previous, maybe 'outdated', or certainly less fashionable ways of conceiving space. This goes for concepts as diverse as geometric spaces (as we show in Box 13.3) and local attachments to 'place'. They each have their problems, but they may remain meaningful to some groups or communities, and continue to deserve the critical but sensitive attention of cultural geographers. Rob Imrie's (2003) work on architects' conceptions of the human body is an excellent case in point. He shows that most practising architects use rather limited notions of both space and the human body in their work. The body remains an absent or assumed component of their building designs; where the body *is* imagined, it is figured as a 'normal' body of geometrical proportions, and able to perform assumed tasks that many people cannot (such as walking, climbing stairs, reaching windows). Rather than dismiss these conceptions as simply an extension of geometric notions of space that ignore social difference, however, Imrie (2003) argues that geographers need to *engage* with architects in productive conversations that will help them to challenge and overcome the difficulties associated with their assumptions.

Reece Jones (2009) argues similarly for the concept of 'borders'. With talk of flow, mobility and spatial relations, the spatial notion of the border has of late been of less central concern to geographers. Rather, he argues, we should re-focus on borders, categories and boundaries – which are both still hard to avoid in daily life, and from which academics have found it difficult to move on in writing about societies, spaces and the divisions within them. This is especially important given what he calls "the xenophobic and exclusionary categorisation of the present era" (Jones 2009: 186) in, for instance, anti-terrorism policies in and between many countries. For him, it is impossible to get beyond boundaries fully, and rather more important to use some of the conceptions of flow and relationality to explore how boundaries are constructed, controlled, maintained and resisted (Jones 2009). So, first, we would suggest to you not to assume that the order of appearance of different spatial motifs in this chapter assumes any 'lineage' or sense of 'real-world importance'. The question 'where next?' can be turned back on the history of thinking about space in cultural geography, as well as forwards.

There is a second way of breaking down the question 'where next for space and place in cultural geography?' That is, to foster further, creative ways of conceiving space and place that map onto emergent spatial processes and, for want of a better phrase, 'ways of doing' space and place. John Law – whose work we cited above – has been a particularly creative thinker in this regard. For instance, in his work on technoscience (Law and Mol 2001), he moves beyond spaces of fluidity to consider 'fire' spaces – those spaces that are renewed by destruction, or by the momentary breakdown of relations, or momentary ripping-apart of a network. In the 'natural' world, trees, shrubs and whole ecosystems may be renewed by the seemingly destructive process of fire, in fact enabling their *continuity* – clearing a space for their continued existence. Absence – of relations, of continuity – matters just as much in the constitution of spaces as presence (Harrison 2007; Kraftl 2010).

Elsewhere, Thrift's (2006) documents many kinds of spatiality. As he says: "[s]pace comes in many guises: points, planes, parabolas, blots, blurs and blackouts" (Thrift 2006: 141). In essence, he argues – as we have just suggested – that there are as many ways of conceiving space as there are peoples, worlds and practices. He explains:

> Truth to tell, all these things [ways of thinking space] exist – and none – as part of the tuning of local variant systems . . . In a world without levels, the words [for space] are necessarily approximations of the right size of the world that chime with the finding that the fundamental fact of human communication is its variability [and] that are a part of how we learn to environ the world and how the world learns to environ us.
>
> (Thrift 2006: 141)

Hence, while we must not forget previous and perhaps unpopular ways of thinking space and place, Thrift encourages us to keep thinking creatively about space, and not allow our notions of space to become assumed in the service of dominant political or cultural orders (also Lefebvre 1991 – see Section 13.3).

Third, though, Thrift's longer quotation contains a curious, parenthetical observation: "all these things exist – *and none*" (Thrift 2006: 141, emphasis added). Thrift appears to intimate that, paradoxically, spaces exist and do not exist. Perhaps this is the substance of Law and Mol's (2001) treatment of 'fire' spaces, indicated above. Or perhaps it suggests that spaces are always and only the product of the human imagination, and of sensory perception, and that therefore, in the phenomenological sense we have summarised in this chapter, do not precede our being in the world. But it also suggests that we can take Marston *et al.*'s (2005) call for 'geography without scale' a little further: to consider a geography *without* space and place. What would a cultural geography without any 'key motifs' of space or place look like? Would it still be a geography? Can 'culture' exist without notions of space and place?

As cultural geographers, we should not be too quick to dispense entirely with notions of space and place. Rather, as we suggested throughout the introductory chapter (Chapter 1), we would urge you – and all cultural geographers – to think about why and how conceptions of space and place are being theorised, in more and more diverse ways, and to think about how these theories of space and place work in the contexts of the examples you are studying or researching. What else, what more, and what difference do notions of space and place make? And are notions of space and place another way to *explain* cultural-geographical phenomena (like landscapes), are they actually emergent *from* those phenomena, and/or what can be *gained* by academics, policy-makers and publics alike from continuing to talk of 'space' and 'place'?

Summary

- 'Traditional' conceptions of space figure 'it' as a noun, an abstract container, and something that can be mapped using geometric coordinates. 'Traditional' conceptions of place see 'it' equally as a noun - somewhere local, named and of deep meaning to the people who live there. Space is often opposed to place, and both are often opposed to time. However, cultural geographers have criticised these definitions and divisions, and have articulated many alternative ideas, which we have described as 'key motifs'.
- Space and place are frequently thought of as 'social constructions'. Rather than being neutral, natural containers for human action, spaces are created, designed, controlled and maintained, often to serve the interests of the dominant group(s) in society.
- Scale may also be seen as a social construction. Many cultural geographers have sought to criticise the notion that the local is 'opposed' to the global scale, and that places are always 'local'. Rather, they see places as the result of ongoing practices - social constructions - at different spatial scales. Places happen as the result of connections of different distances and durations. More recently, some geographers have begun to question whether we can do without scale altogether in geography, although this debate continues.
- Ideas about scale also link to questions of 'flow' and 'mobility'. Many cultural geographers see places, texts, images, cultural objects and identities as the result of their incessant movement at and across different scales. The idea of fluidity helps us to understand not only what cultural artefacts 'mean' as they flow between different cultural contexts, but also the kinds of responsibility that people in particular places have for people in other places connected by those flows.
- The last section of this chapter sought to open out some questions about where cultural geographers might go next in conceiving space and place. In our view, cultural geographers should keep an eye both on 'traditional' and more 'creative' ways of thinking about space and place, while reflecting on why and how ideas about space and place are useful.

Some key readings

Clifford, N., Holloway, S., Rice, S. and Valentine, G. (2009) *Key Concepts in Geography,* Sage, London.

One of several accessible introductions to philosophical approaches to geography. This text is helpful, because approaches to cultural geography appear alongside approaches to other parts of geography, including physical geography and doing research in geography. Note the massive breadth of approaches used by geographers, but note that, often, each approach comes with its own unique brand of 'space' and/or 'place'! Other excellent texts include Hubbard *et al.* (2002, 2008), Johnston and Sidaway (2004) and Nayak and Jeffrey (2011).

Cresswell, T. (2004) *Place: A short introduction,* Blackwell, Oxford.

Perhaps the most authoritative book on place ever written. Readable and accessible, it both explores the historical/philosophical lineage of the term 'place', and provides a sensitive and detailed treatment of some of the key thinkers on place discussed in this chapter.

Massey, D. (2005) *For Space,* Sage, London.

Although more challenging than Cresswell's (2004) introduction, this book represents the result of two decades of thinking about space by a prominent geographer. Massey's ideas about space – stretching back to the early 1980s – have inspired countless studies in cultural geography, feminist geography and beyond. This book contains readable and wide-ranging examples, together with personal insight from Massey's life and work in London and South America, and ends with an important political challenge to readers to take seriously the implications of living in a world of complex relations and flows.

Law, J. and Mol, A. (2001) Situating technoscience: an inquiry into spatialities. *Environment and Planning D: Society and Space* **19,** 609–621.

Marston, S., Jones, J. III and Woodward, K. (2005) Human geography without scale. *Transactions of the Institute of British Geographers* **30,** 416–432.

Thrift, N. (2006) Space. *Theory, Culture and Society* **23,** 139–146.

Three journal articles that say very different things about space and place, and which have been discussed in some depth in this chapter. They are intended not as introductions, but as entry points into some of the details of what has been sketched out in this chapter. They are provocative, and should set you thinking about what space and place mean. They all use lively and engaging examples (especially Law and Mol), and all make arguments that are designed to challenge ways of thinking about space and place that are certainly more familiar, and have perhaps become a little too comfortable.

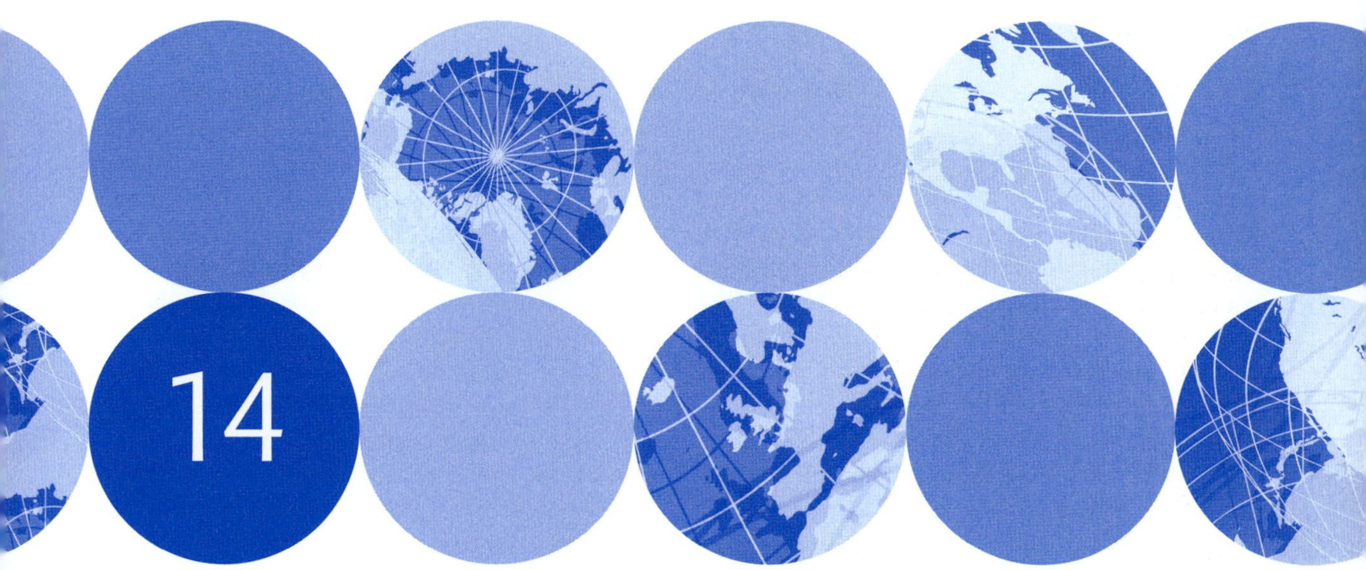

Conclusion: telling stories about cultural geographies

> **Chapter Map**
> - Introduction
> - *We* love cultural geography . . . still? A personal reflection on cultural geographies
> - What next for cultural geographies . . . where will cultural geographies take *you*?

14.1 Introduction

In the introductory chapter of this book, we shied away from providing a neat, singular, snappy definition of 'cultural geography'. To some readers, this will be frustrating; to others, it may seem like a classic case of academic frippery, an over-exaggeration of the depth, complexity, contradictoriness and all-round, bewildering variety of research that somehow goes on under the banner of 'cultural geography'.

For a student of the **new cultural geography** (see Chapter 1) in the 1990s, it might appear, upon reading this book, that cultural geography has lost its way, its sense of vitality, its sense of self-importance, or its radical, 'fashionable', cutting edge, which was the result of the situation of cultural geography on the margins of the discipline for several years (Crang 2010; Cresswell 2010). Alternatively, it might simply appear that cultural geography encompasses everything, and therefore lacks a distinct identity. How can it be possible for one textbook to cover issues as diverse as monumental sculpture, waiting at a bus stop, Japanese fictional writing about landscapes, and the cultural politics of subcultures? Could it, then, be argued that 'cultural geography' is about everything and, therefore, nothing at all? (see Box 1.4)

The above represent just some of the questions and critiques that have been levelled at cultural geography in the recent past. Yet we side with Crang, Cresswell, Wylie and others who – in a special section of the journal *Cultural Geographies* (2010) – reflected upon the many pasts and futures of 'cultural geography', and all found things to like (even to love). Most notably, Crang (2010: 191) found "a welter of wonderful cultural geographic stuff" – a small percentage of which has found its way into the pages of this book, because it reflects some of the diversity of what *we* (the authors) love about cultural geography, as well as that work which has been influential within geography and beyond. *We* still love cultural geography, in spite of the bewildering array of research that goes on, in spite of the fact that it can be hard to see where cultural geography starts and other (sub)disciplines (such as social geography or performance studies) end, and in spite of the sometimes vicious critiques that have been levelled at the subdiscipline.

If, as Price (2010) argues, one interprets the word 'culture' to signify the stories we tell about ourselves, then cultural geographies can, in one way, be understood as a series of critical but creative stories that have been told about the world. Significantly, these are stories that can be told in many ways – in words, images, performances, installations, archives, exhibitions, and so much more besides (also Wylie 2010). They are, in other words, studies of how meaning, narrative and representation are not simply free-floating, but somehow *take place*. Stories are situated (in particular places, at particular times); stories travel (with migrants, via the Internet, with commodities); stories give places meanings and vice versa, because it is through the gradual building-up of stories that people feel they belong (or do not belong) to particular localities.

This book, then, is one more story. But it is in making this observation that it is so hard to provide a neat, singular conclusion to the 'story' of cultural geography thus far. For any story is just that – just one attempt to narrate the world that will only ever be partial, that will gloss over what some people will think to be 'facts' where others find them 'fictions', and which is time- and place-specific. As we have repeatedly argued in this text, one key contribution of cultural geographers has been to de-naturalise *any* and *all* stories about the world. Whether showing that age or gender identities are socially constructed through labels or performances in public spaces, or whether showing that banal, **everyday**, **emotional** experiences really *matter*, cultural geographers have shown that the meanings people attach to the world are rarely, if ever, 'natural' (including 'nature' itself).

Indeed, in a similar way, cultural geographers have often been at the forefront of a move to make *visible* those aspects of cultural life that are often hidden from view, whether through design or through accident. Let us not forget that the early North American cultural geographers (the **Berkeley School**) advocated a way of looking at and recognising landscapes with a depth and richness that had previously been lacking in many studies. This imperative – although utterly different in its expression – continues through contemporary **non-representational** and **actor-network** theories (see Chapters 9 and 10), which, quite simply, seek to expose and witness those aspects of the world that tended to elude previous geographical research.

A final insight of many – if not all – cultural geographies is that stories *are* stories because they are never finished. Since they do not simply reflect 'facts' about the world, they can always be revised, updated or retold with a different spin. As we argued in Chapter 13, **space and place** are concepts that refer not to dead coordinates, but to living, lively sets of relations that never hold still. Cultural geographies are – like the stories they try to capture – always open to further interpretation.

If this book is just one more story about the stories that cultural geographers tell, then this is why it is as difficult to write a neat, singular conclusion as it was to find a neat, singular definition in the introductory chapter. The previous paragraphs do, we hope, provide you with a sense of some of the (very) broad impulses that, in our view, have underpinned the study of cultural geographies over the last 150 years or so. It is with these thoughts in mind that this concluding chapter offers two reflections, both of which are knowingly partial stories, but both of which are designed to prompt you to critically consider the possible futures of cultural geography.

We begin with a personal reflection on writing a book about cultural geography, and on being 'cultural geographers', in the second decade of the twenty-first century. We then consider some possible futures for cultural geography, inspired by a selection of insights from across the cultural materials, texts, practices and theories that have appeared in this book.

14.2 *We* love cultural geography . . . still? A personal reflection on cultural geographies

We (the authors) *still* love cultural geography for the three reasons we outlined at the very beginning of this book, and which this book has exemplified in many different ways through Chapters 2 to 13.

- Cultural geography – in its various forms – presents us with a rich body of research and evidence-based theory about geographies of cultural practices and politics in diverse contexts (see Part 1).
- Cultural geography offers a major resource of concepts and in-depth research exploring the geographical importance of cultural materials, media, texts and representations (see Part 2).
- Cultural geography has been, and continues to be, one of the most theoretically adventurous, open-minded and avant-garde subdisciplines within human geography. Through the work of cultural geographers, many lines of social and cultural theory have become important within the wider discipline of human geography (see Part 3).

In doing these things, the work of cultural geographers past and present has inspired us in our own research (predominantly with children and young people), occasionally prompting us to tell stories about the world – and even ourselves – in some quite unusual ways (see Horton and Kraftl 2006b; Kraftl and Horton 2007).

Yet we conclude this book with a nagging feeling. That is, that neither of us feels as if we belong either to the **Berkeley School** approach, or to the group of researchers collected under the label of the **new cultural geographies**. Indeed, we find it hard to locate ourselves at all within the story of cultural geographies. This reflects both our geographical location (we both live in the UK) and the time when we studied geography (we both began our undergraduate degrees in the late 1990s and completed our PhDs in 2004). If anything, we belong to a group of geographers who have increasingly come to label their work 'non-representational geographies', but are wary both of this label and of labels per se, as well as being acutely aware of the continuities with previous approaches to cultural-geographic research (on which more below). To be sure, we have sought to foreground the importance of a series of ideas and case studies that have been of central significance to non-representational geographies – in particular through the considerable space we have afforded to poststructural and feminist theories of **everydayness, embodiment, emotion/affect, materiality** and **space/place,** which we discussed in Chapters 9 to 13. These are ideas that have inspired us, and which (when we wrote the book) were still considered cutting-edge in cultural geography.

At the same time, the story we have told about cultural geography in this book has been partial, in so many ways. First, thanks in part to our own inability to read and understand more than a couple of languages other than English, this book is foremost a story about Anglophone cultural geographies. We have drawn on a range of case studies from geographical contexts both within and beyond the English-speaking world, but this partiality is actually less an outcome of our linguistic prowess, and more a result of the simple fact that a large proportion of the researchers who call themselves 'cultural geographers' live and work in the English-speaking world.

It is worth noting that there are exciting developments in non-Anglophone countries such as Brazil and Germany (Crang 2010), and there will be countless other examples in other geographical contexts, too. It sounds rather glib to say it, but there is considerably more scope for (truly two-way) cross-fertilisation between Anglophone cultural geographies and those taking place elsewhere, if only so that cultural geographers do not end up mirroring the power relations between places that they seek so hard to critique. That is, as Noxolo *et al.* (2008) argue, that as much as postcolonial analyses (particularly) have sought to uncover the voices of those people and places marginalised by colonial processes, the majority of *research* about those people and places is still done by English-speaking academics from a small selection of Western countries, who impose their theories and research methods upon others.

Our first reflection is, then, that it is unavoidable to write a book about cultural geographies that privileges

Anglophone cultural geographies, not only because of our own locatedness but also, and predominantly, because that is where the majority of research going by that name has – and this is problematic – been done. But, we wonder, in what ways could cultural geographies be opened up and challenged by those cultural media, texts, materials, practices and theories that originate outside the Anglophone world? Indeed, given that research – like identities, commodities and places – is a result of multiple processes originating in many different places (see Chapter 13 on Massey's conception of 'place-beyond-place'), we wonder how Anglophone cultural geographies might, somehow, be decentred.

The second reason why this book is partial is that we do not consider ourselves to be 'just' cultural geographers. Some of the research we do straddles research on social geographies, cultural studies, political geographies, even environmental geographies. Significantly, we have cited many geographers in this book who may not consider themselves to be only 'cultural geographers', and we know that several of the authors whose work we cite would not call themselves 'cultural geographers' at all. Rather, some would say they were 'social geographers', some would say they were simply 'geographers', and some of them do work that straddles many disciplinary boundaries – from performance studies to literary theory, and from sociology to cultural studies.

In some ways, it matters not what 'we' call ourselves, but rather that we love cultural geography because of the kinds of possibility it has afforded for researchers of all persuasions in thinking, doing and writing exciting and different stories about the world. As Cresswell (2010: 169) puts it, in the first years of the twenty-first century, "it seems as though 'we' [cultural geographers] have been successful in insisting on the centrality of the broadly conceived 'cultural' for all of human geography".

This achievement poses both difficulties and opportunities. It poses difficulties, because it leads to fears that the central message and purposes of 'cultural geographies' as defined in any formal sense might become diluted, too 'mainstreamed', and thereby lose their cutting edge. It also leads us – the authors – to a sense that we do not feel quite at home 'in' cultural geography, both because we do not feel truly part of any particular 'school' (such as the **new cultural geography**) and precisely because the success of cultural-geographic theories has been such that they permeate so many areas of geographical research.

We reflect on this feeling by wondering, on the one hand, whether other 'cultural geographers' feel the same. Although you would have to ask them, we suspect that many do. On the other hand, and perhaps more seriously, we urge you not to worry too much about having an authoritative view of what cultural geography 'is' and 'is not'. Rather, it is far more important that the practice of cultural geography is – and specifically makes space for – the development of one's own voice in the stories that one tells. Whether this means writing from the body or writing poetry (Chapter 6), whether it means writing memoirs, travelogues or spatial histories, or whether it means engaging in forms of performance art (Chapter 7), the practice of cultural geography should, perhaps, always be "at once critical *and* creative, at once scholarly *and* story-like" (Wylie 2010: 212, our emphases).

As we insisted at the beginning of this chapter, we take Wylie's point to mean that cultural geographers have often been so radical, inspiring and thought-provoking because they have constantly sought to tell new stories, in new ways, about those aspects of the world that remain hidden, whether by design or accident. If you choose to do an essay or piece of research that draws on cultural geography – and we hope that you do – then we would urge you to bear in mind that cultural-geographical approaches *do* allow for creativity and (a certain measure of) personality, as much as they should be rigorous and critical.

A third reason why the story presented in this book is partial is that it has been written at a particular historical moment. The chapters in Part 3, in particular, tell the story of a specific set of cultural theories that have gained prominence in cultural geography at different times (some from the 1970s onwards), but all of which have been of particular interest to cultural geographers doing research in the early twenty-first century. On the one hand, we have provided a sense of how 'earlier' schools of thought remain current – of how the Berkeley School's work on landscape resonates with contemporary approaches (Chapter 5), or of how the notion of materiality can be traced back to Marxian theorising in geography, and beyond (Chapter 10). This is because, although we studied cultural geographies only relatively recently, we did so at a kind of transitional period (if you like), when ideas about 'representation' in the new cultural geography were challenged by feminist and poststructuralist geographers (during the late 1990s and early 2000s). Having had the fortune of being taught by some of the key cultural geographers from each 'era' (if you will), this has served for us both as a constant reminder not to view a more recent approach as somehow replacing an earlier one. This is why we have tried not to structure chapters in terms of the chronological appearance of different theories, but thematically (as we do in Chapter 4, on **architectural geographies**). This, we hope, gives a sense of both the continuities and the changes that have characterised cultural-geographic research on some enduring objects of study (such as buildings).

On the other hand, while we have framed cultural geography in the terms of some key phrases that have been fashionable in the twenty-first century (such as **affect**), it is the case that much of human geography – indeed, much of cultural geography – still goes on regardless of these theoretical twists and turns. Just as much as some of the academics we cite in this book may not see themselves as 'cultural geographers', so scores of cultural geographers will not identify with (and/or would actively resist) some of the insights of non-representational geographies.

When reflecting on cultural geography – and, perhaps, embarking on an essay or research project that draws on some aspect(s) of cultural geography – we encourage you to reflect further on both of these observations. That is, on the one hand, do not feel that you should affiliate yourself only with the most recent, fashionable, 'sexy' theories in cultural geography, for it could well be that an 'older' approach (or a combination) will be more appropriate to the question you are trying to answer. And, on the other hand, remember that cultural geographers are a diverse group of researchers, each with their own voice and each taking different theoretical approaches to their work; the more you can expose yourself to, and remain sufficiently critical of, a *range* of cultural geographers' work on your chosen topic, the better. There will never be a 'last word' on any given story – and this book is certainly no exception.

Finally, the story this book tells is partial precisely because it cannot provide the 'last word'. Cultural geographies will move on, as will the worlds (and the stories) that cultural geographers try to explore. This means that trying to identify any key 'trends' for the 'future' of cultural geography is a hard task indeed. Nonetheless, we feel that to *ask* the question 'What next for cultural geographies?' is an important task. Asking the question in this way is less an attempt to issue any definitive statement about the direction of the subdiscipline than a way to prompt you to think about what else, and what more, cultural geographies could be or do.

We turn to this question in the next section. However, before you read on, we would urge you to remember two things that we have been at pains to highlight in this section. First, that the questions we raise in the next section are, in part, a reflection of our own positionalities – of the time- and place-specific contexts from which we are writing and doing research as 'cultural geographers', some of which we have discussed above. This acknowledgement of positionality is something of which cultural geographers have become increasingly aware over time, and is an insight that has affected research practices within and beyond cultural geography, as well as the writing of books like this. To this end, in Box 14.1, we reframe our discussion of our own positionality with regard to this book as a series of questions for you to consider before you move on. Second, that, notwithstanding the position from which *we* are writing, there are several key issues that currently concern many cultural geographers, not just us. However, once again, none of these cultural geographers has had the last word on these issues – in fact, far from it – and so if you are looking for an original research question or problem, you could do worse than consider some of the questions we raise in the next section.

Box 14.1
Four sets of questions about 'our' cultural geographies

1. If the stories about cultural geographies that we have told in this book are quite Anglo-centric, what are the prospects for 'decentring' English-speaking researchers, research, theories and examples from cultural geographies? What are the practical steps that could be taken - from collaborative projects with international exchange students whose first language is not English, to using your own proficiency in a language other than English to read examples from cultural geography outside the English-speaking world?
2. If cultural geography spills over into a range of other areas of research, how much does this *really* matter, from your vantage point? If you are studying at a university, how much does this matter from the point of view of the academic staff who are teaching you - do they see themselves as (cultural) geographers? What other fields of research might you read up on after reading this book - social geography, cultural theory, literary theory, architectural history . . . ?
3. If the authors wrote this book at a particular moment in the history of cultural geography, how do you feel about studying cultural geography today? Which aspects of cultural geography really chime with your own experiences? Can you identify more or less with the ideas and examples used by cultural geographers than, say, social geographers or economic geographers?
4. If we cannot have the 'last word' on cultural geography, then where might the story/ies of cultural geography be heading next? And - whether studying for an exam and looking for an original example, or embarking upon some original research for a dissertation - where will cultural geographies take you? We look in more detail at this last set of questions in the next section of this chapter.

14.3 What next for cultural geographies . . . where will cultural geographies take *you*?

In this section, we explore a number of what we want to call 'tensions' in cultural geography. These tensions refer to some of the more important debates and ideas that we have raised during the course of this book. Some of these tensions have troubled cultural geographers for many years, but are ongoing concerns. Others have arisen more recently. These tensions – and we choose just five (Box 14.2) – by no means represent all of the key debates in contemporary cultural-geographic research, but they represent a cross-section of the debates that are likely to shape the future of the subdiscipline. More importantly for a book like this, these are tensions that we believe will provoke *you* into further thought only about not the future of cultural geographies, but also about where cultural geographies could take *your* own work. It may be that you read this section as a kind of summary of some of the major subdisciplinary twists and turns that we have attended to in this book. Or it may be that you read it to prompt a research question for a dissertation or piece of coursework. Either way, you may find it helpful to pick and choose particular parts of this section that appeal to whatever you want to get out of (and hopefully put back into!) cultural geographies. After briefly discussing each 'tension', we include questions designed

Box 14.2
Five 'tensions' for the future of cultural geographies . . . and for *you* to consider

1. Relevance, usefulness and social problems
2. The 'boundaries' of cultural geography
3. What is the 'culture' in *cultural* geographies?
4. Representation/non-representation: or, accounting for 'missing' cultural geographies
5. What are the 'geographies' in cultural *geographies*?

specifically to support your own reflection about where cultural geographies could take *you*.

Tension 1: Relevance, usefulness and social problems

On reading some of the chapters in this book (especially in Part 2), you may have wondered *why* cultural geographers study particular things. Why, for instance, have cultural geographers studied children's sleep, as we have done (Kraftl and Horton 2008); or what people do in cafés (Laurier and Philo 2006a, 2006b); or their own experiences of walking along a coastal path (Wylie 2005)?

While acknowledged as intellectually stimulating and theoretically avant-garde, some research in cultural geography (not necessarily the three examples just cited) has been charged with being ephemeral, irrelevant to 'real-world' concerns, a mere diversion that satisfies a particular author's own idiosyncratic interests (Peck 1999; Martin 2001). However, in our view, this sets up a rather problematic divide between 'useful' research and anything else (Horton and Kraftl 2005). On the one hand, it is impossible to say in advance whether any piece of research will be policy-relevant, or solve a particular social problem, or support local communities in a participatory way. Sometimes the seemingly most obscure piece of research can be picked up because it just happens to resonate with a particular organisation or group of people at a particular time. On the other hand, what is *wrong* with doing research out of personal or intellectual curiosity, as long as it is original, contemporary, and drives forward research agendas or theoretical ideas that could be applied in a range of (sometimes 'useful') contexts?

It is also important to remember that some of the major approaches to cultural geography – especially the new cultural geography – were underpinned by a belief that research on landscape and identity *did* address 'real-world' social problems, such as social injustice, unequal power relations, and the oppression of people on the basis of their identity or beliefs (Cresswell 2010). As we discuss in Chapter 5, new-cultural-geographic approaches to **landscape** explored how class-based inequalities could be embedded into landscapes (whether painted, built or material) in such a way that they powerfully reinforced those inequalities – often by hiding or naturalising them. Chapters 2 and 3 of this book, in particular, provide further examples of how cultural geographers have directly addressed a series of pressing social concerns in a range of geographical and historical contexts, with direct reference to the **production** and **consumption** of cultural spaces.

For your reflection

- To what extent should *cultural geographers* try to tackle – or even solve – 'real-world' social problems?
- And – from where you are reading this – which social problems should they address?

Tension 2: The 'boundaries' of cultural geography

In Section 14.1, we argued from our own perspective that it is virtually impossible – if not undesirable – to set any hard and fast boundaries around where cultural geographies end and other forms of research and practice begin. Here, we raise a specific (kind of) tension around how cultural geographers relate to and work with those 'other forms of research and practice'. As Wylie (2010) argues, cultural geographers' work has been increasingly and enthusiastically taken up by researchers and practitioners in other disciplines – by architects, sociologists, performance artists, literary theorists and dancers, to name but a few. Significantly, there has been mounting interest in formal and informal collaborations between cultural geographers and these 'others' – whether in co-curating an exhibition with a museum on the basis of a piece of archival research, or working with artists to create a piece of public art – on the latter, see Morris's (2011) article on 'sustainable' public art in the Gorbals area of Glasgow; see also examples of walking as performance art in Chapter 7, on **performed geographies**. The 'Cultural Geographies in Practice' section of the journal of *Cultural Geographies* is a great place to look for further examples of such collaborations – as well as of cultural geographers themselves engaging in artistic, performative endeavours.

Building on Tension 1 (above), it could be argued – and you might agree – that such endeavours are 'not geography'; that they distract further from addressing 'real-world' social issues; or that they might dilute the central aims, theories and research methodologies that underpin cultural geographies. But we would not agree with these observations. At the very least, such collaborations make cultural geographies a subdiscipline that is more exciting, more open-minded, and more accepting of non-traditional ways of doing research. In effect, such collaborations as those listed above – while in many cases still in their infancy at the time of writing this book – could actually tell us *more* about some of the key concerns that have troubled cultural geographers for decades (such as how geographers who themselves are experimenting with writing poetry can tell us still more about how people feel about landscapes) and in more recent times (such as how dance practitioners can help non-representational geographers to find new ways to think and write about everyday performances).

For your reflection

- What, for cultural geography, are the potential advantages and disadvantages of collaborating with 'non-cultural-geographers'?
- In what ways could *you* work with artists, dancers, musicians, writers, sportspeople or others . . . and/or in what ways could you (with suitable care and reflection) draw on your *own* interests or expertise in areas such as this to inform a piece of coursework or project work?
- How could *you* do or write the stories of cultural geography in novel ways?

Tension 3: What is the 'culture' in *cultural* geographies?

As we explored in Chapter 1, the question of 'What is culture?' has troubled cultural geographers – and cultural theorists – for decades. This book provides a broad overview of a range of different examples of culture – texts, materials, processes, performances, ideas, beliefs, identities, and much more besides (see Box 1.4 for a more specific, but still non-exhaustive list). At this juncture, we should like to suggest just three further ways to think about the kinds of 'culture' that cultural geographers might study in the future, and with which you might choose to engage.

First, one may always question what elements of human life should be labelled 'culture'. While the debates about 'high' (formal, elite, sponsored, often 'artistic') culture and 'low' (popular, everyday, vernacular, often 'non-artistic') culture have to some extent abated in cultural geography, it remains as important as ever to *question* what is meant by the 'cultural' in cultural geography (Mitchell 2000). In so doing, the intention should not be to police the boundaries of *what* should be studied but, rather, how, when and especially *where* certain forms of life take on the spectre of 'culture'.

Second, as we argued in Section 14.1, the notion of culture refers to stories (in the broadest sense) that are never finished. Popular cultures move on: S Club 7 was a phenomenally successful pop group in the UK during the early 2000s, which spawned an enormous range of merchandise, TV shows and dance moves, especially targeted at and enthusiastically taken up by children (and, to a lesser degree, British university students); a few years later, the group were all but forgotten as the next 'big thing' emerged (Horton 2010). In other words, culture is often something emergent, unexpected and ephemeral as much as it is 'traditional', long-standing, or engrained in the popular imagination. Given the sheer number of possible stories vying for attention (both now and in the past), and the speed at which they appear and disappear, questions then become how to choose – why study this or that particular pop band, or social networking site, or artist, or material object? And, *how* to study cultural artefacts that matter one moment and then disappear next? These are questions with which cultural geographers will continue to grapple.

Third, we began this book by thinking about different definitions of 'culture'. One definition – which you may have skipped past at the time – referred to the 'cultures' that scientists, such as bacteriologists, take with a pipette when they require samples for a medical examination or biological experiment. Yet this definition may now be of more interest than ever to cultural geographers because, increasingly, they are turning their attention to how 'non-human' matters (often termed 'nature') are part and parcel of what we call 'culture' (Chapter 10). Just as pressingly, as the theorist Nikolas Rose argues, with advances in mapping the human genome and contemporary biotechnologies that can actually *alter* and even *(re)create* – not just 'mend' – bits and pieces of the human body, medical experts are increasingly intervening in such a way that "life itself has become open to politics" (Rose 2007: 15). He says that life itself in many contemporary societies has been reduced to the "molecular level" (Rose 2007: 12): it is at this submicroscopic scale that some of the most important decisions are made about whose illness gets treated and whose does not, about what forms of life matter more than others (often along the old identity lines of age, income, gender and race), about who is responsible for the health of society (increasingly, individual citizens rather than national governments). For Rose, these are new ways of doing life itself that recombine at least two of the definitions of culture we discussed in Chapter 1 – as a scientific practice and as a way of life – into what Rose (2007: 12) calls "styles of thought".

Very rapidly, then, these styles of thought are becoming definitional of aspects of contemporary cultures – from new identities forged through pressure groups who lobby for research into a particular illness, to activism around genetically modified crops, animals and human tissues. Of course, Rose identifies just one (albeit important) way in which cultures are being (re)defined in new ways: there are many other contemporary styles of thought, sometimes related to the molecular styles that Rose writes about. In this way, cultural geographers have begun to show how emergent styles of thought impact upon how we define and deal with 'big social problems' (yes, that term again) – from the futures of climate change and the 'war on terror' (Anderson 2010) to those of obesity (Evans 2010). Thus, as we have already argued, the very definition of culture, and its involvement in such styles of thought, remains a live, pressing task for the future of cultural geographies as much as a question for their emergence in the past (Figure 1.1).

For your reflection

- What are the cultures that matter to you, or to people living where you are reading this book?
- In the context of where you live, how are elements of what cultural geographers study – such as texts, materials, performances, identities – involved in emergent attempts to control 'life itself'?

Tension 4: Representation/non-representation: or, accounting for 'missing' cultural geographies

We have considered in many places in this book how cultural geographers have grappled with questions of representation. Think back to how some new cultural geographers viewed paintings, **landscapes** and **buildings** as 'texts' (Chapters 4 and 5) that, with the right tools, could be read so that the intentions of the architects or sponsors could be uncovered. This technique brought enormous insights into cultural geographers' understanding of how seemingly innocent 'artistic' works or everyday landscapes could reinforce and even create some of the deepest divisions within societies.

Or think back to how other work by new cultural geographers during the 1990s sought to represent those social and cultural groups who had traditionally been sidelined in geography – women, children, people with dis/abilities, gays and lesbians, to name but a few (Chapter 8). This work brought to the fore a whole series of important debates about representation – not only about the gains to be made from seeing the world from an-other's vantage point, but about the dangers of privileged academics (even if they were not white, male, middle-class) speaking on behalf of others with (often) considerably less power.

Or think about the whole range of ways in which cultural geographers have sought to move beyond representation – through attention to **emotion**, **affect**, **bodily performances**, **materiality** and the like (see Part 3). On the one hand, these non-representational geographies have heralded a range of exciting ways to witness the world, many of which have still not been exhausted. On the other hand, non-representational geographies have given rise to new debates about representation, as "some [cultural geographers] have deliberated the methodological and interpretive paradoxes of the approach, since academic research, reflection, and writing are to a certain degree inherently representational" (DeLyser and Rogers 2010: 187). These kinds of deliberation have led to new controversies – about, for instance, the differences between emotion and affect (Chapter 11) and the words cultural geographers use to describe these related but distinct features of life (Pile 2010).

These debates about representation and non-representation will continue, and, as the stories and 'styles of thought' that constitute culture themselves move on (see Tension 3), will, we are sure, be constantly reinterpreted. However, we are also convinced that one of the central and enduring features of cultural-geographic research will continue – that is, to uncover and subject to analysis that which is hidden or missing from either popular or academic view, either by design or by default. Remember, too, that some parts of life can be 'hidden in plain view' – that is, those cultural practices that seem so obvious and **everyday** that they pass most of us by (Chapter 9); or those identity groups that may seem 'mainstream' or be positioned at the 'centre' and therefore tend to be ignored (Hopkins and Pain 2007; see also Chapter 8). Cultural geographies have been so diverse and insightful because they seek to understand both the obvious and the truly hidden, both the ordinary and the extraordinary, and to question what happens when any or all of these facets of life come into contact with one another (Kraftl 2009).

For your reflection

- Having read this book (or at least some of it), what do *you* feel are the most important implications of what cultural geographers have had to say about (non) representation?
- From your vantage point, what are the hidden people, places, texts, materials, performances or beliefs that could be uncovered or analysed by cultural geographers?

Tension 5: What are the 'geographies' in cultural *geographies*?

The previous chapter of this book (Chapter 13) ended by asking you to reflect upon the relative merits of different ways of thinking about space and place. If you have not already, you may at least like to read the conclusion of that chapter before continuing. We are convinced – like most cultural geographers – that space and place matter, profoundly, to those forms of life that come, somehow, to be labelled 'culture' (Tension 3). This book could, in one way, be seen broadly as a series of overlapping, layered, competing-yet-complementary stories about the multiple ways in which cultures *take place*. That term – 'taking place' – implies that cultural geographers now understand space as something complex, lively, processual, dynamic and, like the stories that constitute cultures themselves, never finished (Massey 2005). This is not to say that 'older' forms of cultural geography are defunct: *every* approach, from those of the late-nineteenth/early twentieth-century North American and European pioneers onwards (see Figure 1.2), has somehow added to an ever-burgeoning series of insights into how space and place make culture, and vice versa. While the tensions between different approaches to cultural geography have sometimes become extremely embittered (especially during the emergence of the new cultural geography), we must always see these differences as productive conversations. Usually, these conversations happen not just for the sake of high theoretical debate, but because they correspond to real-world concerns: for instance, the incorporation of Marxian approaches in some versions of new cultural geography reflected concerns from the 1970s onwards with shifting class relations in the places in which **new cultural geographers** were working.

Thus it is important not to see the different conceptions of space and place used by cultural geographers as mere metaphors or interchangeable frames of reference for studying (basically) the same things. Instead – as the brief example of Marxian approaches in new cultural geography suggests – conceptions of space and place refer to and should help us to explain the complexities of "material and symbolic life" (Gulsom and Symes 2007: 3; see also Smith and Katz 1993). This is why, to repeat a point we made in Chapter 13, we encourage you to think about *how* theories of space and place apply (or do not apply) across the whole, complex range of cultural geographies that we have presented throughout this book. Doubtless, new ways of theorising space and place will be developed by cultural geographers, but it is always important not to see them as impenetrable, incontestable, immovable '-isms' but, rather, to think critically about what those conceptions of space and place can *do*.

For your reflection

Try not to shy away from theories of space and place; rather, see what they can do for *you*. In particular, if you are embarking on an essay or a piece of project work, why not, after doing some further reading, 'try out' a couple of theories to see which one (or more) helps you to explain a particular case study, or offers a particularly useful way to work out your methodology, or to analyse some data that you have collected, or to generalise about some particular item you have seen in the news?

Summary

- Inevitably, this book reflects the positionality of the authors as much as it does some of the major contemporary (and historical) trends in cultural geographies. However, it is important to recognise that an acknowledgement of positionality has been a key contribution of cultural geography to wider geographical research.
- There are many ongoing 'tensions' in cultural-geographic research, most of which will probably never be resolved. We picked just five key tensions, which we think might both inform the future of cultural geographies, and prompt you to consider how *you* can work on, from and with cultural geographies. To summarise, we suggested that you consider:
  - how, when and where cultural geographies might be 'useful' (and not to worry *too* much about this), or address particular social problems, even if that 'relevance' might not be immediately apparent;
  - what the 'boundaries' of cultural geographies are – both in terms of how cultural-geographic research intersects with work done by other geographers, and in terms of potentially open-minded, creative collaborations with academics and practitioners from other disciplines (such as performance art or architecture);
  - what we mean by the word 'culture' in cultural geographies – not so much in terms of defining *what* cultural geography is or what should be studied, but in terms of an ongoing question as to which emergent forms of life (whether popular cultural fads or biotechnologies) become labelled as 'culture', and how they might be studied by geographers;
  - what can be gained from ongoing questions about (non)representation – in particular, in terms of cultural geographers' long-standing mission to uncover and subject to analysis those people, places, materials, texts or performances that are 'hidden' – even if they are hidden in plain view;
  - what we mean by the 'geographies' in cultural geographies – most significantly, in continuing to pose questions about how culture 'takes' place: that is, in maintaining a decades-old tradition of interrogating how the many stories that we tell about culture are contingent upon complex, dynamic processes of place-making, scaling, boundary-setting and mobility (and vice versa).

References

Adams, A. (1994) Competing communities in the 'great bog of Europe': identity and seventeenth century Dutch landscape painting. In Mitchell, W. (ed.) *Landscape and Power,* The University of Chicago Press, Chicago, 35–76.

Adams, P. (2009) *Geographies of Media and Communication: A critical introduction,* Wiley-Blackwell, Oxford.

Adey, P. (2007) 'May I have your attention': airport geographies of spectatorship, position, and (im)mobility. *Environment and Planning D: Society and Space,* **25,** 515–536.

Adey, P. (2008) Airports, mobility, and the calculative architecture of affective control. *Geoforum,* **39,** 438–451.

Adey, P. (2009) *Mobility,* Routledge, London.

Adey, P., Budd, L. and Hubbard, P. (2007) Flying lessons: exploring the social and cultural geographies of global air travel. *Progress in Human Geography,* **31,** 773–791.

Adorno, T. and Horkheimer, M. (1979[1944]) *Dialectic of Enlightenment,* Verso, London.

Ahmed, S. (2004) On collective feelings, or the impressions left by others. *Theory, Culture and Society,* **20,** 25–42.

Al-Hindi, K. and Stadder, C. (1997) The hidden histories and geographies of neotraditional town planning: the case of Seaside, Florida. *Environment and Planning D: Society and Space,* **15,** 349–372.

Allen, M. (2008) *Cleansing the City: Sanitary geographies in Victorian London,* Ohio University Press, Athens, OH.

Amoore, L. (2006) Biometric borders: governing mobilities in the war on terror. *Political Geography,* **25,** 336–351.

Amoore, L. and de Goede, M. (2008) Transactions after 9/11: the banal face of the preemptive strike. *Transactions of the Institute of British Geographers,* **33,** 173–185.

Anderson, B. (2002) A principle of hope: recorded music, listening practices and the immanence of utopia. *Geografiska Annaler B,* **84,** 211–227.

Anderson, B. (2004) Time stilled, space slowed: how boredom matters. *Geoforum,* **35,** 739–754.

Anderson, B. (2006) Becoming and being hopeful: towards a theory of affect. *Environment and Planning D: Society and Space,* **24,** 733–752.

Anderson, B. (2009a) Emotional geography. In Gregory, D., Johnston, R., Pratt, G., Watts, M. and Whatmore, S. (eds) *The Dictionary of Human Geography,* Wiley-Blackwell, Chichester, 188–189.

Anderson, B. (2009b) Affect. In Gregory, D., Johnston, R., Pratt, G., Watts, M. and Whatmore, S. (eds) *The Dictionary of Human Geography,* Wiley-Blackwell, Chichester, 8–9.

Anderson, B. (2010) Preemption, precaution, preparedness: anticipatory action and future geographies. *Progress in Human Geography,* **34,** 777–798.

Anderson, B. and Harrison, P. (2006) Questioning affect and emotion. *Area,* **38,** 333–335.

Anderson, B. and Harrison, P. (eds) (2010) *Taking-Place: Non-representational theories and geography,* Ashgate, Farnham.

Anderson, B. and Tolia-Kelly, D. (2004) Matter(s) in social and cultural geography. *Geoforum,* **35,** 669–674.

Anderson, K. (1988) Cultural hegemony and the race definition process in Vancouver's Chinatown: 1880–1980. *Environment and Planning D: Society and Space,* **6,** 127–149.

Anderson, K. (1999) Introduction. In Anderson, K. and Gale, F. (eds) *Cultural Geographies,* Longman, New York, NY, 1–24.

Anderson, K. and Gale, F. (eds) (1999) *Cultural Geographies,* Longman, New York, NY.

Anderson, K. and Smith, S. (2001) Editorial: emotional geographies. *Transactions of the Institute of British Geographers,* **26,** 7–10.

Anderson, K., Domosh, M., Pile, S. and Thrift, N. (2003a) A rough guide. In Anderson, K., Domosh, M., Pile, S. and Thrift, N. (eds) *Handbook of Cultural Geography,* Sage, London, 1–35.

Anderson, K., Domosh, M., Pile, S. and Thrift, N. (2003b) Preface. In Anderson, K., Domosh, M., Pile, S. and Thrift, N. (eds) *Handbook of Cultural Geography,* Sage, London, xviii–xix.

Andrews, G., Sudwell, M. and Sparks, A. (2005) Towards a geography of fitness: an ethnographic case study of the gym in British bodybuilding culture. *Social Science and Medicine,* **60,** 877–891.

Angus, T., Cook, I. and Evans, J. (2001) A manifesto for cyborg pedagogy. *International Research in Geographical and Environmental Education,* **10,** 195–201.

Ansell, N. (2005) *Children, Youth and Development.* Routledge, London.

Ansell, N., Hajdu, F., Robson, E., van Blerk, L. and Marandet, E. (2012) Youth policy, neoliberalism and transnational governmentality: a case study of Lesotho and Malawi. In Kraftl, P., Horton, J. and Tucker, F. (eds) *Critical Geographies of Childhood and Youth: Policy and practice,* Policy Press, Bristol, 43–60.

Aoki, D. (2004) True love stories. *Cultural Studies - Critical Methodologies,* **1,** 97–111.

Appadurai, A. (1986) Introduction: commodities and the politics of value. In Appadurai, A. (ed.) *The Social Life of Things,* Cambridge University Press, Cambridge, 3–63.

Arthritis Care (2012) *Living with Arthritis,* Arthritis Care, London.

Askins, K. (2009) 'That's just what I do': placing emotion in academic activism. *Emotion, Society and Space,* **2,** 4–13.

Attfield, J. (2000) *Wild Things: The material culture of everyday life,* Berg, Oxford.

Augé, M. (1995) *Non-Places: Introduction to an anthropology of super-modernity,* Verso, London.

Bale, J. (2003) *Sports Geography,* Taylor & Francis, London.

Barnes, T. (2005) Culture: economy. In Cloke, P. and Johnston, R. (eds) *Spaces of Geographical Thought,* Sage, London, 61–80.

Barnes, T. and Duncan, J. (1992) *Writing Worlds: Discourse, text and metaphor in the representation of landscape,* Routledge, London.

Barnett, C. (1998) Cultural twists and turns. *Environment and Planning D: Society and Space,* **16,** 631–634.

Bartky, S. (1990) *Femininity and Domination: Studies in the phenomenology of oppression,* Routledge, London.

BBC (2011) People urged to say they are Cornish on census. Available at http://www.bbc.co.uk/news/uk-england-cornwall-12809366

Bell, D. and Valentine, G. (1995) *Mapping Desire: Geographies of sexuality,* Taylor & Francis, London.

Bell, D. and Valentine, G. (1996) *Consuming Geographies: We are where we eat,* Routledge, London.

Bendelow, G. and Williams, S. (eds) (1997) *Emotions in Social Life: Critical themes and contemporary issues,* Routledge, London.

Benjamin, W. (1999) *Arcades Project,* Harvard University Press, Boston, MA.

Bennett, J. (2010) *Vibrant Matter: A political ecology of things,* Duke University Press, Durham, NC.

Berger, J. (1972) *Ways of Seeing,* Penguin, Harmondsworth.

Bingham, N. (1996) Object-ions: from technological determinism towards geographies of relations. *Environment and Planning D: Society and Space,* **14,** 635–657.

Bissell, D. (2008) Comfortable bodies: sedentary affects. *Environment and Planning A,* **40,** 1697–1712.

Blanchot, M. (1993 [1969]) Everyday speech. In Hanson, S. (ed.) *The Infinite Conversation,* University of Minnesota Press, Minneapolis, MN, 238–245.

Blum, V. and Nast, H. (1996) Where's the difference? The heterosexualization of alterity in Henri Lefebvre and Jacques Lacan. *Environment and Planning D: Society and Space,* **14,** 559–580.

Blumen, O. (2007) The performative landscape of going-to-work: on the edge of a Jewish ultraorthodox neighborhood. *Environment and Planning D: Society and Space,* **25,** 803–831.

Blunt, A. (1999) Imperial geographies of home: British women in India, 1886–1925. *Transactions of the Institute of British Geographers,* **24,** 421–440.

Blunt, A. and Wills, J. (2004) *Dissident Geographies: An introduction to radical ideas and practice,* Pearson, Harlow.

Boal, F. (2002) Belfast: walls within. *Political Geography,* **21,** 687–694.

Bondi, L. (2005) The place of emotions in research. In Davidson, J., Bondi, L. and Smith, M. (eds) *Emotional Geographies,* Ashgate, Aldershot, 231–246.

Bondi, L., Davidson, J. and Smith, M. (2005) Introduction: geography's 'emotional turn'. In Davidson, J., Bondi, L. and Smith, M. (eds) *Emotional Geographies,* Ashgate, Aldershot, 1–18.

Bonnett, A. (1992) Art, ideology and everyday space: subversive tendencies from Dada to postmodernism. *Environment and Planning D: Society and Space,* **10,** 69–86.

Borden, I., Kerr, J., Rendell, J. and Pivaro, A. (2001) *The Unknown City: Contesting architecture and social space,* MIT Press, Cambridge, MA.

Bordo, S. (1993) *Unbearable Weight: Feminism, Western culture and the body,* University of California Press, Berkeley, CA.

Bourdieu, P. (1977) *Outline of a Theory of Practice,* Cambridge University Press, Cambridge.

Bourdieu, P. (1984) *Distinction: A social critique of the judgement of taste,* Harvard University Press, Boston, MA.

Bowker, G. and Star, S. (1999) *Sorting Things Out: Classification and its consequences,* MIT Press, Cambridge, MA.

Braudel, F. and Reynolds, S. (1975 [1948]) *The Mediterranean and the Mediterranean World in the Age of Phillip II,* Fontana, London.

Brickell, K. and Datta, A. (2011) *Translocal Geographies: Spaces, places, connections,* Ashgate, Farnham.

Bridge, G. and Smith, A. (2003) Intimate encounters: culture – economy – commodity. *Environment and Planning D: Society and Space,* **21,** 257–268.

Brown, G. (2007) Mutinous eruptions: autonomous spaces of radical queer activism. *Environment and Planning A,* **39,** 2685–2698.

Brown, G. (2009) Thinking beyond homonormativity: performative explorations of diverse gay economies. *Environment and Planning A,* **41,** 1496–1510.

Brown, G. and Pickerill, J. (2009) Space for emotion in the spaces of activism. *Emotion, Space and Society,* **2,** 24–35.

Brown, M. (2012) Gender and sexuality I: intersectional anxieties. *Progress in Human Geography,* **36,** 541–550.

Browne, K. (2009) Womyn's separatist spaces: rethinking spaces of difference and exclusion. *Transactions of the Institute of British Geographers,* **34,** 541–556.

Brusseau, J. (1998) *Isolated Experiences,* State University of New York Press, Albany, NY.

Bryman, A. (1999) *The Disneyization of Society,* Sage, London.

Bull, M. (2000) *Sounding Out the City: Personal stereos and the management of everyday life*, Berg, Oxford.

Butler, J. (1990a) Performative acts and gender constitution: an essay in phenomenology and feminist theory. In Case, S. (ed.) *Performing Feminisms: Feminist critical theory and theatre,* Johns Hopkins University Press, Baltimore, MD, 270–282.

Butler, J. (1990b) *Gender Trouble: Feminism and the subversion of identity,* Routledge, New York, NY.

Butler, R. and Parr, H. (eds) (2004) *Mind and Body Spaces: Geographies of illness, impairment and disability,* Routledge, London.

Butler, T. (2006) A walk of art: the potential of the sound walk as practice in cultural geography. *Social and Cultural Geography,* **7,** 889–908.

Butz, K. and Besio, D. (2004a) The value of autoethnography for field research in transcultural settings. *The Professional Geographer,* **56,** 350–360.

Butz, K. and Besio, D. (2004b) Autoethnography: a limited endorsement. *The Professional Geographer,* **56,** 432–438.

CAFOD [Catholic Overseas Development Agency] (2009) *Working Conditions in the Electronics Industry,* CAFOD, London.

Callard, F. (2004) Doreen Massey. In Hubbard, P., Kitchin, R. and Valentine, G. (eds) *Key Thinkers on Space and Place,* Sage, London, 219–225.

Callon, M. (1986) Some elements of a sociology of translation: domestication of the scallops and the fishermen of St Brieuc Bay. In Law, J. (ed.) *Power, Action and Belief: A new sociology of knowledge,* Routledge, London, 196–227.

Carney, G. (1998) Music geography. *Journal of Cultural Geography,* **18,** 1–10.

Carpena-Méndez, F. (2007) 'Our lives are like a sock inside-out': children's work and youth identity in neoliberal rural Mexico. In Panelli, R., Punch, S. and Robson, E. (eds) *Global Perspectives on Rural Childhood and Youth: Young rural lives,* Routledge, London, 41–56.

Castells, M. (2000) *The Rise of the Network Society,* Blackwell, Oxford.

CCCE [Center for Communication and Civic Engagement] (2011) Culture jamming and meme-based communication, available at http://depts.washington.edu/ccce/polcommcampaigns/CultureJamming.htm

Chaney, D. (2002) *Cultural Change and Everyday Life,* Palgrave, Basingstoke.

Chatterton, P. (2003) *Urban Nightscapes: Youth cultures, pleasure spaces and corporate power,* Routledge, London.

Childs, I. (1991) Japanese perception of nature in the novel *Snow Country. Journal of Cultural Geography,* **11,** 1–19.

Claeys, G. and Sargent, L. (1999) *The Utopia Reader,* New York University Press, New York, NY.

Clark, N., Massey, D. and Sarre, P. (eds) (2008) *Material Geographies: A world in the making,* Sage, London.

Clarke, D., Doel, M. and Housiaux, K. (2003) *The Consumption Reader,* Routledge, London.

Clay, G. (1994) *Real Places: An unconventional guide to America's generic landscape,* University of Chicago Press, Chicago, IL.

Clifford, N., Holloway, S., Rice, S. and Valentine, G. (2009) *Key Concepts in Geography,* Sage, London.

Cloke, P. and Jones, O. (2004) Turning in the graveyard: trees and the hybrid geographies of dwelling, monitoring and resistance in a Bristol cemetery. *Cultural Geographies,* **11,** 313–341.

Cloke, P., Philo, C. and Sadler, D. (1991) *Approaching Human Geography: An introduction to contemporary theoretical debates,* Paul Chapman, London.

Cloke, P., Cooke, P., Cursons, J., Milbourne, P. and Widdowfield, R. (2000) Ethics, reflexivity and research: encounters with homeless people. *Ethics, Place and Environment,* **3,** 133–154.

Cloke, P., Cook, I., Crang, M., Goodwin, M., Painter, J. and Philo, C. (2004) *Practising Human Geography,* Sage, London.

Cloke, P., Crang, P. and Goodwin, M. (eds) (2005) *Introducing Human Geographies,* Arnold, London.

Cohen, L. (1997) *Glass, Paper, Beans: Revelations on the nature and value of ordinary things,* Doubleday, New York, NY.

Cohen, S. (1967) *Folk Devils and Moral Panics,* Paladin, London.

Colls, R. (2012) Feminism, bodily difference and non-representational geographies. *Transactions of the Institute of British Geographers,* **37,** 430–445.

Connell, J. and Gibson, C. (2004) World music: deterritorializing place and identity. *Progress in Human Geography,* **28,** 342–361.

Connolly, J. and Prothero, A. (2008) Green consumption: life-politics, risk and contradictions. *Journal of Consumer Culture,* **8,** 117–145.

Conradson, D. (2003a) Geographies of care: spaces, practices, experiences. *Social and Cultural Geography,* **4,** 451–454.

Conradson, D. (2003b) Spaces of care in the city: the place of a community drop-in centre. *Social and Cultural Geography,* **4,** 507–525.

Cook, I. (2004) Follow the thing: papaya. *Antipode,* **36,** 642–664.

Cook, I. (2005) Commodities: the DNA of capitalism. Available at: http://followthethings.files.wordpress.com/2010/07/commodities-dna.pdf

Cook, I., Crouch, D., Naylor, S. and Ryan, J. (2000) Foreword. In Cook, I., Crouch, D., Naylor, S. and Ryan, J. (eds) *Cultural Turns, Geographical Turns,* Prentice Hall, Harlow, xi–xii.

Cook, I., Evans, J., Griffiths, H., Mayblin, L., Payne, B. and Roberts, D. (2007) Made in...? Appreciating the everyday geographies of connected lives. *Teaching Geography,* Summer, 80–83.

Cosgrove, D. (1984) *Social Formation and Symbolic Landscape,* Croom Helm, Beckenham.

Cosgrove, D. (1985) Prospect, perspective and the evolution of the landscape idea. *Transactions of the Institute of British Geographers,* **10,** 45–62.

Cosgrove, D. (1997) *Social Formation and Symbolic Landscape,* University of Wisconsin Press, Madison, WI.

Cosgrove, D. and Daniels, S. (eds) (1988) *The Iconography of Landscape: Essays on the symbolic representation, design, and use of past environments,* Cambridge University Press, Cambridge.

Crampton, K. (2011) Cartographic calculations of territory. *Progress in Human Geography,* **35,** 92–103.

Crane, D. (1992) *The Production of Culture: Media and the urban arts,* Sage, London.

Crang, P. (1994) It's showtime: on the workplace geographies of display in a restaurant in southeast England. *Environment and Planning D: Society and Space,* **12,** 675–704.

Crang, M. (1998) *Cultural Geography,* Routledge, London.

Crang, M., Crang, P. and May, J. (1999) *Virtual Geographies: bodies, spaces and relations,* Routledge, London.

Crang, P. (2010) Cultural geography: after a fashion. *Cultural Geographies,* **17,** 191–201.

Cresswell, T. (1996) *In Place/Out of Place: Geography, ideology and transgression,* University of Minnesota Press, Minneapolis, MN.

Cresswell, T. (2004) *Place: A short introduction,* Blackwell, Oxford.

Cresswell, T. (2006) *On the Move: Mobility in the modern Western world,* Taylor & Francis, London.

Cresswell, T. (2010) New cultural geography – an unfinished project? *Cultural Geographies,* **17,** 169–174.

Crewe, L. (2000) Geographies of retailing and consumption. *Progress in Human Geography,* **24,** 274–290.

Crouch, D. (2003) Spatialities and the feeling of doing. *Social and Cultural Geography,* **2,** 61–75.

Daniels, S. (1993) *Fields of Vision,* Princeton University Press, Princeton, NJ.

Daniels, S. and Nash, C. (2004) Lifepaths: geography and biography. *Journal of Historical Geography,* **30,** 449–458.

Dant, T. (1999) *Material Culture in the Social World: Values, activities, lifestyles,* Open University Press, Buckingham.

Datta, A. and Brickell, K. (2009) 'We have a little bit more finesse as a nation': constructing the Polish worker in London's building sites. *Antipode,* **41,** 439–464.

Davidson, J. (2003) 'Putting on a face': Sartre, Goffman and agoraphobic anxiety in social space. *Environment and Planning D: Society and Space,* **21,** 107–122.

Davis, M. (1990) *City of Quartz: Excavating the future in Los Angeles,* Verso, London.

Dear, M. (2000) *The Postmodern Urban Condition,* Blackwell, Oxford.

de Certeau, M. (1984) *The Practice of Everyday Life,* University of California Press, Berkeley, CA.

DEEWR [Australian Department of Education, Employment and Workplace Relations] (2010) *Building the Education Revolution: Overview,* available online at: http://www.deewr.gov.au/Schooling/BuildingTheEducationRevolution/Pages/default.aspx

de Leeuw, S. (2009) 'If anything is to be done with the Indian, we must catch him very young': colonial constructions of Aboriginal children and the geographies of Indian residential schooling in British Columbia, Canada. *Children's Geographies,* **7,** 107–122.

DeLyser, D. and Rogers, B. (2010) Meaning and methods in cultural geography: practicing the scholarship of teaching. *Cultural Geographies,* **17,** 185–190.

den Besten, O., Horton, J., Adey, P. and Kraftl, P. (2011) Claiming events of school redesign: materialising the promise of Building Schools for the Future. *Social and Cultural Geography,* **12,** 9–26.

Denevan, W. and Mathewson, K. (eds) (2009) *Carl Sauer on Culture and Landscape: Readings and commentaries,* Louisiana State University Press, Baton Rouge, LA.

Dennis, R. (1989) The geography of Victorian values: philanthropic housing in London, 1840–1900. *Journal of Historical Geography,* **15,** 40–54.

de Propris, L., Chapain, C., Cooke, P., MacNeill, S. and Mateos-Garcia, J. (2009) *The Geography of Creativity,* NESTA, London.

Dery, M. (1999) *Culture Jamming: Hacking, slashing and sniping in the empire of signs,* Open Magazine Pamphlet Series, Westfield.

Desforges, L. (2004) Bananas and Citizens, available at: http://www.exchange-values.org/

DeSilvey, C. (2007) Salvage memory: constellating material histories on a hardscrabble homestead. *Cultural Geographies,* **14,** 401–424.

Dewsbury, J. (2000) Performativity and the event: enacting a philosophy of difference. *Environment and Planning D: Society and Space,* **18,** 473–496.

Dewsbury, J. and Naylor, S. (2002) Practising geographical knowledge: fields, bodies and dissemination. *Area,* **34,** 253–250.

Dewsbury, J., Harrison, P., Rose, M. and Wylie, J. (2002) Introduction: enacting geographies. *Geoforum,* **33,** 437–440.

DfES [UK Department for Education and Skills] (2003) *Building Schools for the Future: Consultation on a new approach to capital investment,* DfES, London.

DHS [Department of Homeland Security] (2001) *President's Address to a Joint Session of Congress and the American People, United States Capitol, Washington, D.C., 20 September 2001 (9:00 p.m. EDT), DHS, Washington DC,* available at: http://www.state.gov/documents/organization/10308.pdf

Doel, M. (1994) Writing difference. *Environment and Planning A,* **26,** 1015–1020.

Doel, M. (1999) *Poststructuralist Geographies: The diabolical art of spatial science,* Edinburgh University Press, Edinburgh.

Doel, M. (2005) Deconstruction and geography: settling the account. *Antipode,* **37,** 246–249.

Domosh, M. (1989) A method for interpreting landscape: a case study of the New York World Building. *Area,* **21,** 347–355.

Domosh, M. (2002) A 'civilized' commerce: gender, 'race,' and empire at the 1893 Chicago Exposition. *Cultural Geographies,* **9,** 183–203.

Dovey, K. (1999) *Framing Places: Mediating power in built form,* Routledge, London.

du Gay, P. (1997) Introduction. In du Gay, P., Hall, S., Janes, L., Mackay, P. and Negus, K. (eds) *Doing Cultural Studies: The story of the Sony Walkman,* Sage, London, 1–7.

Duncan, J. and Ley, D. (1993) *Place/Culture/Representation,* Routledge, London.

Edgell, S. and Hetherington, K. (1996) Introduction: consumption matters. In Edgell, S., Hetherington, K. and Warde, A. (eds) *Consumption Matters,* Blackwell, Oxford, 1–10.

Ekinsmyth, C. and Shurmer-Smith, P. (2002) Humanistic and behaviouralist geography. In Shurmer-Smith, P. (ed.) *Doing Cultural Geography,* Sage, London, 19–28.

Eldridge, J. (1993) News, truth and power. In Eldridge, J. (ed.) *Getting the Message: News, truth and power,* Routledge, London, 3–28.

Elliott, A. (2001) *Concepts of the Self,* Polity Press, Cambridge.

Engels, F. (1987[1845]) *The Condition of the Working Class in England,* Penguin, Harmondsworth.

Evans, B. (2006) 'Gluttony or sloth': critical geographies of bodies and morality in (anti) obesity policy. *Area,* **38,** 259–267.

Evans, B. (2010) Anticipating fatness: childhood, affect and the pre-emptive 'war on obesity'. *Transactions of the Institute of British Geographers,* **35,** 21–38.

Evans, B. and Colls, R. (2009) Measuring fatness, governing bodies: the spatialities of the Body Mass Index (BMI) in anti-obesity politics. *Antipode,* **41,** 1051–1083.

Evans, B. and Honeyford, E. (2012) Brighter futures, greener lives: children and young people in UK sustainable development policy. In Kraftl, P., Horton, J. and Tucker, F. (eds) *Critical Geographies of Childhood and Youth,* Policy Press, Bristol, 61–78.

Eyles, J. (1989) The geography of everyday life. In Gregory, D. and Walford, R. (eds) *Horizons in Human Geography,* Macmillan, London, 102–117.

Featherstone, D. (2008) *Resistance, Space and Political Identities: The making of counter-global networks,* Wiley, London.

Featherstone, D., Thrift, N. and Urry, J. (eds) (2005) *Automobilities,* Sage, London.

Featherstone, M. (1991) *Consumer Culture and Postmodernism,* Sage, London.

Featherstone, M. (2000) Body modification: an introduction. In Featherstone, M. (ed.) *Body Modification,* Sage, London, 1–15.

Featherstone, M., Hepworth, M. and Turner B. (eds) (1991) *The Body: Social process and cultural theory,* Sage, London.

Fine, B. and Leopold, E. (1993) *The World of Consumption,* Routledge, London.

Fisher, P., Wood, J. and Cheng, T. (2004) Where is Helvellyn? Multiscale morphometry and the mountains of the English Lake District. *Transactions of the Institute of British Geographers,* **29,** 106–128.

Fiske, J. (1989) *Understanding Popular Culture,* Routledge, London.

Florida, R. and Jackson, S. (2010) Sonic city: the evolving economic geography of the music industry. *Journal of Planning Education and Research,* **29,** 310–321.

Fotheringham, A., Charlton, M. and Brunsdon, C. (2000) *Quantitative Geography: Perspectives on spatial data analysis,* Sage, London.

Foucault, M. (1972) *The Archaeology of Knowledge,* Tavistock, London.

Foucault, M. (1991[1975]) *Discipline and Punish: The birth of the prison,* Penguin, Harmondsworth.

Frost, L. (2001) *Young Women and the Body: A feminist sociology,* Palgrave, Basingstoke.

Fuller, D., Askins, K., Mowl, G., Jeffries, M.J. and Lambert, D. (2008) Mywalks: fieldwork and living geographies. *Teaching Geography,* **33**, 80–83.

Gagen, E. (2004) Making America flesh: physicality and nationhood in early-twentieth century physical education reform. *Cultural Geographies,* **11,** 417–442.

Gardiner, M. (2004) Everyday utopianism: Lefebvre and his critics. *Cultural Studies,* **18,** 228–254.

Gatens, M. (1983) A critique of the sex/gender distinction. In Allen, J. and Patten, P. (eds) *Beyond Marxism? Interventions after Marx,* Intervention, Sydney, 143–161.

Gelder, K. (2005) The field of subcultural studies. In Gelder, K. (ed.)*The Subcultures Reader,* Routledge, London, 1–16,

Gelder, K. and Jacobs, J. (1998) *Uncanny Australia: Sacredness and identity in a postcolonial nation,* University of Melbourne Press, Melbourne.

Gertler, M. (2003) A cultural economic geography of production. In Anderson, K., Domosh, M., Pile, S. and Thrift, N. (eds) *Handbook of Cultural Geography,* Sage, London, 131–146.

Gibson, C. and Homan, S. (2004) Urban redevelopment, live music and public space. *International Journal of Cultural Policy,* **10,** 67–84.

Gibson-Graham, J.K. (2006) *A Postcapitalist Politics,* University of Minnesota Press, Minneapolis, MN.

Gibson-Graham, J.K. (2008) Diverse economies: performative practices for 'other worlds'. *Progress in Human Geography,* **32,** 613–632.

Giddens, A. (1991) *Modernity and Self-Identity: Self and society in the late modern age,* Stanford University Press, Stanford, CT.

Goffman, E. (1963) *Stigma: Notes on the management of spoiled identity,* Spectrum, New York, NY.

Golledge, R. (1993) Geography and the disabled: a survey with special reference to vision impaired and blind populations. *Transactions of the Institute of British Geographers,* **18,** 63–85.

Goodwin, M. (2004) Recovering the future: a post-disciplinary perspective on geography and political economy. In Cloke, P., Crang, P. and Goodwin, M. (eds) *Envisioning Human Geography,* Arnold, London, 65–80.

Goss, J. (1988) The built environment and social theory: towards an architectural geography. *Professional Geographer,* **40,** 392–403.

Goss, J. (1993) The 'Magic of the Mall': an analysis of form, function, and meaning in the contemporary retail built environment. *Annals of the Association of American Geographers,* **83,** 18–47.

Gowans, G. (2003) Imperial geographies of home: Memsahibs and Miss-sahibs in India and Britain, 1915–1947. *Cultural Geographies,* **10,** 424–441.

Granger, R. and Hamilton, C. (2011) *Breaking New Ground: Spatial mapping of the creative economy,* Institute for Creative Enterprise, Coventry.

Graves-Brown, P. (2002) Introduction. In Graves-Brown, P. (ed.) *Matter, Materiality and Modern Culture,* Routledge, London, 1–9.

Greco, M. and Stenner, P. (2008) Introduction: emotion and social science. In Greco, M. and Stenner, P. (eds) *Emotions: A social science reader,* Routledge, London, 1–24.

Greene, V. (2005) Dealing with diversity: Milwaukee's multiethnic festivals and urban identity, 1840–1940. *Journal of Urban History,* **31,** 820–849.

Greenhough, B. (2006) Decontextualised? Dissociated? Detached? Mapping the networks of bio-informatic exchange. *Environment and Planning A,* **38,** 445–463.

Greenhough, B. and Roe, E. (2006) Towards a geography of bodily biotechnologies. *Environment and Planning A,* **38,** 416–422.

Gregory, D. (1994) *Geographical Imaginations,* Blackwell, Oxford.

Gregory, D. (2009) Regional geography. In Gregory, D., Johnston, R., Pratt, G., Watts, M. and Whatmore, S. (eds) *The Dictionary of Human Geography,* Wiley-Blackwell, Chichester, 632–636.

Gregson, N. and Rose, G. (2000) Taking Butler elsewhere: performativities, spatialities and subjectivities. *Environment and Planning D: Society and Space,* **18,** 433–452.

Gregson, N., Kothari, U., Cream, J., Dwyer, C., Holloway, S., Maddrell, A. and Rose, G. (1997) Gender in feminist geography. In Women and Geography Study Group (eds) *Feminist Geography: Explorations in diversity and difference,* Longman, London, 49–85.

Griffiths, H., Cook, I. and Evans, J. (2009) *Making the Connection: Mobile phone geographies,* available at: http://makingtheconnectionresources.wordpress.com/

Griggs, G. (2009) 'Just a sport made up in a car park': the soft landscape of Ultimate Frisbee. *Social and Cultural Geography,* **10,** 757–770.

Gruffudd, P. (1995) Remaking Wales: nation-building and the geographical imagination. *Political Geography,* **14,** 219–239.

Gruffudd, P. (1996) The countryside as educator: schools, rurality and citizenship in inter-war Wales. *Journal of Historical Geography,* **22,** 412–423.

Gruffudd, P. (2001) 'Science and the stuff of life': modernist health centres in 1930s London. *Journal of Historical Geography,* **27,** 395–416.

Gulson, K. and Symes, C. (2007) *Spatial Theories of Education: Policy and geography matters,* Routledge, London.

Haegerstrand, T. (1978) A note on the quality of life-times. In Carlstein, T., Parkes, D. and Thrift, N. (eds) *Making Sense of Time, Volume 2: Human activity and time geography,* Wiley, Chichester, 214–224.

Hagen, J. (2008) Parades, public space, and propaganda: the Nazi culture parades in Munich. *Geografiska Annaler, Series B,* **90,** 349–367.

Hagen, J. and Ostergren, R. (2006) Spectacle, architecture and place at the Nuremberg Party Rallies: projecting a Nazi vision of past, present and future. *Cultural Geographies,* **13,** 157–181.

Hague, E. and Mercer, J. (1998) Geographical memory and urban identity in Scotland: Raith Rovers FC and Kirkcaldy. *Geography,* **83,** 105–116.

Hall, S. (1980) Encoding/decoding. In Hall, S., Hobson, D., Lowe, A. and Willis, P. (eds) *Culture, Media, Language,* Routledge, London, 128–138.

Hall, S. (1990) Cultural identity and diaspora. In: Rutherford, J. (ed.) *Identity: Community, culture, difference,* Lawrence & Wishart, London, 222–237.

Hall, S. (1997) The work of representation. In Hall, S. (ed.) *Representation: Cultural representations and signifying practices,* Sage, London, 13–64.

Hall, S. and Jefferson, T. (eds) (1976) *Resistance Through Rituals: Youth subcultures in post-war Britain,* Centre for Contemporary Cultural Studies, Birmingham.

Hall, T. and Hubbard, P. (1998) *The Entrepreneurial City: Geographies of politics, regime and representation,* John Wiley, Chichester.

Hallam, E. and Hockey, J. (2001) *Death, Memory and Material Culture,* Berg, Oxford.

Hannigan, J. (1998) *Fantasy City: Pleasure and profit in the postmodern metropolis,* Routledge, London.

Haraway, D. (1991) A cyborg manifesto: science, technology and socialist-feminism in the late twentieth-century. In Haraway, D. (ed.) *Simians, Cyborgs and Women: The reinvention of nature,* Routledge, London, 149–181.

Harley, J. (2002) *The New Nature of Maps: Essays in the history of cartography,* Johns Hopkins University Press, Baltimore, MD.

Harrell, J. (1994) The poetics of deconstruction: death metal rock. *Popular Music and Society,* **18,** 91–107.

Harrison, P. (2000) Making sense: embodiment and the sensibilities of the everyday. *Environment and Planning D: Society and Space,* **18,** 497–517.

Harrison, P. (2002) The Caesura: remarks on Wittgenstein's interruption of theory, or, why practices elude explanation. *Geoforum,* **33,** 487–503.

Harrison, P. (2007) 'How shall I say it?' Relating the nonrelational. *Environment and Planning A,* **39,** 590–608.

Harrison, P. (2009) In the absence of practice. *Environment and Planning D: Society and Space,* **27,** 987–1009.

Hartley, J. (1994) *Understanding News,* Routledge, London.

Hartshorne, R. (1959) *Perspective on the Nature of Geography,* Association of American Geographers/Rand McNally & Co., Chicago, IL.

Harvey, D. (1973) *Social Justice and the City,* Arnold, London.

Harvey, D. (1989) *The Condition of Postmodernity: An enquiry into the origins of cultural change,* Blackwell, Oxford.

Harvey, D. (1990) Between space and time: reflections on the geographical imagination. *Annals of the Association of American Geographers,* **80,** 418–434.

Harvey, D. (1996) *Justice, Nature and the Geography of Difference,* Blackwell, Oxford.

Harvey, D. (2000) *Spaces of Hope,* Edinburgh University Press, Edinburgh.

Hastings, A. (1999) Discourse and urban change. *Urban Studies,* **36,** 7–12.

Hayes, B. (2005) *Infrastructure: A field guide to the industrial landscape,* W. W. Norton, New York, NY.

Hebdige, D. (1979) *Subculture: The meaning of style,* Methuen, London.

Henry, N. and Pinch, S. (2000) Spatialising knowledge: placing the knowledge community of Motor Sport Valley. *Geoforum,* **31,** 191–208.

Herman, E. and Chomsky, N. (1988) *Manufacturing Consent: The political economy of the mass media,* Pantheon, New York, NY.

Hetherington, K. (1997) Museum topology and the will to connect. *Journal of Material Culture,* **2,** 199–218.

Highmore, B. (2002a) *Everyday Life and Cultural Theory: An introduction,* Routledge, London.

Highmore, B. (ed.) (2002b) *The Everyday Life Reader,* Routledge, London.

Highmore, B. (2004) Homework: routine, social aesthetics and the ambiguity of everyday life. *Cultural Studies,* **18,** 306–327.

Hinchliffe, S. (2008) *Geographies of Nature: Societies, environments, ecologies,* Sage, London.

Hodkinson, P. (2007) Interactive online journals and individualization. *New Media and Society,* **9,** 625–650.

Holloway, L. and Hubbard, P. (2001) *People and Place: The extraordinary geographies of everyday life,* Pearson, Harlow.

Holloway, S. (2003) Outsiders in rural society? Constructions of rurality and nature–society relations in the racialisation of English Gypsy-Travellers, 1869–1934. *Environment and Planning D: Society and Space,* **21,** 695–715.

Holloway, S. (2005) Identity and difference: age, dis/ability and sexuality. In Cloke, P., Crang, P. and Goodwin, M. (eds) *Introducing Human Geographies,* Arnold, London, 400–410.

Holloway, S., Valentine, G. and Bingham, N. (2000) Institutionalising technologies: masculinities, femininities and the heterosexual economy of the IT classroom. *Environment and Planning A,* **32,** 617–633.

Holt, L. (2007) Children's socio-spatial (re)production of disability in primary school playgrounds. *Environment and Planning D: Society and Space,* **25,** 783–802.

Holt-Jensen, A. (2009) *Geography: History and concepts,* Sage, London.

Hones, S. (2008) Text as it happens: literary geography. *Geography Compass,* **2,** 1301–1317.

Hopkins, P. and Hill, M. (2008) Pre-flight experiences and migration stories: the accounts of unaccompanied asylum-seeking children. *Children's Geographies,* **6,** 257–268.

Hopkins, P. and Pain, R. (2007) Geographies of age: thinking relationally. *Area,* **39,** 287–294.

Hopkins, T. and Wallerstein, I. (1986) Commodity chains in the world economy prior to 1880. *Review,* **10,** 157–170.

Hörschelmann, K. and Colls, R. (eds) (2009) *Contested Bodies of Childhood and Youth,* Palgrave Macmillan, Basingstoke.

Horton, J. (2008) A 'sense of failure'? Everydayness and research ethics. *Children's Geographies,* **6,** 363–383.

Horton, J. (2010) 'The best thing ever': how children's popular culture matters. *Social and Cultural Geography,* **11,** 377–398.

Horton, J. (2012) 'Got my shoes, got my Pokémon': spaces of children's popular culture. *Geoforum,* **43,** 4–13.

Horton, J. and Kraftl, P. (2005) For more-than-usefulness: six overlapping points about children's geographies. *Children's Geographies,* **3,** 131–143.

Horton, J. and Kraftl, P. (2006a) What else? Some more ways of thinking about and doing children's geographies. *Children's Geographies,* **4,** 69–95.

Horton, J. and Kraftl, P. (2006b) Not just growing up, but going on: children's geographies as becomings; materials, spacings, bodies, situations. *Children's Geographies,* **4,** 259–276.

Horton, J. and Kraftl, P. (2009) Time for bed! Rituals, practices and affects in children's bed-time routines. In Colls, R. and Hörschelmann, K. (eds) *Contested Bodies of Childhood and Youth,* Palgrave Macmillan, Basingstoke, 215–231.

Horton, J. and Kraftl, P. (2011) Tears and laughter at a Sure Start Centre: preschool geographies, policy contexts. In Holt, L. (ed.) *Geographies of Children, Youth and Families: An international perspective,* Routledge, London, 235–249.

Horton, J. and Kraftl, P. (2012a) Clearing out a cupboard: memories and materialities. In Jones, O. and Garde-Hansen, J. (eds) *Geography and Memory,* Palgrave Macmillan, Basingstoke.

Horton, J. and Kraftl, P. (2012b) School building redesign: everyday spaces, transformational policy discourses. In Waters, J., Brooks, R. and Fuller, A. (eds) *Changing Spaces of Education: New perspectives on the nature of learning,* Routledge, London, 114–134.

Howell, P. (2004a) Sexuality, sovereignty and space: law, government and the geography of prostitution in colonial Gibraltar. *Social History,* **29,** 444–464.

Howell, P. (2004b) Race, space and the regulation of prostitution in Colonial Hong Kong. *Urban History,* **31,** 229–248.

Howell, P. (2009) *Geographies of Regulation: Policing prostitution in nineteenth-century Britain and the Empire,* Cambridge University Press, Cambridge.

Howell, P., Beckingham, D. and Moore, F. (2008) Managed zones for sex workers in Liverpool: contemporary proposals, Victorian parallels. *Transactions of the Institute of British Geographers,* **33,** 233–250.

Howson, A. (2004) *The Body in Society: An introduction,* Polity Press, Cambridge.

Hubbard, P., Kitchin, R., Bartley, B. and Fuller, D. (2002) *Thinking Geographically: Space, theory and contemporary human geography,* Continuum, London.

Hubbard, P., Kitchin, R. and Valentine, G. (2008) *Key Texts in Human Geography,* Sage, London.

Hudson, R. (2006) Regions and place: music, identity and place. *Progress in Human Geography,* **30,** 626–634.

Imrie, R. (2001) Barriered and bounded places and the spatialities of disability. *Urban Studies,* **38,** 231-237.

Imrie, R. (2003) Architects' conceptions of the human body. *Environment and Planning D: Society and Space,* **21,** 47–65.

Ingold, T. (2000) *Perceptions of the Environment: Essays in livelihood, dwelling and skill,* Routledge, London.

Iveson, K. (2007) *Publics and the City,* Blackwell, Oxford.

Jabry, A. (2002) *Children in Disasters: After the cameras have gone,* Plan UK, London.

Jackson, P. (1989) *Maps of Meaning,* Routledge, London.

Jackson, P. (2003) Mapping culture. In Rogers, A. and Viles, H. (eds) *The Student's Companion to Geography,* Blackwell, Oxford, 133–137.

Jackson, P. (2005) Identities. In Cloke, P., Crang, P. and Goodwin, M. (eds) *Introducing Human Geographies,* Arnold, London, 391–399.

Jackson, P., Russell, P. and Ward, N. (2004) *Commodity Chains and the Politics of Food,* available at www.consume.bbk.ac.uk/working_papers/jackson.doc

Jacobs, J. (2006) A geography of big things. *Cultural Geographies,* **13,** 1–27.

Jacobs, J. and Smith, S. (2008) Guest editorial: living room: rematerialising home. *Environment and Planning A,* **40,** 515–519.

Jacobs, J., Cairns, S. and Strebel, I. (2007) 'A tall storey…but, a fact just the same': the Red Road high-rise as a black box. *Urban Studies,* **44,** 609–629.

James, A. and James, A.L. (2004) *Constructing Childhood: Theory, policy and practice,* Palgrave Macmillan, Basingstoke.

Jayne, M., Holloway, S. and Valentine, G. (2006) Drunk and disorderly: alcohol, urban life and public space. *Progress in Human Geography,* **30,** 452–468.

Jazeel, T. (2005) The world is sound? Geography, musicology and British-Asian soundscapes. *Area,* **37,** 233–241.

Jenkins, L. (2002) Geography and architecture: 11, Rue de Conservatoire and the permeability of buildings. *Space and Culture,* **5,** 222–236.

Johnston, R. (1997) *Geography and Geographers: Anglo-American human geography since 1945,* 5th edn, Arnold, London.

Johnston, R. and Sidaway, J. (2004) *Geography and Geographers: Anglo-American human geography since 1945,* 6th edn, Hodder Arnold, London.

Jones, O. (1995) Lay discourses of the rural: developments and implications for rural studies. *Journal of Rural Studies,* **11,** 35–49.

Jones, O. (2003) 'Endlessly revisited and forever gone': on memory, reverie and emotional imagination in doing children's geographies. *Children's Geographies,* **1,** 25–36.

Jones, R. (2009) Categories, borders and boundaries. *Progress in Human Geography,* **33,** 174–189.

Jupp, E. (2007) Participation, local knowledge and empowerment: researching public spaces with young people. *Environment and Planning A,* **39,** 2832–2844.

Jupp, E. (2008) The feeling of participation: everyday spaces and urban change. *Geoforum,* **39,** 331–343.

Kahn-Harris, K. (2004) Unspectacular subculture? Transgression and mundanity and the global extreme metal scene. In Bennett, A. and Kahn-Harris, K. (eds) *After Subculture: Critical studies in contemporary youth culture,* Palgrave Macmillan, Basingstoke, 107–118.

Karsten, L. and Pel, E. (2000) Skateboarders exploring public space: ollies, obstacles and conflicts. *Journal of Housing and the Built Environment,* **15,** 327–340.

Kawabata, Y. (2011[1956]) *Snow Country,* Penguin, Harmondsworth.

Kearns, G. (2010) Geography, geopolitics and Empire. *Transactions of the Institute of British Geographers,* **25,** 187–203.

Keith, M. and Pile, S. (1993) *Place and the Politics of Identity,* Routledge, London.

Kenny, J. (1992) Portland's Comprehensive Plan as text: the Fred Meyer case and the politics of reading. In Barnes, T. and Duncan, J. (eds) *Writing Worlds: Discourse, text and metaphor in the representation of landscape,* Routledge, London, 176–192.

King, A. (1984) *The Bungalow,* Routledge, London.

King, A. (2004) *Spaces of Global Cultures,* Routledge, London.

Kitchin, R. and Kneale, J. (2001) Science fiction or future fact? Exploring imaginative geographies of the new millennium. *Progress in Human Geography,* **25,** 19–35.

Kleinman, A. (1988) *Illness Narratives,* Basic Books, New York, NY.

Kniffen, F. (1965) Folk housing: key to diffusion. *Annals of the Association of American Geographers,* **55,** 549–577.

Kniffen, F. and Glassie, H. (1966) Building in wood in the eastern United States. *Geographical Review,* **56,** 40–66.

Kobayashi, A. (2004) Critical 'race' approaches to cultural geography. In Duncan, J., Johnson, N. and Schein, R. (eds) *The Companion to Cultural Geography,* Blackwell, Oxford, 238–249.

Kolb, A. (2010) *Dance and Politics,* Peter Lang, Oxford.

Kong, L. (1995a) Popular music in geographical analyses. *Progress in Human Geography,* **19,** 183–198.

Kong, L. (1995b) Music and cultural politics: ideology and resistance in Singapore. *Transactions of the Institute of British Geographers,* **20,** 447–459.

Kopytoff, I. (1986) The cultural biography of things: commoditisation as process. In Appadurai, A. (ed.) *The Social Life of Things,* Cambridge University Press, Cambridge, 64–94.

Koskela, H. and Pain, R. (2000) Revisiting fear and place: women's fear of attack and the built environment. *Geoforum,* **31,** 269–280.

Kraftl, P. (2006a) Building an idea: the material construction of an ideal childhood. *Transactions of the Institute of British Geographers,* **31,** 488–504.

Kraftl, P. (2006b) Ecological buildings as performed art: Nant-y-Cwm Steiner School, Pembrokeshire. *Social and Cultural Geography,* **7,** 927–948.

Kraftl, P. (2007) Utopia, performativity and the unhomely. *Environment and Planning D: Society and Space,* **25,** 120–143.

Kraftl, P. (2009) Living in an artwork: the extraordinary geographies of everyday life at the Hundertwasser-Haus, Vienna. *Cultural Geographies,* **16,** 111–134.

Kraftl, P. (2010) Architectural movements, utopian moments: (in)coherent renderings of the Hundertwasser-Haus, Vienna. *Geografiska Annaler, Series B: Human Geography,* **92,** 327–345.

Kraftl, P. (2012) Utopian promise or burdensome responsibility? A critical analysis of the UK Government's Building Schools for the Future Policy. *Antipode,* **44,** 847–870.

Kraftl, P. and Adey, P. (2008) Architecture/affect/dwelling. *Annals of the Association of American Geographers,* **98,** 213–231.

Kraftl, P. and Horton, J. (2007) 'The Health Event': everyday, affective politics of participation. *Geoforum,* **38,** 1012–1027.

Kraftl, P. and Horton, J. (2008) Spaces of every-night life: geographies of sleep, sleeping and sleepiness. *Progress in Human Geography,* **31,** 509–524.

Kraftl, P., Horton, J. and Tucker, F. (2012a) Introduction. In Kraftl, P., Horton, J. and Tucker, F. (eds) *Critical Geographies of Childhood and Youth: Contemporary policy and practice,* Policy Press, Bristol, 1–24.

Kraftl, P., Horton, J. and Tucker, F. (eds) (2012b) *Critical Geographies of Childhood and Youth: Contemporary policy and practice,* Policy Press, Bristol.

Kytö, M. (2011) We are the rebellious voice of the terraces, we are the Çarşı: constructing a football supporter group through sound. *Soccer and Society,* **12,** 77–93.

Kumar, K. (1991) *Utopianism,* Open University Press, Milton Keynes.

Lande, B. (2007) Breathing like a soldier: culture incarnate. *Sociological Review,* **55,** 95–108.

Lanegran, D. (2007) Cultural geography and place: introduction. In Moseley, W., Lanegran, D. and Pandit, K. (eds) *The Introductory Reader in Human Geography,* Blackwell, Oxford, 181–183.

Lash, S. (2001) Technological forms of life. *Theory, Culture and Society,* **18,** 105–120.

Lash, S. and Lury, C. (2007) *Global Culture Industry,* Polity Press, Cambridge.

Latham, A. (2003) The possibilities of performance. *Environment and Planning A,* **35,** 1901–1906.

Latour, B. (1999) *Pandora's Hope: Essays on the reality of science studies,* Harvard University Press, Cambridge, MA.

Latour, B. (2000) The Berlin key or how to do words with things. In Graves-Brown, P. (ed.) *Matter, Materiality and Modern Culture,* Routledge, London, 10–21.

Latour, B. and Woolgar, S. (1979) *Laboratory Life: The construction of scientific facts,* 1st edn, Sage, London.

Latour, B. and Woolgar, S. (1987) *Laboratory Life: The construction of scientific facts,* 2nd edn, Princeton University Press, Princeton, NJ.

Laurier, E. (2003) Searching for a parking space, available at http://web2.ges.gla.ac.uk/~elaurier/texts/park_search.pdf

Laurier, E. and Parr, H. (2000) Emotions and interviewing. *Ethics, Place and Environment,* **3,** 61–102.

Laurier, E. and Philo, C. (2003) The region in the boot: mobilising lone subjects and multiple objects. *Environment and Planning D: Society and Space,* **21,** 85–106.

Laurier, E. and Philo, C. (2006a) Cold shoulders and napkins handed: gestures of responsibility. *Transactions of the Institute of British Geographers,* **31,** 193–208.

Laurier, E. and Philo, C. (2006b) Possible geographies: a passing encounter in a café. *Area,* **38,** 353–363.

Law, J. (1992) *Notes on the Theory of the Actor Network: Ordering, strategy and heterogeneity,* Centre for Science Studies, Lancaster, available at: http://www.lancs.ac.uk/sociology/papers/law-notes-on-ant.pdf

Law, J. (2003) Making a mess with method, available at http://www.lancs.ac.uk/fass/sociology/papers/law-making-a-mess-with-method.pdf

Law, J. and da Costa Marques, I. (2000) Invisibility. In Pile, S. and Thrift, N. (eds) *City A–Z,* Routledge, London, 119–121.

Law, J. and Mol, A. (2001) Situating technoscience: an inquiry into spatialities. *Environment and Planning D: Society and Space,* **19,** 609–621.

Lees, L. (2001) Towards a critical geography of architecture: the case of an ersatz colosseum. *Ecumene,* **8,** 51–86.

Lefebvre, H. (1988) Toward a Leftist cultural politics: remarks occasioned by the centenary of Marx's death. In Nelson, C. and Grossberg, L. (eds) *Marxism and the Interpretation of Culture,* University of Illinois Press, Urbana, IL, 75–88.

Lefebvre, H. (1991) *The Production of Space,* Wiley, London.

Legg, S. (2003) Gendered politics and nationalised homes: women and the anticolonial struggle in Delhi, 1930–47. *Gender, Place and Culture,* **10,** 7–28.

Leonard, T. (2010) Artist spends 700 hours sitting in performance art record. *Daily Telegraph,* 1 June.

Lerup, L. (1977) *Building the Unfinished: Architecture and human action*, Sage, London.

Leslie, D. and Reimer, S. (1999) Spatializing commodity chains. *Progress in Human Geography*, **23**, 401–420.

Ley, D. (1993) Co-operative housing as a moral landscape. In Duncan, J. and Ley, D. (eds) *Place/Culture/Representation*, Routledge, London, 128–148.

Leyshon, A., Matless, D. and Revill, G. (eds) (1998) *The Place of Music*, Guilford Press, New York, NY.

LibCom (2009) Subvertising billboards, available at http://libcom.org/organise/subvertising-billboards

Liu, F. (2012) Politically indifferent nationalists? Chinese youth negotiating political identity in the internet age. *European Journal of Cultural Studies*, **15**, 53–69.

Livingstone, D. (1999) *The Geographical Tradition: Episodes in the history of a contested enterprise*, Wiley, London.

Livingstone, D. (2003) *Putting Science in its Place: Geographies of scientific knowledge*, University of Chicago Press, Chicago, IL.

Llewellyn, M. (2004) 'Urban village' or 'white house': envisioned spaces, experienced places, and everyday life at Kensal House, London in the 1930s. *Environment and Planning D: Society and Space*, **22**, 229–249.

Longhurst, R. (2000) Corporeographies of pregnancy: 'bikini babes'. *Environment and Planning D: Society and Space*, **18**, 453–472.

Longhurst, R. (2001) *Bodies: Exploring fluid boundaries*, Routledge, London.

Lorimer, H. (2003) Telling small stories: spaces of knowledge and the practice of geography. *Transactions of the Institute of British Geographers*, **28**, 197–217.

Lorimer, H. (2005) Cultural geography: the busyness of being 'more-than-representational'. *Progress in Human Geography*, **29**, 83–94.

Lorimer, H. (2007) Cultural geography: worldly shapes, differently arranged. *Progress in Human Geography*, **31**, 89–100.

Lorimer, H. (2008) Cultural geography: nonrepresentational conditions and concerns. *Progress in Human Geography*, **32**, 551–9.

Löwenheim, O. and Gazit, O. (2009) A critique of citizenship tests. *Security Dialogue*, **40**, 145–167.

Lupton, D. (1996) *Food, the Body and the Self*, Sage, London.

Lury, C. (1996) *Consumer Culture*, Polity Press, Cambridge.

MacKay, H. (1997) Introduction. In MacKay, H. (ed.) *Consumption and Everyday Life*, Sage, London, 1–12.

Madanipour, A. (2003) *Public and Private Spaces of the City*, Psychology Press, London.

Maddrell, A. (2008) The 'Map Girls': British women geographers' war work, shifting gender boundaries and reflections on the history of geography. *Transactions of the Institute of British Geographers*, **33**, 127–148.

Maddrell, A. (2009a) *Complex Locations: Women's geographical work in the UK 1850–1970*, Blackwell, Oxford.

Maddrell, A. (2009b) A place for grief and belief: the Witness Cairn at the Isle of Whithorn, Galloway, Scotland. *Social and Cultural Geography*, **10**, 675–693.

Maddrell, A. and Sidaway, J. (2010) Introduction: bringing a spatial lens to death, dying, mourning and remembrance. In Maddrell, A. and Sidaway, J. (eds) *Deathscapes: Spaces for death, dying, mourning and remembrance*, Ashgate, Farnham, 1–16.

Maddrell, A., Strauss, K., Thomas, N. and Wyse, S. (2008) *Careers in UK HE Geography Survey: Choices, status and experience*, Report for the Royal Geographical Society (with IBG), available at: http://wgsg.org.uk/newsite1/pdfs/wgsg_presentation.pdf

Madge, C. and Eshun, G. (2012) 'Now let me share this with you': exploring poetry as a method for postcolonial geography research. *Antipode*, **44**, 1395–1428.

Madge, C., Meek, J., Wellens, J. and Hooley, T. (2009) Facebook, social integration and informal learning at University: 'It is more for socialising and talking to friends about work than for actually doing work'. *Learning, Media and Technology*, **34**, 141–155.

Malbon, B. (1999) *Clubbing: Dancing, ecstasy and vitality*, Routledge, London.

Mansvelt, J. (2005) *Geographies of Consumption*, Sage, London.

Marcus, G. (1989) *Lipstick Traces: A secret history of the twentieth century*, Secker & Warburg, London.

Marshall, W. (2010) Running across the rooves of Empire: parkour and the postcolonial city. *Modern and Contemporary France*, **18**, 157–173.

Marston, S., Jones, J. III and Woodward, K. (2005) Human geography without scale. *Transactions of the Institute of British Geographers*, **30**, 416–432.

Martin, J. (2005) Identity. In Sibley, D., Jackson, P., Atkinson, D. and Washbourne, N. (eds) *Cultural Geography: A critical dictionary of key ideas*, Taurus, London, 97–102.

Martin, R. (2001) Geography and public policy: the case of the missing agenda. *Progress in Human Geography*, **25**, 189–210.

Marx, K. (1959 [1844]) *Economic and Philosophical Manuscripts of 1844*, Progress, Moscow.

Marx, K. (1976[1867]) *Capital, Volume I*, Penguin, Harmondsworth.

Massey, D. (1984) *Spatial Divisions of Labour: Social structures and the geography of production*, Macmillan, London.

Massey, D. (1991) A global sense of place. *Marxism Today,* **38,** 24–29.

Massey, D. (1994) *Space, Place and Gender,* Polity Press, Cambridge.

Massey, D. (2005) *For Space,* Sage, London.

Massumi, B. (1996) The autonomy of affect. In Patton, P. (ed.) *Deleuze: A critical reader,* Blackwell, Oxford, 217–239.

Matheson, D. (2005) *Media Discourses: Analysing media texts,* Open University Press, Maidenhead.

Matthews, H., Taylor, M., Percy-Smith, B. and Limb, M. (2000) The unacceptable flaneur: the shopping mall as a teenage hangout. *Childhood,* **7,** 279–294.

Mathewson, K. (2000) Cultural landscapes and ecology III: foraging/farming, food, festivities. *Progress in Human Geography,* **24,** 457–474.

Matless, D. (1993) Appropriate geography: Patrick Abercrombie and the energy of the world. *Journal of Design History,* **6,** 167–177.

Matless, D. (1998) *Landscape and Englishness,* Reaktion, London.

May, J., Wills, J., Datta, K., Evans, Y., Herbert, J. and McIlwaine, C. (2007) Keeping London working: global cities, the British state, and London's new migrant division of labour. *Transactions of the Institute of British Geographers,* **37,** 151–167.

McCormack, D. (2003) An event of geographical ethics in spaces of affect. *Transactions of the Institute of British Geographers,* **28,** 488–507.

McCormack, D. (2004) Drawing out the lines of the event. *Cultural Geographies,* **11,** 211–220.

McCormack, D. (2006) For the love of pipes and cables: a response to Deborah Thien. *Area,* **38,** 330–332.

McCracken, G. (1988) *Culture and Consumption: New approaches to the symbolic character of consumer goods and activities,* Indiana University Press, Bloomington, IN.

McDowell, L. (1992) Doing gender: feminism, feminists and research methods in human geography. *Transactions of the Institute of British Geographers,* **17,** 399–416.

McDowell, L. (1994) The transformation of cultural geography. In Gregory, D., Martin, R. and Smith, G. (eds) *Human Geography: Society, space and social science,* Macmillan, London, 146–173.

McDowell, L. (2003) Masculine identities and low paid work: young men in urban labour markets. *International Journal of Urban and Regional Research,* **27,** 828–848.

McEwan, C. and Blunt, A. (2002) Introducing postcolonial geographies. In Blunt, A. and McEwan, C. (eds) *Postcolonial Geographies,* Continuum, London, 1–6.

McIntyre, A. (2002) Women researching their lives: exploring violence and identity in Belfast. *Qualitative Research,* **2,** 387–409.

McRobbie, A. (1991) Introduction. In McRobbie, A. and Gerber, J. (eds) *Feminism and Youth Culture: From 'Jackie' to 'Just 17',* Macmillan, London, 1–15.

McRobbie, A. (2000) *Feminism and Youth Culture,* 2nd edn, Palgrave, Basingstoke.

Meinig, D. (1979) *The Interpretation of Ordinary Landscapes,* Oxford University Press, Oxford.

Melina, L. (2008) Backstage with The Hot Flashes: performing gender, performing age, performing Rock 'n' Roll. *Qualitative Inquiry,* **14,** 90–110.

Merrifield, A. (1993) Space and place: a Lefebvrian reconciliation. *Transactions of the Institute of British Geographers,* **18,** 516–531.

Merriman, P. (2006) 'A new look at the English landscape': landscape architecture, movement and the aesthetics of motorways in early post war Britain. *Cultural Geographies,* **13,** 78–105.

Merriman P. (2007) *Driving Spaces: A cultural-historical geography of England's M1 motorway,* Wiley-Blackwell, Oxford.

Merriman, P. (2010) Architecture/dance: choreographing and inhabiting spaces with Anna and Lawrence Halprin. *Cultural Geographies,* **17,** 427–449.

Merriman, P. and Webster, C. (2009) Travel projects: landscape, art, movement. *Cultural Geographies,* **16,** 525–535.

Miles, S. (1998a) The consuming paradox: a new research agenda for urban consumption. *Urban Studies,* **35,** 1001–1008.

Miles, S. (1998b) *Consumerism: As a way of life,* Sage, London.

Miles, S. and Paddison, R. (1998) Urban consumption: an historiographical note. *Urban Studies,* **35,** 815–824.

Miller, D. (1998) Why some things matter. In Miller, D. (ed.) *Material Cultures: Why some things matter,* UCL Press, London, 3–21.

Miller, D. (2008) *The Comfort of Things,* Polity Press, Cambridge.

Milligan, C. and Wiles, J. (2010) Landscapes of care. *Progress in Human Geography,* **34,** 736–754.

Mills, C. (1993) Myths and meanings of gentrification. In Duncan, J. and Ley, D. (eds) *Place/Culture/Representation,* Routledge, London, 149–170.

Mitchell, D. (1995) There's no such thing as culture: towards a reconceptualization of the idea of culture in geography. *Transactions of the Institute of British Geographers,* **20,** 102–116.

Mitchell, D. (2000) *Cultural Geography: A critical introduction,* Blackwell, Oxford.

Mitchell, W.J.T. (1994) *Landscape and Power,* University of Chicago Press, Chicago, IL.

Monmonier, M. (1996) *How to Lie With Maps,* University of Chicago Press, Chicago, IL.

Monmonier, M. (2005) Lying with maps. *Statistical Science,* **20,** 215–222.

Moreira, T. (2004) Surgical monads: a social topology of the operating room. *Environment and Planning D: Society and Space,* **22,** 53–69.

Morris, N. (2011) The Orchard: cultivating a sustainable public artwork in the Gorbals, Glasgow. *Cultural Geographies,* **18,** 413–420.

Moser, I. and Law, J. (2006) Fluids or flows? Information and qualculation in medical practice. *Information, Technology and People,* **19,** 55–73.

Moser, S. (2008) Personality: a new positionality? *Area,* **40,** 383–392.

Moss, P. and Dyck, I. (2003) Embodying social geography. In Anderson, K., Domosh, M., Pile, S. and Thrift, N. (eds) *Handbook of Cultural Geography,* Sage, London, 58–73.

Mould, O. (2009) Parkour, the city, the event. *Environment and Planning D: Society and Space,* **27,** 738–750.

Moylan, T. (1986) *Demand the Impossible: Science fiction and the utopian imagination,* Taylor and Francis, London.

Muggleton, D. (2000) *Inside Subculture: The postmodern meaning of style,* Berg, Oxford.

Murdoch, J. and Pratt, A. (1993) Rural studies: modernism, postmodernism and the 'post-rural'. *Journal of Rural Studies,* **9,** 411–427.

Nash, C. (2000) Performativity in practice: some recent work in cultural geography. *Progress in Human Geography,* **24,** 653–664.

Nayak, A. and Jeffrey, A. (2011) *Geographical Thought: An introduction to ideas in human geography,* Pearson, Harlow.

Negus, K. (1997) The production of culture. In du Gay, P. (ed.) *Production of Culture/Cultures of Production,* Sage, London, 67–104.

Nettleton, S. and Watson, J. (eds) (1998) *The Body in Everyday Life,* Routledge, London.

Nicholls, W. (2009) Place, networks, space: theorising the geographies of social movements. *Transactions of the Institute of British Geographers,* **34,** 78–93.

Nolan, N. (2003) The ins and outs of skateboarding and transgression in public space in Newcastle, Australia. *Australian Geographer,* **34,** 311–327.

Nold, C. (2009) Introduction: emotional cartography. In Nold, C. (ed.) *Emotional Cartography: Technologies of the self,* available at: http://emotionalcartography.net/EmotionalCartographyLow.pdf

No Tesco (2011) No Tesco in Stokes Croft: why don't we want a Tesco? Available at http://notesco.wordpress.com/thecampaign/why-dont-we-want-a-tesco/

Noxolo, P., Raghuram, P. and Madge, C. (2008) 'Geography is pregnant' and 'Geography's milk is flowing': metaphors for a postcolonial discipline? *Environment and Planning D: Society and Space,* **26,** 146–168.

O'Connor, J. and Wynne, D. (1992) The uses and abuses of popular culture: cultural policy and popular culture. *Leisure and Society,* **14,** 465–483.

Ogborn, M. (2005) Mapping words. *New Formations,* **57,** 145–149.

Okely, J. (2007) Fieldwork embodied. *Sociological Review,* **55,** 65–79.

Olwig, K. (1996) Recovering the substantive nature of landscape. *Annals of the Association of American Geographers,* **86,** 630–653.

ONS [Office for National Statistics] (2011) *Ethnicity and National Identity in England and Wales 2011,* ONS, London.

Openshaw, S. (1991) A view on the GIS crisis in geography: or, using GIS to put Humpty-Dumpty back together again. *Environment and Planning A,* **23,** 621–628.

Openshaw, S. (1996) Fuzzy logic as a new scientific paradigm for doing geography. *Environment and Planning A,* **28,** 761–768.

Oswin, N. (2008) Critical geographies and the uses of sexuality. *Progress in Human Geography,* **32,** 89–103.

Pain, R. (2009) Globalized fear? Towards an emotional geopolitics. *Progress in Human Geography,* **33,** 466–486.

Pain, R. and Smith, S. (eds) (2008) *Fear: Critical geopolitics and everyday life,* Ashgate, Farnham.

Palmer, S. (2007) *Toxic Childhoods: How the modern world is damaging our children and what we can do about it,* Orion, London.

Parr, H. (2005) Emotional geographies. In Cloke, P., Crang, P. and Goodwin, M. (eds) *Introducing Human Geographies,* Arnold, London, 472–485.

Parr, H. (2008) *Mental Health and Social Space: Towards inclusionary geographies?* Wiley, Chichester.

Parviainen, J. (2010) Choreographing resistances: spatial-kinaesthetic intelligence and bodily knowledge as political tools in activist work. *Mobilities,* **5,** 311–329.

Paterson, M. (2005) Affecting touch: towards a felt phenomenology of therapeutic touch. In Davidson, J., Bondi, L. and Smith, M. (eds) *Emotional Geographies,* Ashgate, Aldershot, 161–176.

Paterson, M. (2006) *Consumption and Everyday Life,* Routledge, London.

Paterson, M. (2011) More-than-visual approaches to architecture: vision, touch, technique. *Social and Cultural Geography,* **12,** 263–281.

Peck, J. (1999) Editorial: grey geography. *Transactions of the Institute of British Geographers,* **24,** 131–135.

Peet, R. (2007) *Geography of Power: The making of global economic policy,* Zed Press, London.

Pels, D., Hetherington, K. and Vandenberghe, F. (2002) The status of the object: performances, mediations and techniques. *Theory, Culture and Society,* **19,** 1–21.

Perec, G. (1999[1974]) *Species of Spaces,* Penguin, Harmondsworth.

Petrov, A. (1995) The sound of suburbia (death metal). *American Book Review,* **16,** 5.

Phillips, M. (2002) The production, symbolisation and socialisation of gentrification: a case study of a Berkshire village. *Transactions of the Institute of British Geographers,* **27,** 282–308.

Philo, C. (ed.) (1991a) *New Words, New Worlds: Reconceptualising social and cultural geography,* Cambrian, Aberystwyth.

Philo, C. (1991b) Introduction, acknowledgements and brief thoughts on older words and older worlds. In Philo, C. (ed.) *New Words, New Worlds: Reconceptualising social and cultural geography,* Cambrian, Aberystwyth, 1–13.

Philo, C. (1992) Neglected rural others: a review. *Journal of Rural Studies,* **8,** 193–207.

Philo, C. (1994) In the same ballpark? Looking in on the new sports geography. In Bale, J. (ed.) *Community, Landscape and Identity: Horizons in a geography of sports,* Keele University Occasional Monograph, Keele, 1–18.

Philo, C. (2000) More words, more worlds: reflections on the 'cultural turn' in human geography. In Cook, I., Crouch, D., Naylor, S. and Ryan, J. (eds) *Cultural Turns, Geographical Turns,* Prentice Hall, Harlow, 25–53.

Philo, C. (2003) 'To go back up the side hill': memories, imaginations and reveries of childhood. *Children's Geographies,* **1,** 3–24.

Philo, C. (2004) *A Geographical History of Institutional Provision for the Insane from Medieval Times to the 1860s in England and Wales: The space reserved for insanity,* Edwin Mellen Press, Lampeter.

Philo, C. and Wilbert, C. (2000) Animal spaces, beastly places: an introduction. In Philo, C. and Wilbert, C. (eds) *Animal Spaces, Beastly Places: New geographies of human–animal relations,* Routledge, London, 1–36.

Pickerel, W., Jorgenson, H. and Bennett, L. (2002) Culture jams and meme warfare: Kalle Lasn, Adbusters and media activism, available at http://depts.washington.edu/ccce/assets/documents/pdf/culturejamsandmemewarfare.pdf

Pickerill, J. (2003) *Cyberprotest: Environmental activism on-line,* Manchester University Press, Manchester.

Pickerill, J. (2009) Finding common ground? Spaces of dialogue and the negotiation of indigenous interests in environmental campaigns in Australia. *Geoforum,* **40,** 66–79.

Pickerill, J. and Chatterton, P. (2006) Notes towards autonomous geographies: creation, resistance and self-management as survival tactics. *Progress in Human Geography,* **30,** 1–17.

Pike, J. (2008) Foucault, space and primary school dining rooms. *Children's Geographies,* **6,** 413–422.

Pile, S. (1996) *The Body and the City: Psychoanalysis, space and subjectivity,* Routledge, London.

Pile, S. (1999) Introduction: opposition, political identities and spaces of resistance. In Keith, M. and Pile, S. (eds) *Geographies of Resistance,* Routledge, London, 1–32.

Pile, S. (2010) Emotions and affect in recent human geography. *Transactions of the Institute of British Geographers,* **35,** 5–20.

Pinder, D. (2005a) *Visions of the City: Utopianism, power and politics in twentieth-century urbanism,* Edinburgh University Press, Edinburgh.

Pinder, D. (2005b) Arts of urban exploration. *Cultural Geographies,* **12,** 383–411.

Platt, R. (1962) The rise of cultural geography in America. In Wagner, P. and Mikesell, M. (eds) *Readings in Cultural Geography,* University of Chicago Press, Chicago, IL, 35–43.

Ploszajska, T. (1994) Moral landscapes and manipulated space: class, gender and space in Victorian reformatory schools. *Journal of Historical Geography,* **20,** 413–429.

Ploszajska, T. (1996) Constructing the subject: geographical models in English Schools, 1870–1944. *Journal of Historical Geography,* **22,** 388–398.

Porteous, J. (1985) Literature and humanist geography. *Area,* **17,** 117–122.

Powell, R. (2002) The Sirens' voices? Field practices and dialogue in geography. *Area,* **34,** 261–272.

Pratt, A. (2004) Mapping the cultural industries. In Power, D. and Scott, A. (eds) *Cultural Industries and Production of Culture,* Routledge, London, 19–36.

Pratt, M. (2007) *Imperial Eyes: Travel writing and transculturation,* 2nd edn, Routledge, London.

Pred, A. (1991) Spectacular articulations of modernity: the Stockholm Exhibition of 1897. *Geografiska Annaler B,* **73,** 45–84.

Pred, A. (2000) Intersections. In Pile, S. and Thrift, N. (eds) *City A–Z,* Routledge, London, 117–118.

Price, P. (2010) Cultural geography and the stories we tell ourselves. *Cultural Geographies,* **17,** 203–210.

Putnam, R. (2000) *Bowling Alone,* Simon & Schuster, New York.

Queneau, R. (1981 [1947]) *Exercises in Style,* Oneworld, London.

Relph, E. (1976) *Place and Placelessness,* Pion, London.

Revill, G. (2000) Music and the politics of sound: nationalism, citizenship, and auditory space. *Environment and Planning D: Society and Space,* **18,** 597–613.

Revill, G. (2004) Performing French folk music: dance, authenticity and nonrepresentational theory. *Cultural Geographies,* **11,** 199–209.

Rich, A. (1986) *Blood, Bread and Poetry: Selected prose 1979–85,* Norton, London.

Rigg, J. (2007) *An Everyday Geography of the Global South,* Routledge, London.

Ritzer, G. (1993) *The McDonaldization of Society,* Pine Forge Press, Thousand Oaks, CA.

Rodaway, P. (1994) *Sensuous Geographies: Body, sense, and place,* Routledge, London.

Roe, E. (2006) Things becoming food and the embodied, material practices of an organic food consumer. *Sociologia Ruralis,* **46,** 104–121.

Rose, G. (1991) On being ambivalent: women and feminisms in geography. In Philo, C. (ed.) *New Words, New Worlds: Reconceptualising social and cultural geography,* Cambrian, Aberystwyth, 156–163.

Rose, G. (1993) *Feminism and Geography: The limits of geographical knowledge,* University of Minnesota Press, Minneapolis, MN.

Rose, G. (2003) Family photographs and domestic spacings: a case study. *Transactions of the Institute of British Geographers,* **28,** 5–18.

Rose, G. (2010) *Doing Family Photography: The domestic, the public and the politics of sentiment,* Ashgate, Farnham.

Rose, G., Degen, M. and Basdas, B. (2010) More on 'big things': building events and feelings. *Transactions of the Institute of British Geographers,* **35,** 334–349.

Rose, N. (2007) *The Politics of Life-Itself: Biomedicine, power, and subjectivity in the twenty-first century,* Princeton University Press, Princeton, NJ.

Sack, R. (1997) *Homo Geographicus,* Johns Hopkins University Press, Baltimore, MD.

Said, E. (1978) *Orientalism,* Penguin, Harmondsworth.

Saldanha, A. (2002) Music, space, identity: geographies of youth culture in Bangalore. *Cultural Studies,* **16,** 337–350.

Sarup, M. (1996) *Identity, Culture and the Postmodern World,* Edinburgh University Press, Edinburgh.

Saunders, A. (2010) Literary geography: reforging the connections. *Progress in Human Geography,* **34,** 436.

Sauer, C. (1925) *The Morphology of Landscape.* University of California Press, Berkeley, CA.

Sauer, C. (1934) The distribution of Aboriginal tribes and languages in northwestern Mexico. *Ibero-Americana,* **5,** 94–100.

Sauer, C. (1950) Grassland climax, fire, and man. *Journal of Range Management,* **3,** 16–21.

Sauer, C. (1956) The agency of man on the earth. In Williams, T. (ed.) *Man's Role in Changing the Face of the Earth,* University of Chicago Press, Chicago, IL, 49–69.

Sauer, C. (1962[1931]) Cultural geography. In Wagner, P. and Mikesell, M. (eds) *Readings in Cultural Geography,* University of Chicago Press, Chicago, IL, 30–34.

Sauer, C. (1965) Cultural factors in plant domestication in the New World. *Euphytica,* **14,** 301–306.

Saville, S. (2008) Playing with fear: parkour and the mobility of emotion. *Social and Cultural Geography,* **9,** 891–914.

Schiffer, M. (1992) *Technological Perspectives on Behavioral Change,* University of Arizona Press, Tucson, AZ.

Schollhammer, K. (2008) A walk in the invisible city. *Knowledge, Technology and Policy,* **21,** 143–148.

Scott, S. and Morgan, D. (1993) *Body Matters: Essays on the sociology of the body,* Falmer, Basingstoke.

Seidler, V. (1994) *Unreasonable Men: Masculinity and social theory,* Routledge, London.

Seigworth, G. (2000) Banality for cultural studies. *Cultural Studies,* **14,** 227–268.

Seigworth, G. and Gardiner, M. (2004) Rethinking everyday life. *Cultural Studies,* **18,** 139–159.

Seiter, E., Borchers, H., Kreutzner, G. and Warth, E. (eds) (1989) *Remote Control: Television, audiences and cultural power,* Routledge, London.

Sennett, R. and Cobb, A. (1977) *The Hidden Injuries of Class,* W. W. Norton, New York, NY.

Sharp, J. (2000) Towards a critical analysis of fictive geographies. *Area,* **32,** 327–334.

Sheringham, M. (2006) *Everyday life: Theories and practices from Surrealism to the present,* Oxford University Press, Oxford.

Shields, R. (1989) Social spatialisation and the built environment: the case of the West Edmonton Mall. *Environment and Planning D: Society and Space,* **7,** 147–164.

Shields, R. (1999) *Lefebvre, Love and Struggle: Spatial dialectics,* Routledge, London.

Shilling, C. (1993) *The Body and Social Theory,* 1st edn, Sage, London.

Shilling, C. (2003) *The Body and Social Theory,* 2nd edn, Sage, London.

Shilling, C. (2007) *Embodying Sociology,* Blackwell, Oxford.

Short, J. (1991) *Imagined Country: Environment, culture and society,* Routledge, London.

Shotter, J. (1993) *The Cultural Politics of Everyday Life,* Open University Press, Buckingham.

Shotter, J. (1997) Wittgenstein in practice: from 'The Way of Theory' to a 'Social Poetics', available at http://www.focusing.org/apm_papers/shotter.html

Shurmer-Smith, P. (2002a) Introduction. In Shurmer-Smith, P. (ed.) *Doing Cultural Geography,* Sage, London, 1–7.

Shurmer-Smith, P. (ed.) (2002b) *Doing Cultural Geography,* Sage, London.

Shurmer-Smith, P. and Hannam, K. (1994) *Worlds of Desire, Realms of Power: A cultural geography,* Edward Arnold, London.

Sibley, D. (1995) *Geographies of Exclusion: Society and difference in the West,* Routledge, London.

Sibley, D. (1999) Creating geographies of difference. In Massey, D., Allen, J. and Sarre, P. (eds) *Human Geography Today,* Wiley, Chichester, 115–128.

Skelton, T. and Valentine, G. (1998) *Cool Places: Geographies of youth cultures,* Routledge, London.

Skelton, T. and Valentine, G. (2003) 'It feels like being Deaf is normal': an exploration into the complexities of defining D/deafness and young D/deaf people's identities. *Canadian Geographer,* **47,** 451–466.

Slymovics, S. (1998) *The Object of Memory: Arabs and Jews narrate the Palestinian village,* University of Pennsylvania Press, Philadelphia, PA.

Smith, D. (1989) *The Everyday World as Problematic: A feminist sociology,* University Press of New England, Lebanon, NH.

Smith, D.M. (2000) *Moral Geographies: Ethics in a world of difference,* Edinburgh University Press, Edinburgh.

Smith, N. and Katz, C. (1993) Grounding metaphor: towards a spatialized politics. In Keith, M. and Pile, S. (eds) *Place and the Politics of Identity,* Routledge, London, 67–83.

Snyder, J. (1994) Territorial photography. In Mitchell, W. (1994) *Landscape and Power,* University of Chicago Press, Chicago, IL, 175–201.

Söderström, O. (2005) Representation. In Sibley, D., Jackson, P., Atkinson, D. and Washbourne, N. (eds) *Cultural Geography: A critical dictionary of key ideas,* Taurus, London, 11–15.

Soja, E. (1996) *Thirdspace: Journeys to Los Angeles and other real-and-imagined places,* Wiley, London.

Sorkin, M. (1992) *Variations on a Theme Park: The new American city and the end of public space,* Hill and Wang, Boston.

Sorre, M. (1962[1948]) The role of historical explanation in human geography. In Wagner, P. and Mikesell, M. (eds) *Readings in Cultural Geography,* University of Chicago Press, Chicago, IL, 44–47.

Spacehijackers (2009) *A–Z Retail Tricks to Make You Shop,* available at http://www.spacehijackers.org/html/ideas/archipsy/tricks.html

Spinney, J. (2006) A place of sense: a kinaesthetic ethnography of cyclists on Mont Ventoux. *Environment and Planning D: Society and Space,* **24,** 709–732.

Stallybrass, P. and White, A. (1986) *The Politics and Poetics of Transgression,* Methuen, London.

Stanton-Jones, K. (1992) *An Introduction to Dance Music Therapy in Psychiatry,* Routledge, London.

Steffen, A. (2009) Bright green, light green, dark green, gray: the new environmental spectrum, available at http://www.worldchanging.com/archives/009499.html

Stenning A. (2005) Post-socialism and the changing geographies of the everyday in Poland. *Transactions of the Institute of British Geographers,* **30,** 113–127.

Stewart, K. (2007) *Ordinary Affects,* Duke University Press, Durham, NC.

Strohmayer, U. (1996) Pictoral symbolism in the age of innocence: material geographies at the Paris World's Exposition of 1937. *Ecumene,* **3,** 282–304.

Tallontire, A., Rentsendorj, E. and Blowfield, M. (2001) *Ethical Consumers and Ethical Trade: A review of current literature,* Natural Resources Institute, Greenwich.

Taylor, P. (1982) A materialist framework for political geography. *Transactions of the Institute of British Geographers,* **7,** 15–34.

Thien, D. (2004) Love's travels and traces: the impossible politics of Luce Irigaray. In Sharp, J., Browne, K. and Thien, D. (eds) *Women and Geography Study Group: Gender and Geography Reconsidered,* 43–48, available at: http://wgsg.org.uk/newsite1/?page_id=34

Thien, D. (2005) After or beyond feeling? A consideration of affect and emotion in geography. *Area,* **37,** 450–456.

Thrift, N. (1991) Over-wordy worlds? Thoughts and worries. In Philo, C. (ed.) *New Words, New Worlds: Reconceptualising social and cultural geography,* Cambrian, Aberystwyth, 144–148.

Thrift, N. (1994) Inhuman geographies: landscapes of speed, light and power. In Cloke, P., Doel, M., Matless, D., Phillips, M. and Thrift, N. (eds) *Writing the Rural: Five cultural geographies,* Paul Chapman, London, 191–248.

Thrift, N. (1997) The still point: resistance, expressive embodiment and dance. In Pile, S. and Keith, M. (eds) *Geographies of Resistance,* Routledge, London, 124–154.

Thrift, N. (1999) Steps to an ecology of place. In Massey, D., Allen, J. and Sarre, D. (eds) *Human Geography Today,* Blackwell, Oxford, 295–321.

Thrift, N. (2000a) Afterwords. *Environment and Planning D: Society and Space,* **18,** 213–225.

Thrift, N. (2000b) Introduction: dead or alive? In Cook, I., Crouch, D., Naylor, S. and Ryan, J. (eds) *Cultural Turns, Geographical Turns,* Prentice Hall, Harlow, 1–6.

Thrift, N. (2000c) Non-representational theory. In Johnston, R., Gregory, D., Pratt, G. and Watts, M. (eds) *The Dictionary of Human Geography,* Blackwell, Oxford, 556.

Thrift, N. (2004) Intensities of feeling: towards a spatial politics of affect. *Geografiska Annaler B,* **86,** 57–78.

Thrift, N. (2006) Space. *Theory, Culture and Society,* **23,** 139–146.

Thrift, N. and Dewsbury, J. (2000) Dead geographies – and how to make them live. *Environment and Planning D: Society and Space,* **18,** 411–432.

Till, K. (1993) Neotraditional towns and urban villages: the cultural production of a geography of 'otherness'. *Environment and Planning D: Society and Space,* **11,** 709–732.

Titscher, S., Meyer, M., Wodak, R. and Vetter, E. (2000) *Methods of Text and Discourse Analysis,* Sage, London.

Tolia-Kelly, D. (2004a) Locating processes of identification: studying the precipitates of re-memory through artefacts in the British Asian home. *Transactions of the Institute of British Geographers,* **29,** 314–329.

Tolia-Kelly, D. (2004b) Materialising post-colonial geographies: examining the textural landscapes of migration in the South Asian home. *Geoforum,* **35,** 675–688.

Tolia-Kelly, D. (2006) Affect – an ethnocentric encounter? Exploring the 'universalist' imperative of emotional/affectual geographies. *Area,* **38,** 213–217.

Tonts, M. (2005) Competitive sport and social capital in rural Australia. *Journal of Rural Studies,* **21,** 137–149.

Tuan, Y.-F. (1974) *Topophilia: A study of environmental perception, attitudes, and values,* Columbia University Press, New York.

Tuan, Y.-F. (1976) Humanistic geography. *Annals of the Association of American Geographers,* **66,** 266–276.

Tuan, Y.-F. (1977) *Space and Place: The perspective of experience,* University of Minnesota Press, Minneapolis, MN.

Turner, B. (1984) *The Body and Society: An introduction,* 1st edn, Sage, London.

Turner, B. (1991) The secret history of the body in social theory. in Turner, B. (ed.) *The Body: Social processes and cultural theory,* Sage, London, 12–18.

Turner, B. (1996) *The Body and Society: Explorations in social theory,* 2nd edn, Sage, London.

Turner, B. (2008) *The Body and Society: Explorations in social theory,* 3rd edn, Sage, London.

Turner, S. and Manderson, D. (2007) Socialisation in a space of law: student performativity at 'Coffee House' in a university law faculty. *Environment and Planning D: Society and Space,* **25,** 761–782.

UN (2008) *Creative Economy Report 2008,* United Nations, New York, NY.

UN (2011) *Creative Economy Report 2011,* United Nations, New York, NY.

Urry, J. (1990) *The Tourist Gaze,* Sage, London.

Urry, J. (1995) *Consuming Places,* Routledge, London.

Urry, J. (2007) *Mobilities,* Polity Press, Cambridge.

Valentine, G. (1989) The geography of women's fear. *Area,* **21,** 385–390.

Valentine, G. (1996) Angels and devils: moral landscapes of childhood. *Environment and Planning D: Society and Space,* **14,** 581–599.

Valentine, G. (1999) A corporeal geography of consumption. *Environment and Planning D: Society and Space,* **17,** 328–351.

Valentine, G. (2001) *Social Geographies: Space and society,* Pearson, Harlow.

Valentine, G. (2003) Boundary crossings: transitions from childhood to adulthood. *Children's Geographies,* **1,** 37–52.

Valentine, G., Skelton, T. and Chambers, D. (1998) Cool places: an introduction to youth and youth cultures. In Skelton, T. and Valentine, G. (eds) *Cool Places: Geographies of youth cultures,* Routledge, London, 1–34.

van Blerk, L. and Ansell, N. (2006) Moving in the wake of AIDS: children's experiences of migration in Southern Africa. *Environment and Planning D: Society and Space,* **24,** 449–471.

Venkatesh, A. and Meamber, L. (2006) Arts and aesthetics: marketing and cultural production. *Marketing Theory,* **6,** 11–39.

Vidal de la Blache, P. (1908) *On the Geographical Interpretation of Landscapes,* University of Quebec, Chicoutimi.

Vidal de la Blache, P. (1913) *Characteristic Features of Geography,* University of Quebec, Chicoutimi.

Vidal de la Blache, P. (1926) *Principles of Human Geography,* Henry Holt, Baltimore, MD.

Von Thünen, J. (1826) *Die Isolierte Staat in Beziehung auf Landwirtshaft und Nationalökonomie,* Pergamon Press, New York, NY.

Wagner, P. and Mikesell, M. (eds) (1962) *Readings in Cultural Geography,* University of Chicago Press, Chicago, IL.

Wagnleitner, R. (1994) *Coca-Colonization and the Cold War,* University of North Carolina Press, Chapel Hill, NC.

Waterman, S. (1998) Place, culture and identity: summer music in Upper Galilee. *Transactions of the Institute of British Geographers,* **23,** 253–267.

Watson, S. (2006) *City Publics: The (dis)enchantments of modern encounters,* Taylor & Francis, London.

Weed, M. (2007) The pub as a virtual football fandom venue: an alternative to being there? *Soccer and Society,* **8,** 399–414.

WGSG [Women and Geography Study Group] (1997) *Feminist Geography: Explorations in diversity and difference,* Longman, London, 86–111.

Whatmore, S. (2002) *Hybrid Geographies: Natures, cultures,* spaces, Sage, London.

Whatmore, S. (2006) Materialist returns: cultural geography in and for a more-than-human world. *Cultural Geographies,* **13,** 600–609.

White, B. and Day, F. (1997) Country music radio and American culture regions. *Journal of Cultural Geography,* **16,** 21–35.

Williams, B. and Barlow, S. (1998) Falling out with my shadow: lay perceptions of the body in the context of arthritis. In Nettleton, S. and Watson, J. (eds) (1998) *The Body in Everyday Life,* Routledge, London, 124–141.

Williams, R. (1976) *Keywords,* Oxford University Press, Oxford.

Williams, R. (1979) *Politics and Letters: Interviews with New Left Review,* New Left Review, London.

Williams, S. (2001) *Emotion and Social Theory: Corporeal reflections of the (ir)rational,* Sage, London.

Williams, S. and Bendelow, G. (1998) *The Lived Body: Sociological themes, embodied issues,* Routledge, London.

Williams-Ellis, C. (1928) *England and the Octopus,* Golden Dragon, Portmeirion.

Willis, P. (1990) *Common Culture: Symbolic work at play in the everyday cultures of the young,* Open University Press, Basingstoke.

Wilson, A. (1991) *The Culture of Nature,* Between the Lines, Toronto.

Withers, C. (2006) Eighteenth-century geography: texts, practices, sites. *Progress in Human Geography,* **30,** 711–729.

Widdowfield, R. (2000) The place of emotions in academic research. *Area,* **32,** 199–208.

Wodak, R. (1996) *Disorders of Discourse,* Longman, London.

Woodside, S. (2001) *Every Joke is a Tiny Revolution: Culture jamming and the role of humour,* University of Amsterdam, Amsterdam.

Wooldridge, S. and Morgan, R. (1937) *The Physical Basis of Geography: An outline of geomorphology,* Longmans Green, London.

Wylie, J. (2002) An essay on ascending Glastonbury Tor. *Geoforum,* **33,** 441–454.

Wylie, J.W. (2005) A single day's walking: narrating self and landscape on the Southwest Coast Path. *Transactions of the Institute of British Geographers,* **30,** 234–247.

Wylie, J. (2007) *Landscape,* Routledge, London.

Wylie, J. (2009) Landscape, absence and the geographies of love. *Transactions of the Institute of British Geographers,* **34,** 275–289.

Wylie, J. (2010) Cultural geographies of the future, or looking rosy and feeling blue. *Cultural Geographies,* **17,** 211–217.

Yarwood, J. and Shaw, J. (2009) 'N-gauging' geographies: craft consumption, indoor leisure and model railways. *Area,* **42,** 425–433.

Young, I. (1990) *Throwing Like a Girl and Other Essays in Feminist Philosophy and Social Theory,* Indiana University Press, Bloomington, IN.

Young, R. (2003) *Postcolonialism: A very short introduction,* Oxford University Press, Oxford.

Zelinsky, W. (1958) The New England connecting barn. *The Geographical Review,* **48,** 540–553.

Zelinsky, W. (1973) *The Cultural Geography of the United States,* Prentice Hall, Englewood Cliffs, NJ.

Zukin, S. (1995) *The Cultures of Cities,* Blackwell, Oxford.

Zukin, S. (2003) *Point of Purchase: How shopping changed American culture,* Routledge, London.

Index